Smart Innovation, Systems and Technologies

Volume 388

The Smart Innovation, Systems and Technologies book series encompasses the topics of knowledge, intelligence, innovation and sustainability. The aim of the series is to make available a platform for the publication of books on all aspects of single and multi-disciplinary research on these themes in order to make the latest results available in a readily-accessible form. Volumes on interdisciplinary research combining two or more of these areas is particularly sought.

The series covers systems and paradigms that employ knowledge and intelligence in a broad sense. Its scope is systems having embedded knowledge and intelligence, which may be applied to the solution of world problems in industry, the environment and the community. It also focusses on the knowledge-transfer methodologies and innovation strategies employed to make this happen effectively. The combination of intelligent systems tools and a broad range of applications introduces a need for a synergy of disciplines from science, technology, business and the humanities. The series will include conference proceedings, edited collections, monographs, handbooks, reference books, and other relevant types of book in areas of science and technology where smart systems and technologies can offer innovative solutions.

High quality content is an essential feature for all book proposals accepted for the series. It is expected that editors of all accepted volumes will ensure that contributions are subjected to an appropriate level of reviewing process and adhere to KES quality principles.

Indexed by SCOPUS, EI Compendex, INSPEC, WTI Frankfurt eG, zbMATH, Japanese Science and Technology Agency (JST), SCImago, DBLP.

All books published in the series are submitted for consideration in Web of Science.

Roumen Kountchev (Deceased) · Srikanta Patnaik ·
Wenfeng Wang · Roumiana Kountcheva
Editors

AI Methods and Applications in 3D Technologies

Proceedings of 3DWCAI 2023

Springer

Editors
Roumen Kountchev (Deceased)
Technical University of Sofia
Sofia, Bulgaria

Wenfeng Wang
School of Electrical
Shanghai Institute of Technology
Shanghai, China

Srikanta Patnaik
Interscience Institute of Management
and Technology
Bhubaneswar, Odisha, India

Roumiana Kountcheva
TK Engineering
Sofia, Bulgaria

ISSN 2190-3018 ISSN 2190-3026 (electronic)
Smart Innovation, Systems and Technologies
ISBN 978-981-97-2143-6 ISBN 978-981-97-2144-3 (eBook)
https://doi.org/10.1007/978-981-97-2144-3

This Springer imprint is published by the registered company Springer Nature Singapore Pte Ltd.
The registered company address is: 152 Beach Road, #21-01/04 Gateway East, Singapore 189721, Singapore

Paper in this product is recyclable.

Preface

This book is Volume I of the Proceedings of the Second World Conference on Intelligent and 3D Technologies (WCI3DT2023): *AI Methods and Applications in 3D Technologies.*

The conference took part during May 26–28, 2023, in Shanghai, China.

The aim of WCI3DT 2023 was to provide a wide forum to academicians and practitioners where they to present latest scientific results and to exchange ideas in the area of Artificial Intelligence and Deep Learning and their applications in augmented reality and 3D technologies. The conference organizers were focused on the establishment of an effective platform for all participants to introduce their work to scientists, engineers, and students from all over the world.

After reviewing, 61 papers were accepted for presentation and publication in the conference proceedings, of which 30 are in this volume. The selected works present contemporary research works aimed at the area of multidimensional signal processing, and based on contemporary methods and applications in 3D technologies, such as the multi-terrain motion control method for quadruped robot based on reinforcement learning, the method for detection of basal units beneath the Antarctic ice, the study on the applications of AI algorithms in medicine, the creation of an intelligent recommendation algorithm for music website, the creation of a 3D scene simulation optimization model, a special approach for gesture recognition in complex background, various contemporary applications based on AI and DL algorithms, and other interesting ideas. The chapters are arranged in three groups, which cover different parts of the related scientific areas:

- AI-Based Approaches (11 chapters);
- 3D Technologies and DL (9 chapters);
- Intelligent Methods and Applications (10 chapters).

In memory of Prof. Dr. Roumen Kountchev, Dr. Sc., *Best Paper Award* was established by the Organizing Committee. Due to the high number of very good works, the selection was extremely difficult. Two papers were selected as "*winner paper*", and published in this volume as Chap. 1 and Chap. 2, respectively.

The winner papers are

"Extended Prototypical Network for Few-shot Learning", with authors: Jingjing Zhang, Lujie Cui, Wenfeng Wang, and Lalit Mohan Patnaik;

and

"Low Carbon Scheduling of Thermal Power Unit Thermal Storage Capacity Based on Particle Swarm Optimization", with authors: Wenbin Cao, Mingkai Wang, Bo Han, Qi Chai, Hongtao Li, and Ying Liu.

The book editors express their special thanks to IRNet International Academic Communication Center who organized this conference in correspondence with their dedication to build platforms for scholars and researchers for better knowledge sharing, together with providing full-scale conference services that meet the standards of the renowned publishing organizations.

We also thank Prof. Lakhmi Jain (Honorary chair), Prof. Wenfeng Wang, Prof. Srikanta Patnaik, Prof. Gang Sun, and Prof. Xudong Jiang (General chairs), Dr. Yonghang Tai, Dr. Shoulin Yin, and Prof. Hang Li (Organizing chairs), and S. R. Dr. Roumiana Kountcheva (International advisory chair).

The editors express their warmest thanks to the excellent Springer team for making this book possible.

Sofia, Bulgaria — Prof. Dr. Roumen Kountchev
Bhubaneswar, India — Prof. Dr. Srikanta Patnaik
Shanghai, China — Prof. Dr. Wenfeng Wang
Sofia, Bulgaria — S. R. Dr. Roumiana Kountcheva
August 2023

Contents

Part I
AI-Based Approaches

Chapter 1
Extended Prototypical Network for Few-Shot Learning

Jingjing Zhang, Lujie Cui, Wenfeng Wang, and Lalit Mohan Patnaik

Abstract This paper proposes a learning method based on extended distance for unsupervised meta-learning. Compared with previous algorithms, the network can also learn new classes and correctly classify them, and each class only requires a few shots for training. Among them, extension distance emerged as a new distance measurement method, which has obvious advantages over the previous Euclidean distance. This paper compares it with the prototypical network based on Euclidean distance and compares and analyzes the experimental results, the experiment that works best is the Miniimagenet dataset which improves accuracy by about 2.36%. The meta-learning method based on metric space is further explored, and experiments are carried out on three data sets and achieved good experimental results on MNIST, miniimagenet, and omniglot data sets.

1.1 Introduction

Extenics is a new discipline founded by Chinese scholars headed by Professor CAI Wen. It utilizes formal models to study the possibility of expansion of things and methods of exploration and innovation, primarily for dealing with contradictions [1]. The basic element theory puts forward the basic element that describes events, matters, and relation elements—"event elements", "matter elements" and "relation elements". In this paper, we regard each class as a matter element, and classes should have corresponding features, and the features also correspond to values, which usually expresses in extensions as (N, C, V). Extension distance mainly describes the distance relationship between point and interval. Furthermore, we can refer to the construction of the elementary correlation function of interval side distance based

J. Zhang · L. Cui · W. Wang (✉)
Shanghai Institute of Technology, Shanghai 201418, China
e-mail: wangwenfeng@sit.edu.cn

W. Wang · L. M. Patnaik
National Institute of Advanced Studies, Indian Institute of Science Campus, Bangalore 560012, India

R. Kountchev (Deceased) et al. (eds.), *AI Methods and Applications in 3D Technologies*, Smart Innovation, Systems and Technologies 388,
https://doi.org/10.1007/978-981-97-2144-3_1

on area distance [2]. According to current studies, three metric-based meta-learning models are Matching Net, Proto Net, and relation Net [3]. The previous Siamese Network is a two-path Neural Network [4]. During training, different pairs of samples are constructed by combination and input into the Network for training. At the top layer, the distance of sample pairs is used to judge whether they belong to the same class and generate corresponding probability distribution. In the prediction stage, the twin network processes each sample pair between the test sample and the support set, and the ultimate prediction result is the category with the highest probability on the support set [5]. Compared with twin networks, Match Network constructs different encoders for the support set and batch set, and the output of the final classifier is the weighted sum of predicted values between the support set and the query set [6]. It can generate labels for unknown categories without changing the network model, and its main innovation is reflected in the modeling process and training process.

Problems existing in metric-based meta-learning: the core of matrix operation learning lies in the setting of the loss function, and the matrix operation is large [7]; High requirements on data sets, in some newly proposed loss functions, may increase a lot in face data sets, but the effect on other tasks will be worse. And in this paper, we aim to ease the problem of large computation of the meta-learning matrix and improve the accuracy of classifications [8]. And we mainly used K-means for clustering division.

The purpose of this study is to apply theoretical knowledge to expansion, study the meta learning algorithm based on metric space optimization problem, formalize samples and quantitative features, and establish matter element model [9], so that things can be displayed more intuitively and formally for analysis and conversion. Another research focus of this paper is to design a suitable extension distance to measure the distance between samples, and then conduct training under the model structure, so that the training and testing effect can achieve the optimal [10]. It is beneficial to improve the effect of classification and recognition by introducing the theory knowledge of extensions and studying more distance measurement methods. We compare the improved prototypical network using extension distance with the prior prototypical network using Euclidean distance and get good experimental results on three data sets. Firstly, the input data requires to be mapped, and then the unlabeled data set is classified by clustering partition through K-means algorithms. Afterwards, by comparing the query dataset and interval, the closer the distance, the higher the similarity.

1.2 Two Stages of Meta-Learning

Stage 1: Training Train Tasks

Meta-learning mainly consists of two stages, the first stage is the inner cycle training task training; The second stage is the outer loop test task training. The partitioning of the data set remains the same, except that the training set also includes the training set and the test set, and the test set is the same. Through the support set training, the model parameters for each sub-task are trained respectively. The query sets of different subtasks are then used to test the performance separately and calculate the loss between the predicted and true values. Finally, the gradient descent method is used to update parameters, so as to find the optimal super-parameter setting. The training process of the training task in stage 1 can be shown in the Fig. 1.1.

There are N tasks sampled during the training task stage as shown in the Fig. 1.1, and each training support task set can be represented as $\{(x_1^{ks}, y_1^{ks}), (x_2^{ks}, y_2^{ks}), \ldots, (x_n^{ks}, y_n^{ks})\}, k = 1, 2, \ldots, N$, n represents the number of samples in the support set. Each training query task set can be represented as $\left\{\left(x_1^{kq}, y_1^{kq}\right), \left(x_2^{kq}, y_2^{kq}\right), \ldots, \left(x_m^{kq}, y_m^{kq}\right)\right\}, k = 1, 2, \ldots, N$, m represents the number of samples in the query set. Wherein φ indicates the hyperparameter to be set, θ represents the parameters of the neural network to be trained. The purpose of meta learning is to enable function f_{φ,θ_k} to automatically train $F_{\theta_k^*}$ in training tasks, then utilizing the prior knowledge of F_{φ^*,θ_k^*} to train the parameters in the model for

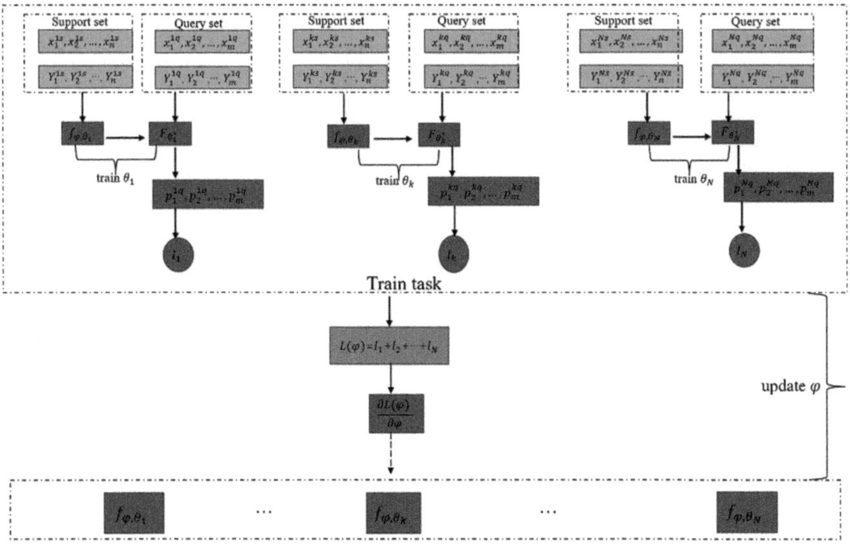

Fig. 1.1 Meta-learning train tasks training parameters

specific tasks in testing tasks。Where $\{p_1^{kq}, p_2^{kq}, \ldots, p_m^{kq}\}$ represents the test label value corresponding to the trained θ_k^* in the training test set sample.

The box above shows the update parameter θ, and the updating process can be specifically described as: the training parameter θ of the support set of the training task, and then the query set in the training task updates the parameter θ, and finally, the partial derivative with respect to φ is obtained:

Stage 2: Test Tasks Update Parameters

The test task is similar to the machine learning process, and the test set is also divided into training set and testing set. The phase 1 training task is to find good parameter Settings and perform the task better in a specific test task. This stage is mainly divided into training data and test data. The training data is mainly responsible for training the parameter θ, and the test data is updated with the known parameter changes, and finally, the output is obtained. In the test task, the parameter update process is shown in Fig. 1.2:

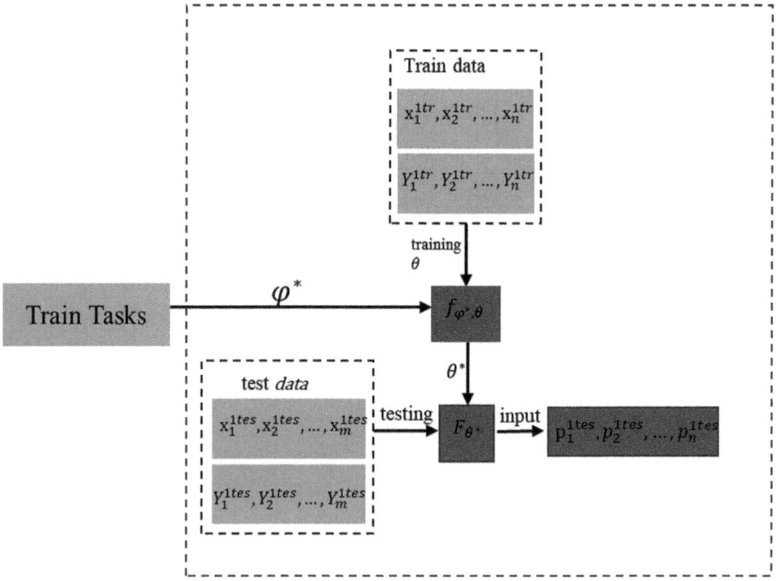

Fig. 1.2 Meta-learning test tasks updates parameters

1.3 Research on Unsupervised Meta-Learning in Metric Space

Notation

The concept is established to formalize the description of the matter element, event element, and relation element 11. They are the logical cells of extensions and are collectively referred to as primitives. This paper mainly introduces the concept of matter elements.

Definition 2.1 With matter O_m as the objects, c_m as the features, O_m as the ordered triad of value v_m about c_m:

$$M = (O_m, c_m, v_m) \tag{1.1}$$

As the basic element of describing matters, they are collectively called one-dimensional matter elements, O_m, c_m, v_m are the three elements. $c_m$andv_m construct the dualistic group of (c_m, v_m) as the feature elements.

Definition 2.2 Let x be any point on the real axis, X = < a, b > is any interval on the real field, called

$$\rho(x, X) = \left|x - \frac{a+b}{2}\right| - \frac{b-a}{2} \tag{1.2}$$

The above formula is the extension distance of the point x and the interval X = < a, b > , where X can be an open, closed, or half-open and half-closed interval.

For any point x_0 in the real axis, there is:

$$\rho(x, X) = \left|x_0 - \frac{a+b}{2}\right| - \frac{b-a}{2} = \begin{cases} a - x_0, x_0 \le \frac{a+b}{2} \\ x_0 - b, x_0 \ge \frac{a+b}{2} \end{cases} \tag{1.3}$$

Extension Distance and Establish the Matter-Element Model

To describe the difference between classes, the distance relationship between the point and interval is defined before establishing the correlation function, which is called extension distance.

Properties 2.1 A given interval $X =< a, b >$, so

(1) The sufficient and necessary conditions for the point $x \in X$ and x $\neq$ a, b are $\rho(x, X) < 0$;

(2) The sufficient and necessary conditions for the point $x \notin X$ and $x \neq a, b$ are $\rho(x, X) > 0$;
(3) The sufficient and necessary conditions for the point x = a or x = b are $\rho(x, X) = 0$;

Step 1: Establish the corresponding matter-element model according to the features or dimensions of examples:

$$X = (x, I, a_i(x)) = \begin{bmatrix} x & I_1 & a_1(x) \\ & I_2 & a_2(x) \\ & \cdots & \cdots \\ & I_m & a_m(x) \end{bmatrix} \tag{1.4}$$

Where $a_j(x)$ the value of jth feature I_j of matter-element X, j = 1,2…,m.

Step 2: Let N dimension vectors $x_j(j = 1, 2, \ldots, N)$ divided into N groups $v_1, v_2, \ldots, v_n$, suppose there are k_1 group from class v_1,…, and there are N_n from class v_n. And find the center of each group to minimize the value function of similarity (or distance) features, where the distance is measured by extension distance.

Step 3: Calculate the central matter element of this class:

$$C = (C_i, I, c_i(w_i)) = \begin{bmatrix} C_i & I_1 & c_1(w_1) \\ & I_2 & c_2(w_2) \\ & \cdots & \cdots \\ & I_m & c_m(w_i) \end{bmatrix} \tag{1.5}$$

Among them $c_i(w_i) = \frac{\sum_{h=1}^{N_i} a_j(x_h)}{N_i}$, $j = 1, 2, \ldots, m; i = 1, 2, .., n$.

Minimum value of corresponding feature of matter-element $v_j^l = \min_{h \in N_i}\{a_j(x_h)\}$, minimum value of corresponding feature of matter-element $v_j^l = \max_{h \in N_i}\{a_j(x_h)\}$, joint domain for $V_j(w_i) = \left[v_j^l, v_j^l\right]$, j = 1,2,..,m;i = 1,2,…,n. Then the matter-element model of this class can be constructed as:

$$W = (w_i, I, V) = \begin{bmatrix} w_i & I_1 & V_1(w_1) \\ & I_2 & V_2(w_2) \\ & \cdots & \cdots \\ & I_m & V_m(w_i) \end{bmatrix} \tag{1.6}$$

In this paper, we define the extension distance function is:

$$d(x, w_i) = \sum_{j=1}^{m} \left[\frac{(a_j(x) - c_j(w_i))^2 - \left(\frac{v_j^u - v_j^l}{2}\right)^2}{2}\right] \tag{1.7}$$

Which satisfies the properties 2.1.

Based Metric Space Meta-Learning Model

The prototypical network is mainly based on the embedding space after nonlinear mapping, clustering each class of the embedding space to gain the mean value of samples as the prototype of every class. And we use a neural network to learn a nonlinear mapping of the input into the embedding space. Classification is mainly about comparing the distance between the query point and the nearest neighbor prototype. First, the data is mapped to an embedded space, then the prototype network can identify new categories that have never been seen during training, each requiring very little information from the support set12. Compared to the Euclidean distance to measure the similarity of query set and prototype, we want to further optimize the Prototypical Networks. In the clustering automatically construct task for unsupervised (CACTUs) [13] prototypical networks for classifications, there gives a support set of m labeled examples and a query set s_q, y_m is the one-hot labels which get from CACTUs:

$S_k = \{(x_1, y_1), (x_2, y_2), \ldots, (x_m, y_m), y_i \in \{1, 2, \ldots, K\}$ is the corresponding set. S_k denotes the set of examples labeled with class k. In this case, the relationship between embedded query points and the class prototype is:

$$p_\phi(y = k|x_i) \propto \exp(-d(f_\varnothing(x_i), c_k)) \tag{1.8}$$

In the embedding space, we can calculate the prototype of the k-th class:

$$c_k = \frac{1}{|S_k|} \sum_{(x_i, y_i) \in S_k} f_\phi(x_i) \tag{1.9}$$

c_k is the prototype, f_ϕ is an embedding function;

Given a distance function, there we use extension distance for prototypical networks to choose the relatively closer prototype, which is the extension distance to produce a distribution over class for a query point x based on a softmax:

$$p_\phi(y = k|x_i) = \frac{\exp(-d(f_\phi(x_i), \mathrm{w}_i))}{\sum_{k\prime} exp(-d(f_\phi(x_i), \mathrm{w}_i))} \tag{1.10}$$

Where we define an extension distance:

$$d(x, \mathrm{w}_i) = \sum_{j=1}^{m} \left[\frac{\left(a_j(x) - c_j(\mathrm{w}_i)\right)^2 - \left(\frac{\mathrm{v}_j^u - \mathrm{v}_j^l}{2}\right)^2}{2} \right] \tag{1.11}$$

Table 1.1 Unsupervised extension prototypical network algorithm

Algorithm 1 Extension Prototypical Network for Few-shot Learning
1: Require: Given a dataset D and clustering partition get $S_k = \{(x_i, y_i)\ y_i \in \{1,2,\ldots,K\}$
2: Random samples $(\{1,2,\ldots,K\}, S_k)$,S_Q is the query set.
Compute the class prototype, c_k is the prototype, f_ϕ is an embedding function:
$$c_k = \frac{1}{\lvert S_k \rvert} \sum_{(x_i,y_i)\in S_k} f_\phi(x_i)$$
3: Compute the probability of query point which belongs to one of the prototypes through Extension distance:
$$p_\phi(y = k \mid x_i) = \frac{\exp(-d(f_\phi(x_i), w_i))}{\sum_{k'} exp(-d(f_\phi(x_i), w_i))}$$
4: $J \leftarrow 0$
$for\ j\ in\ \{1,\ldots,S_k\}\ do$
$for\ (x,y)\ in\ S_Q\ do$
$$J \leftarrow J + \frac{1}{S_k S_Q}(-logp_\phi(y = k \mid x_i))$$

Finally, we can calculate the loss function we can apply

$$J(\phi) = -logp_\phi(y = k \mid x_i) =$$

$$d\left(f_\varnothing(x_i), \mathrm{w}_k\right) + log \sum_{k'} \exp\left(-d\left(f_\varnothing(x_i), \mathrm{w}_{k'}\right)\right)$$

The above is the process in which we combine the entropy weight method to improve the prototypical network. The algorithm is shown in Table 1.1:

1.4 Experiments

Unsupervised Learning for Prototypical Network on MNIST

The MNIST dataset is relatively simple compared with the other two datasets, so the classification accuracy is relatively higher than the other two datasets under the same N-way and K-shot conditions. Extension distance can improve generalization ability, in order to prove this, we carried out the following experiment. As shown in the following figure, the comparison about the prototypical network is under two different metric distances. As shown in Table 1.2, for 5-way and 1-shot, we statistic experimental results and compare Extension distance and Euclidean distance, the Extension distance higher by 1.3% than Euclidean distance. And for 5way-5shot, we can get an accuracy of about 92.38% which is higher than before and the accuracy of Euclidean is 92.26%. What can be further discussed is that although the MNIST data set is simple, our experimental effect is better than before under the condition of 20way and 1shot, but the accuracy is not high as shown in Table 1.2.

Unsupervised Learning for Prototypical Network on Omniglot

As shown in Fig. 1.3, applying Euclidean Distance in the prototype network has a higher training loss than the extension distance at the beginning, although the global trend is similar to the extension Distance. It reflects our method is relatively stable. According to Fig. 1.4, we can see that the situation of train loss is the same as in Fig. 1.3. The accuracy of val accuracy is higher than the prototype network of Euclidean Distance in both the beginning and the end stages. As shown in Fig. 1.5, the initial extended distance in the prototype network has a higher training loss. To observe the accuracy of validation more clearly, as shown in Fig. 1.6, the method of extension is significantly higher than that of Euclidean Distance. As depicted in Table 1.3, we statistic three groups of experiments and the corresponding results of the prototype network that using extension distance are 64.77, 79.32 and 41.99%, while the corresponding results of the prototype network using Euclidean distance are 64.71, 78.99 and 41.74%, respectively.

Unsupervised Learning for Prototypical Network on Miniimagenet

Since this data is more difficult to classification than the other two data sets, we adopt three ways to construct the task for N-way K-shot here, for 5-way,20-shot, we take the cluster number is k = 50, the partition number is p = 10; And for 5-way, 5-shot, we take the cluster number is k = 100, the partition number is p = 50; Finally, for 5-way,2-shot, we take the cluster number is k = 500, the partition number is p = 100. The comparison results are shown in Figs. 1.7 and 1.8. Surprisingly, the effect is better than the above data set. As depicted in Table 1.4, we statistic two groups of experiments and the corresponding results of the prototype network that using extension distance are 50.37 and 41.62%, while the corresponding results of the prototype network using Euclidean distance are 48.92 and 40.37%, respectively. As depicted in Table 1.5, by setting the number of clusters to k = 500, it show better classification when the data set is more difficult.

Table 1.2 Based on different metric distance of few-shot classification accuracies on Mnist; We all take the cluster number is k = 50, partition number is p = 10

Model	Dist	5-way Acc		20-wayAcc
		1-shot	5-shot	1-shot
Prototypical network	Extension	85.17%(↑1.3)	92.38% (↑0.12)	37.92%(↑0.07)
Prototypical network	Euclidean	83.87%%	92.26%	37.85%

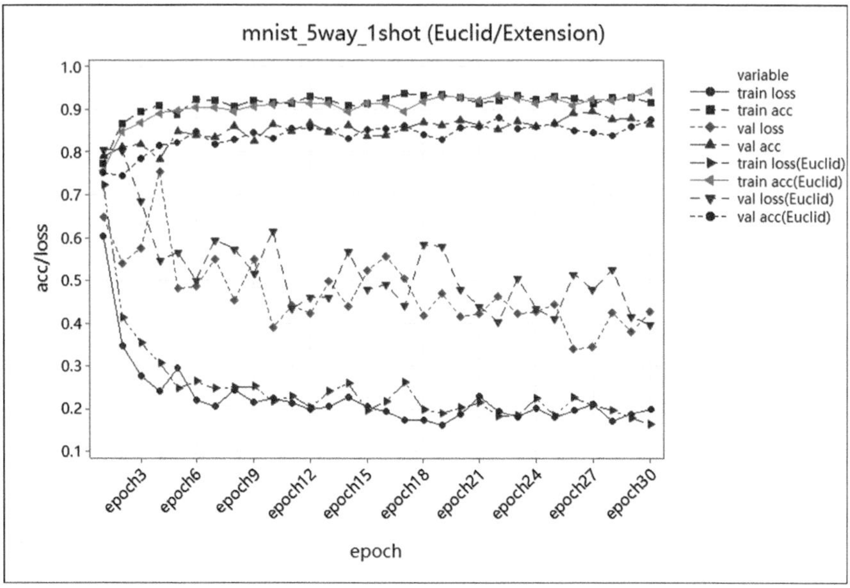

Fig. 1.3 Comparison of two distance ways for the prototypical network with the same MNIST dataset 5-way 1-shot

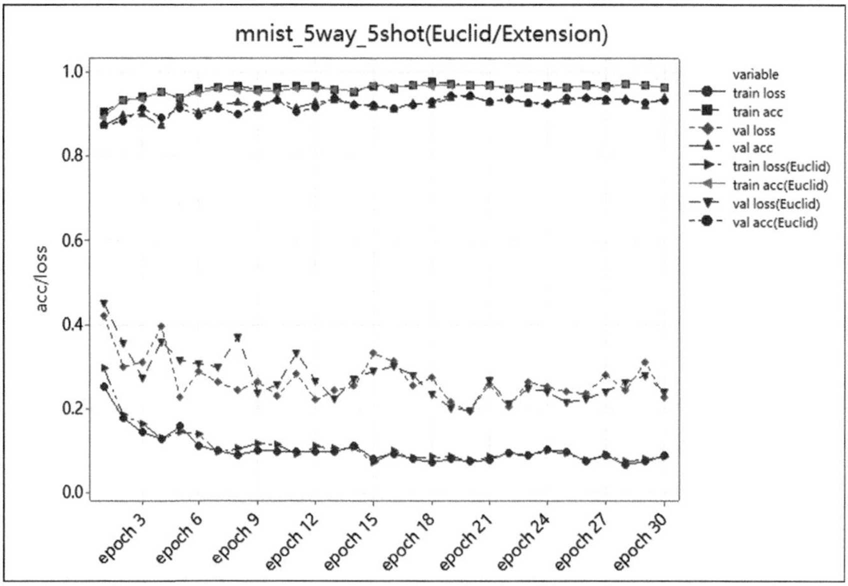

Fig. 1.4 Comparison two distance ways for prototypical network with the same 5-way 5-shot

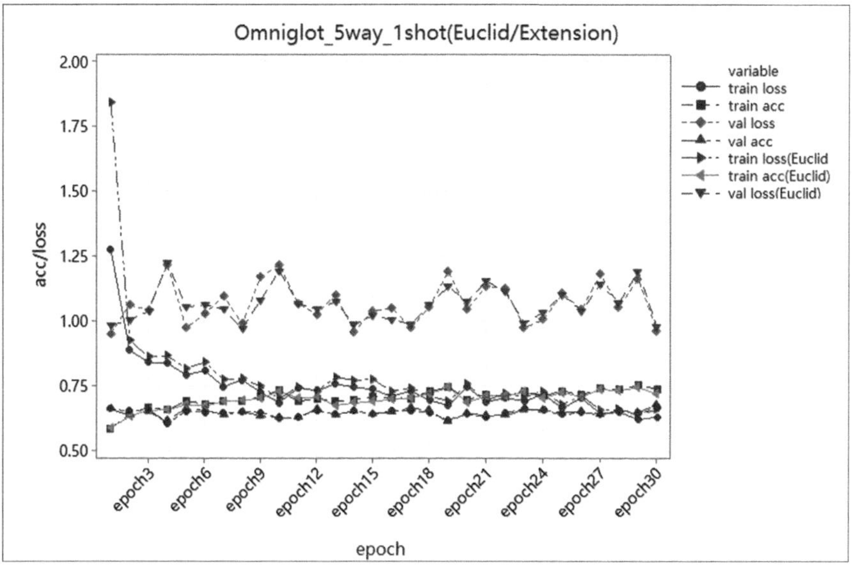

Fig. 1.5 Comparison of two distance ways for the prototypical network with the same Omniglot dataset 5-way 1-shot

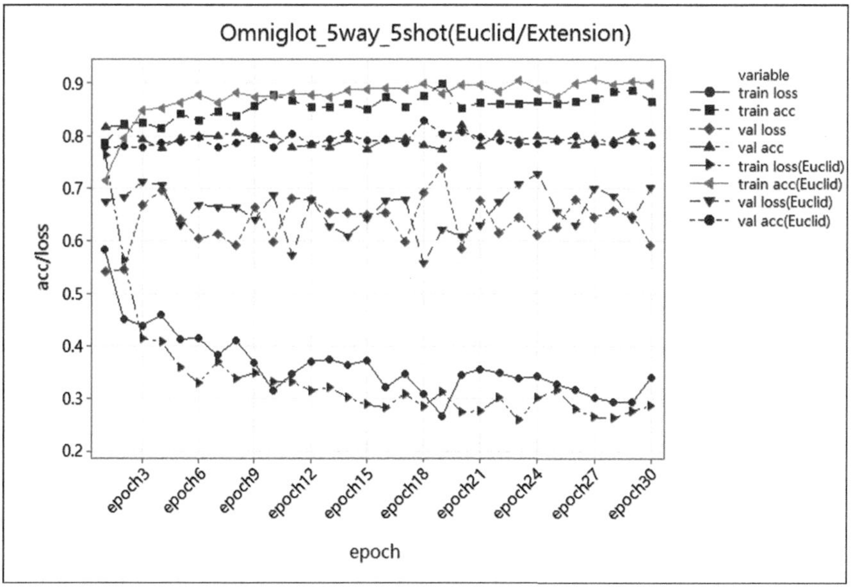

Fig. 1.6 Comparison two distance ways for the prototypical network with the same 5-way 5-shot

Table 1.3 Based on two different metric distances of few-shot classification accuracies on Omniglot; We all take the cluster number as k = 50, partition number is p = 10

Model	Dist	5-way Acc		20-way
		1-shot	5-shot	1-shot
Prototypical network	Extension	64.77%(↑0.07)	79.32%(↑0.33)	41.99%(↑0.25)
Prototypical network	Euclidean	64.71%	78.99%	41.74%

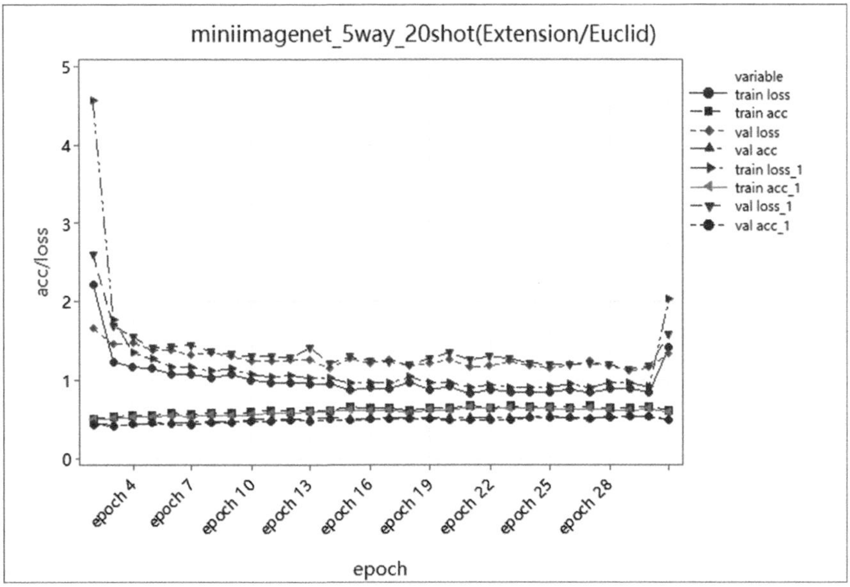

Fig. 1.7 Comparison of two distance ways for prototypical network with the same 5-way 20-shot

1.5 Discussion

The prototypical network is a meta-learning method based on metric space14. It is helpful to improve generalization ability by improving the query point's search for prototypes in embedding space. Compared with the prototype network with label learning, the data set without label learning is more difficult, in which the automatic clustering tasks construction method in unsupervised meta-learning (CACTUS) is adopted. Our main work is to further study the measurement method between the query point and the prototype in the embedded space, to improve the generalization ability of experiments in the query set15. And we demonstrate that based on Extension distance meta-learning we can achieve better performance in learning downstream tasks. At the same time, it can be seen from the experiment that the task construction mechanism is also very important. The paper mainly adopts the K-means algorithm for clustering division16. The definition of distance in extensions is

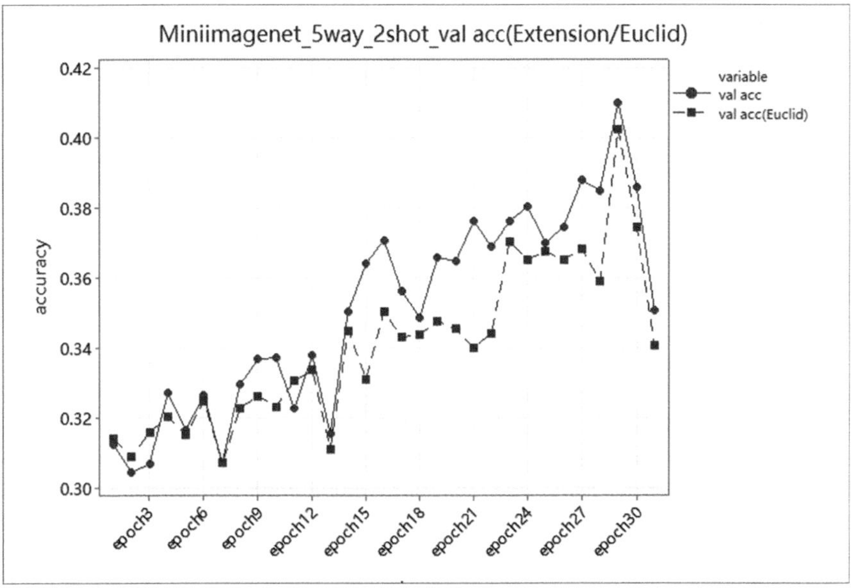

Fig. 1.8 Comparison of two distance ways for prototypical network with the same 5-way 2-shot

Table 1.4 Based on two different metric distances of few-shot classification accuracies on Miniimagenet, we take the cluster number as k = 50, the partition number as p = 10 for 5-way, 20-shot; And we take the cluster number as k = 100, the partition number is p = 50 for 5-way, 5-shot

Model (Miniimagenet)	Dist	5-way Acc	
		20-shot	5-shot
Prototypical network	Extension	50.37%(↑1.45)	41.62%(↑1.25)
Prototypical network	Euclidean	48.92%	40.37%

Table 1.5 Based on two different metric distances of few-shot classification accuracies on Miniimagenet. We take the cluster number as k = 500, the partition number is p = 100 for 5-way, 2-shot

Model (Miniimagenet)	Dist	Train loss	Val loss	Train acc	Val acc
Prototypical network	Extension (5way-2-shot)	2.35	2.54	57.17%(↑2.36)	35.06%(↑1.0)
Prototypical network	Euclidean (5way-2shot)	3.94	3.69	54.81%	34.06%

also called extension distance, which is a new distance measure proposed to make up for the inability to describe the process of qualitative change in classical mathematics. Quantitative change and qualitative change can be visually defined. Understanding the lateral distance is also very meaningful in extensions, side distance can further describe the distribution of the optimal value, also reflected in some papers 17. K-nearest neighbor and K-means clustering algorithm also can learn from the side distance, and extension from the side distance can also find out the initial clustering centers, and optimize the clustering center, introducing the idea of extension distance [18–20].

1.6 Conclusion

This paper introduces a novel distance measurement to evaluate the distance between the prototype and query point in embedding space. The relationship between a point and an interval in extensions can be explained with Extension distance. Introducing the extension from the concept, and designing a new extension distance at the same time, the experimental results show that our method unlike the Euclidean distance is better to measure query point and prototype, which also makes the classification more effective. And in the next work, we can go further explore metric distance meta-learning and extensions based on many kinds of literature, which can study the combination and application of more theoretical knowledge.

Author Contributions J. J., Z was mainly responsible for data collection, code convenience, and paper writing as the core contributor of this paper. W. F., W led the project and conceive the main idea of this research and L. M. P improved the quality of the whole paper.

Funding This research was funded by the High-level Base-building Project for Industrial Technology Innovation (1021GN204005-A06). Lalit Mohan Patnaik would like to thank the National Academy of Sciences India (NASI), Allahabad, India for the support and to the Director, National Institute of Advanced Studies (NIAS), Bengaluru, India for providing the infrastructure facilities.

Data Availability All the data utilized to support the theory and models of the present study are available from the corresponding authors upon request.

Conflicts of Interest The authors declare that there are no conflicts of interest regarding the publication of this article.

References

1. Cai, W.: Introduction of extenics. Syst Eng-Theory & Pract, (1998)
2. Xiang, C.: K-nearest Neighbor algorithm based on extension Distance. J. Hubei Univ. Natl.: Nat. Sci. Ed., (1), 4 (2010)
3. Oriol, V., Charles, B., Tim, L., Daan, W., et al.: Matching networks for one shot learning. In: Advances in neural information processing systems, pp. 3630–3638 (2016)
4. Bertinetto, L., Valmadre, J., Henriques, J. F., et al.: Fully-convolutional siamese networks for object tracking. In: European conference on computer vision. Springer, Cham (2016)
5. Abramson, N., Braverman, D.J., Sebestyen, G.S.: Pattern recognition and machine learning. Publ. Am. Stat. Assoc. **103**(4), 886–887 (2006)

6. Liu, M. Y., Huang, X., Mallya, A., et al.: Few-shot unsupervised image-to-image translation, (2019)
7. Ren, J., Wang, N.: Research on loss function in artificial neural network. J. Gansu Norm. Univ. **23**(02), 65–67 (2018)
8. Vilalta, R., Giraud-Carrier, C., Brazdil, P.: Meta-learning. Data Min. & Knowl. Discov. Handb., (2005)
9. Yang, C., Tang, L.: A review of theories, methods and systems for incompatible problem solving. CAAI Trans. Intell. Syst. **11**(6), 799–806 (2016)
10. Li, B., Zhang, L.: A new method for global optimal value estimation. J. Shandong Inst. Min. & Technol.: Nat. Sci. Ed. **02**, 107–109 (1999)
11. Zhang, Y.: Intelligent control based on extension logic. Nanjing University of Science and Technology, (2002)
12. Banerjee, M., Pal, N.: Unsupervised feature selection with controlled redundancy (UFeSCoR). IEEE Trans. Knowl. Data Eng. **27**(12), 3390–3403 (2015)
13. Hsu, K., Levine, S., Finn, C.: Unsupervised learning via meta-learning. arXiv (2018)
14. Finn, C.: Learning to learn with gradients. University of California, Berkeley (2018), Technical Report No. UCB/EECS-2018–105
15. Smarandache, F.: Generalizations of the distance and dependent function in extenics to 2D, 3D, and n−D. Prog. Phys., (2012)
16. Coates, A., Ng, A.Y.: Learning feature representations with K-means. Lect. Notes Comput. Sci. **7700**, 561–580 (2012)
17. Li, Q., Liu, S.: Elementary correlation function construction based on interval distance and interval side distance. Chinese Society for Artificial Intelligence, (2006)
18. Zhao, Y.: Research on extension method of intelligent conceptual design. Shanghai University, (2005)
19. Liang, G., Xu, W.: On xiaolong.Application of extension theory based on entropy weight in rock mass quality evaluation. Rock Soil Mech., **31**(002), 535–540 (2010)
20. Zhao, Y., Zhu, F., Gui, F., et al.: Improved K-means clustering algorithm based on extension distance. J. Intell. Syst., **15**(2), 8 (2020)

Chapter 2
Low Carbon Scheduling of Thermal Power Unit Thermal Storage Capacity Based on Particle Swarm Optimization

Wenbin Cao, Mingkai Wang, Bo Han, Qi Chai, Hongtao Li, and Ying Liu

Abstract A novel low carbon scheduling method for thermal power unit storage capacity is introduced to address the limitations of traditional approaches. By employing a particle swarm optimization algorithm, this method optimizes the scheduling of thermal storage capacity while considering the operational characteristics of the thermal power units. Through an analysis of the unit's pressure parameters, capacity level, and fuel type, different carbon emission allowances are allocated to each unit based on a baseline. The operation mechanism of the heat storage device is then optimized. The proposed low carbon scheduling model is effectively solved using the particle swarm optimization algorithm. Experimental analysis demonstrates the favorable scheduling performance of this method for thermal power units, highlighting its potential for reducing carbon emissions and achieving efficient operations.

2.1 Introduction

As thermal power units will still occupy the main position in China's power supply structure for a long time, some scholars have studied how to transform existing thermal power units into carbon capture systems and convert them into carbon capture units to reduce carbon emissions and alleviate social and environmental pressures [1, 2]. Compared with traditional thermal power generation units, carbon capture units have more convenient and flexible regulation capabilities. By quickly adjusting the net power generation, the units can track the random fluctuations of wind power.

W. Cao · Q. Chai · H. Li · Y. Liu
State Grid Chaoyang Electric Power Supply Company, Chaoyang, China

M. Wang (✉)
State Grid Liaoning Electric Power Supply Co. LTD, Shenyang, China
e-mail: wangmingkai_ln@outlook.com

B. Han
Shenyang Institute of Computing Technology, Chinese Academy of Sciences, Shenyang, China

R. Kountchev (Deceased) et al. (eds.), *AI Methods and Applications in 3D Technologies*, Smart Innovation, Systems and Technologies 388,
https://doi.org/10.1007/978-981-97-2144-3_2

Regarding carbon capture technology, existing literature has discussed the ability of carbon capture power plants to promote wind power consumption. However, the above research is limited to solar heat or carbon capture with fluctuating power output, and few people have explored the optimal scheduling of the three joint operations. With the further deepening of power reform, it is a trend to consider the carbon emission reduction of emerging renewable energy power plants and traditional thermal power units. Reference [3] studied the impact of carbon emissions trading and green technology investment under the carbon ceiling on manufacturing enterprises' low-carbon manufacturing decisions from the perspective of micro low-carbon economy. Reference [4] introduced an adaptive image swarm optimization algorithm called SA-EHO, which aimed to achieve the desired scheduling model of a microgrid (MG) incorporating electric vehicles and renewable energy sources (RES). However, the aforementioned methods primarily focused on scheduling and did not yield satisfactory low-carbon effects. Thus, to ensure the safe operation of the power grid, there is an urgent need to investigate a low-carbon scheduling approach for the thermal power unit's thermal storage capacity, considering carbon emission rights comprehensively. This study examines the operational characteristics of thermal power units in the context of carbon emission rights and establishes the operating mechanism of thermal storage devices based on this analysis framework. Leveraging the particle swarm optimization algorithm, we construct the objective function and scheduling model for low-carbon scheduling of the thermal power unit's thermal storage capacity, thereby achieving effective low-carbon scheduling in this domain.

2.2 Analysis of the Operating Characteristics of Thermal Power Units Considering Carbon Emission Rights

This paper utilizes the industry benchmark approach to determine the initial allowances, employing the carbon emission trading regulations currently being tested in China. The allocation of carbon emission allowances to various units is based on 11 baseline categories, which are determined by pressure parameters, capacity levels, and fuel types of the primary generating units. The calculation of these allowances is conducted using the equation outlined.

$$E_{i,t} = \delta_i P_t \quad (2.1)$$

Where, $E_{i,t}$ represents the carbon trading allowance of thermal unit i in time period t, δ_i represents the initial carbon emission allowance per unit of electricity of unit i, and P_t represents the output of thermal unit. The carbon dioxide emission of thermal power unit is related to the unit output, and the relevant expression is shown in (2.2).

$$E_{i,t}^C = \delta_i^C P_t \quad (2.2)$$

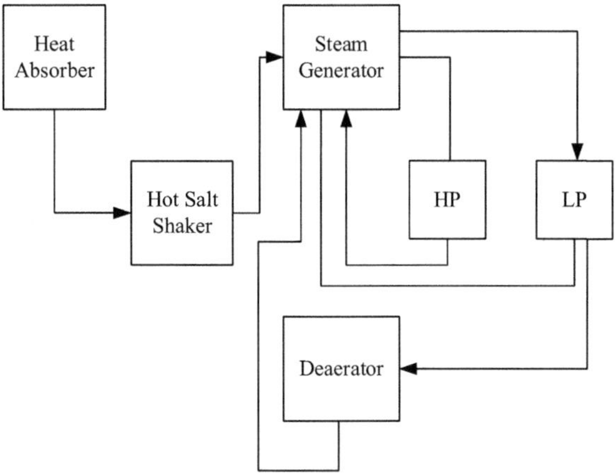

Fig. 2.1 Structure diagram of thermal power unit generating station

Where: $E_{i,t}^{C}$ is the actual system carbon emission of thermal power unit i at time t, and δ_i^C is the actual carbon emission intensity of unit i. Thus, the carbon trading cost of unit i can be calculated by (2.3).

$$C_{i,t}^{C} = C^{CO_2}(E_{i,t}^{C} - E_{i,t}) \tag{2.3}$$

Where: $C_{i,t}^{C}$ is the carbon trading cost of thermal unit i at time t, and C^{CO_2} is the carbon trading price factor.

Taking the most widely used thermal power plant with thermal storage as an example, we introduce the basic structure of thermal power generation [5–7], as shown in the (Fig. 2.1).

The working principle of thermal power plant is as follows: the fixed sun mirror in the concentrating collector system can track the solar radiation in real time and converge the light energy to the collector at the top of the tower. The collector temperature rises and thus heats the heat-conducting fluid, which, in the process of flowing, transfers heat to the power generation unit to produce steam to drive the turbine to generate electricity; at the same time, the redundant heat flows into the heat storage system through the heat-conducting fluid, and the heat between the hot and cold salt tanks is exchanged to realize the change of heat storage and discharge. This satisfies the time-sharing utilization of heat and plays the role of output leveling, which makes the solar thermal power plant excellent scheduling capability. The energy flow process of the thermal power plant is shown in the (Fig. 2.2).

The above diagram clearly shows the energy conversion during the operation of thermal power generation, which shows that thermal power generation realizes the step-by-step conversion of light-thermal energy to electrical energy, and there is storage and release of redundant heat. This unique operating characteristic makes thermal power generation different from other new energy generation [8], which can realize smooth and controllable power output and is conducive to grid dispatching and regulation, which is of positive significance to improving grid peaking pressure.

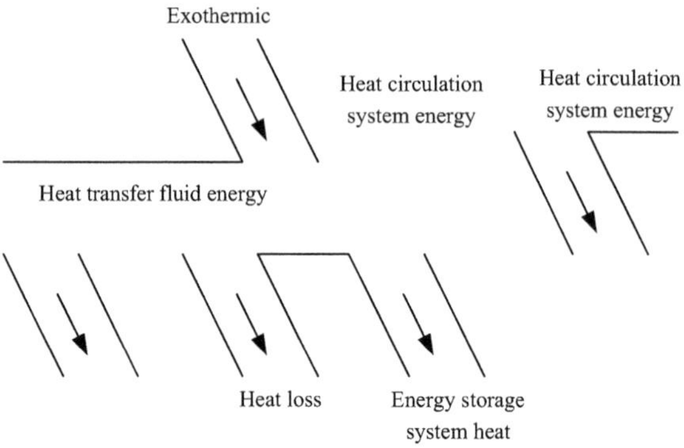

Fig. 2.2 Energy flow process of thermal power unit

The expression of light-to-heat power conversion in the concentrated heat collection system is as follows:

$$P_t^{SF} = \eta_{sh} S_{SF} D_t \tag{2.4}$$

Where, P_t^{SF} represents the thermal power obtained by the heat absorber in time t, η_{sh} represents the solar thermal conversion efficiency, S_{SF} represents the area occupied by the generator set, and D_t represents the direct irradiation index in time t. The thermal energy collected by the heat absorber can either be stored by the thermal conductivity fluid to the thermal storage device or directly supplied to the power generation system for power generation at peak load, while the absorbed thermal energy in the heat absorber can be discarded to ensure the smooth operation of the solar thermal power plant.

$$P_t^{SF} = (P_t^{esp,h} + P_t^{TS,c} - \eta_{dis})\Delta t \tag{2.5}$$

Where, $P_t^{esp,h}$ represents the thermal power of the heat absorber directly used for power generation in section t, $P_t^{TS,c}$ represents the thermal power of the thermal storage device in time t [9, 10], and η_{dis} represents the exothermic efficiency of the thermal storage system. Limited to the material characteristics of the energy storage device itself, heat dissipation cannot be avoided, and the calculation formula is as follows.

$$C_t^{TS,h} = (1 - \eta_{loss})C_{t-1}^{TS,h} \Delta t \eta_{cha} \tag{2.6}$$

Where, $C_t^{TS,h}$ represents the heat storage state of the heat storage device at time t, η_{loss} represents the heat dissipation coefficient, η_{cha} represents the heat storage efficiency of the heat storage system, and Δt represents the time interval.

The equivalent output power of thermal power plant is jointly determined by the power generated by the thermal energy of both the heat collection device and the heat storage device, and the calculation formula is shown.

$$P_t^e = \eta_{h-e} + P_t^{esp,h} \tag{2.7}$$

Where P_t^e represents the equivalent power output of thermal units at time t and η_{h-e} represents the thermoelectric conversion efficiency.

The above steps can complete the analysis of thermal unit operation characteristics and provide help for the subsequent low-carbon dispatch of thermal unit storage capacity.

2.3 Construction of Thermal Storage Device Operation Mechanism

In thermal power heating systems containing thermal storage devices, the thermal storage device is used as a storage and release device for thermal energy, and its thermal storage methods are mainly divided into three types, namely chemical reaction thermal storage, sensible thermal storage and latent thermal storage [11–13]. Compared with the other two heat storage methods, sensible heat storage adopts direct thermal conversion and high heat transfer efficiency, which is a common heat storage method in thermal power plants, and the following figure shows the structure of heat storage device.

The expression of heat storage capacity of heat storage device is:

$$S = cm\Delta T = \rho v\Delta T \tag{2.8}$$

Where, S is the heat storage capacity; c, m are the specific heat capacity and mass of medium, respectively; ρ is the density of medium; V is the volume of heat storage tank; ΔT is the temperature difference between supply and return water. The mathematical model expressions of heat storage and heat release of the storage device are:

$$S_t = S_t - 1 + h_c - h_d \tag{2.9}$$

$$0 \leq S_t \leq S_{\max} \tag{2.10}$$

Where, S_t is the heat storage capacity at time t; h_c and h_d are the heat storage power and heat release rate at time t, respectively; $S_{\max}$ is the maximum heat storage capacity. The constrained expressions for the corresponding heat storage and heat release rates of thermal power unit equipment are as follows.

$$0 \leq h_c \leq h_{c\max} \tag{2.11}$$

$$0 \leq h_d \leq h_{d\max} \tag{2.12}$$

Where $h_{c\max}$ and $h_{d\max}$ are the maximum heat storage power and maximum heat release power for thermal power units, respectively.

According to the above analysis, it can be seen that there is a linear coupling relationship between the power supply of thermal power units and the heating power. Since there is no decoupling measure, the regulation ability of thermal power units is poor, which leads to the limitation of the system wind power consumption space.

2.4 Construction of Low-Carbon Dispatch Objective Function for Thermal Storage Capacity of Thermal Power Units

In order to balance the economy and low carbon in the system generation process, and to consider the impact of thermal power prediction error on the system operation, the optimization objective of this time is to minimize the combined fuel cost, start-up and shutdown cost, carbon trading cost, and the penalty of load loss or wind abandonment brought by wind power uncertainty in the system operation, which is expressed as follows.

$$\min F = \sum_{i=1}^{T} (C^{coal} m_{i,t} + u^{coal} c_i^{fix} + u^{on} s_i^{on} + C_t^{curl}) \tag{2.13}$$

Where T is the number of periods in the dispatch cycle; N is the number of thermal units; C^{coal} is the unit coal price; $m_{i,t}$ is the coal quality consumed by thermal unit i at time t; u^{coal} is the start-up and shutdown status of thermal unit i at time t, $u^{coal} = 1$ indicates that the thermal unit is in operation [14, 15]; otherwise it is 0; c_i^{fix} is the fixed cost of operating thermal unit i; u^{on} indicates the start-up status of thermal unit i at time t, when the unit is changed from shutdown to start-up it is 1 and 0 otherwise; s_i^{on} is the start-up cost of unit i at one start-up; C_t^{curl} is the system load cost at time t.

In (2.13), the losses to system operation caused by thermal power forecast errors are mainly load shedding and wind abandonment penalty costs. The power shortage caused by the positive thermal power forecast error is compensated by the reserved up-regulated reserve, which will cause load shedding if the up-regulated reserve capacity is insufficient; the wind power over-generation caused by the negative wind power forecast error is absorbed by the down-regulated reserve, which will result in wind abandonment if the down-regulated reserve capacity is insufficient. If a normal distribution is used to approximate the probability density of the prediction error, the calculation of load shedding and wind abandonment penalties are shown in the following expressions, respectively.

$$C_t^{curl} = E_{EENST} C^{cural} + \alpha \tag{2.14}$$

Where E_{EENST} represents the value of lost load per unit, C^{cural} represents the expected value of power shortfall, and α represents the thermal power forecast error.

Through the above scheduling objective function, each renewable energy unit gives full play to its own characteristics during a typical day of schedule, so that it can complement each other's power output with its own advantages and disadvantages.

2.5 Construction and Solution of Low-Carbon Dispatch Model for Thermal Storage Capacity of Thermal Power Units

The particle swarm optimization (PSO) algorithm mainly changes the position of particles by the behavior of the population rather than the principle of survival of the fittest, so there is no evolutionary operator. Each potential solution tends to be associated with its movement speed, which is constantly resized and oriented according to the experience of surrounding particles, and is controlled by a learning factor that iteratively searches for the global optimum.

Assume a D-dimensional space with m particles forming a particle swarm, where the position of the i-*th* particle ($i = 1, 2, ...m$) in the D-dimensional space is denoted as $x_i = (x_{i1}, x_{i2}, x_{i3}, ...x_{iD})$. The velocity is denoted as $\mathrm{v}_i = (\mathrm{v}_{i1}, \mathrm{v}_{i2}, \mathrm{v}_{i3}, ...\mathrm{v}_{iD})$, the position of each particle is a potential solution, and the fitness associated with x_i is substituted into the optimization objective function to measure the superiority of the particles. The position of the individual particle i experiencing the historical optimum is recorded in each iteration as $Pbest = (P_{i1}^{best}, P_{i2}^{best}, P_{i3}^{best}, ...P_{iD}^{best})$, and all particles experiencing the optimum position as $gbest = (g_1^{best}, g_2^{best}, g_3^{best}, ..., g_D^{best})$. The velocity and position of particles in the d+1st generation are denoted as:

$$\mathrm{v}_i^{d+1} = \mathrm{w}\mathrm{v}_i^d + cr(Pbest - x_i^d) \tag{2.15}$$

$$x_i^{d+1} = x_i^d + \mathrm{v}_i^{d+1} \tag{2.16}$$

Where, c is the learning factor, which indicates the influence of the particle's historical optimal position on the particle's seeking direction, is a non-negative number, r is a pseudo-random number obeying the uniform human distribution within [0,1], and the two are independent of each other, x_i^d and v_i^d are the velocity and position of the particle after the d-*th* cycle, respectively, and w is the inertia weight, then the constraint expression for the inertia weight is as follows.

$$\mathrm{w} = \mathrm{w}_{\max} - \frac{\mathrm{w}_{\max} - \mathrm{w}_{\min}}{\mathrm{I}_{max}} \cdot I \tag{2.17}$$

Where, $\mathrm{w}_{\max}$ and $\mathrm{w}_{\min}$ are the maximum and minimum values of inertia weights, respectively; I is the current number of bad cycles, and I_{max} is the maximum number of iterations. The value of inertia weights gradually decreases as the number of iterations increases, i.e., the importance of individual particle positions increases, which is conducive to local optimization seeking and accelerates the convergence of the algorithm.

Firstly, the upper and lower limits of heat output are constrained. The upper and lower limits of the unit's heat output are compared with $h_{i,\max}$ and $h_{i,\min}$. If $h \leq h_{i,\min}$, then $h = h_{i,\min}$ is taken, and if $h \geq h_{i,\max}$, then $h = h_{i,\max}$ is taken, so that the unit's heat output will satisfy the upper and lower limits. Then the heating output creep is constrained. The particle updates are adjusted according to the heat output creep constraint starting from the second dispatching period. Finally, a constraint is applied to the heat output power balance. For not satisfying the heating power balance constraint, the feasibility adjustment is carried out, and the expressions for calculating the upward adjustment of unit heating output $\Delta h_{i,up}$, and the downward adjustment of $\Delta h_{i,do}$ under satisfying the heating balance condition are as follows.

$$\begin{cases} \Delta h_{i,up} = \min\{h_{i,\max}, h_i^{t-1} + h_i^t\} \\ \Delta h_{i,do} = h_i^t - \max\{h_{i,\max}, h_i^{t-1}\} \end{cases} \tag{2.18}$$

In addition, the power supply output constraint is also required. After adjusting the heat supply output to meet the constraint, the power supply output constraint needs to be calculated based on the determined heat supply output. The upper and lower limits of power supply output are calculated to adjust the power output that does not satisfy the constraints. The creeping situation is judged and adjusted one by one according to the unit's power supply output creeping constraints. Similarly, after the constraint adjustment according to (2.15), the electric boiler's electricity consumption converts all infeasible solutions into feasible solutions. The calculation of this improved particle swarm self-adjusting optimization algorithm has a substantially higher convergence speed than the basic particle swarm algorithm. The specific flow is shown in the (Fig. 2.3).

The aforementioned steps encompass the establishment and resolution of the thermal storage capacity dispatching model for thermal power units. By integrating the analysis of thermal power unit operation characteristics and the construction of an objective function, a low-carbon dispatching method for thermal storage capacity, taking into account carbon emission rights, has been successfully developed.

2.6 Experimental Verification

Experiment Preparation

The experiment uses MATLAB software to build a simulation experiment environment, as shown in Fig. 2.4.

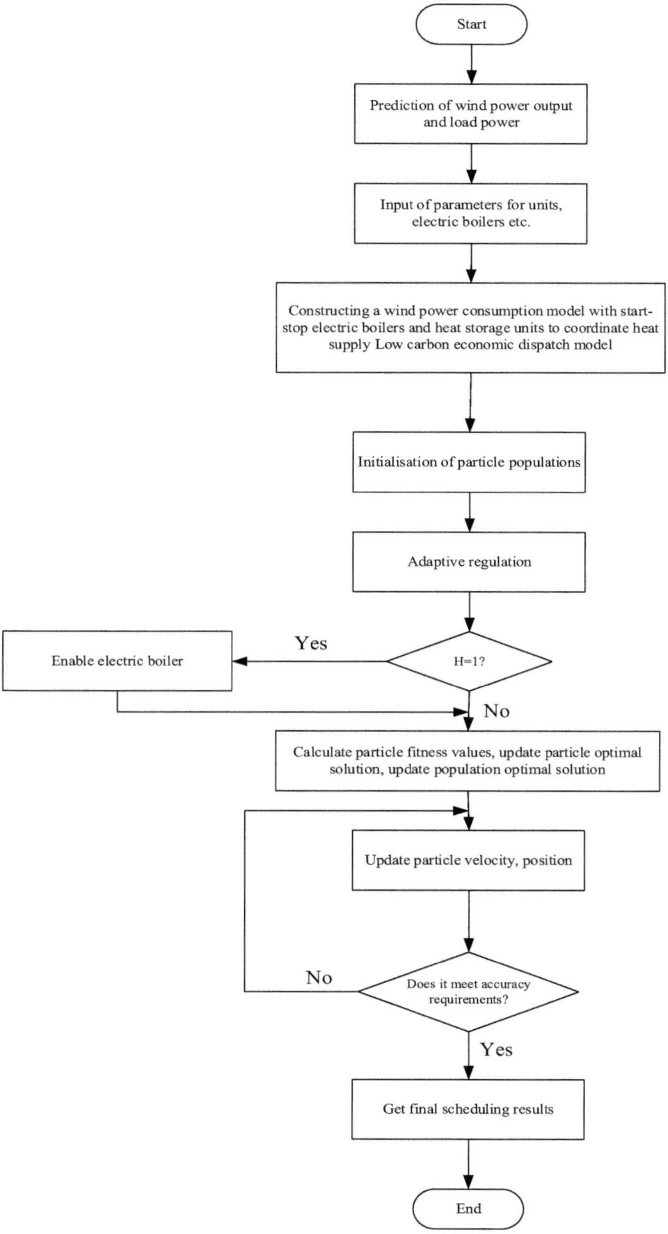

Fig. 2.3 Flow chart of the improved thermal storage capacity scheduling algorithm for thermal power units

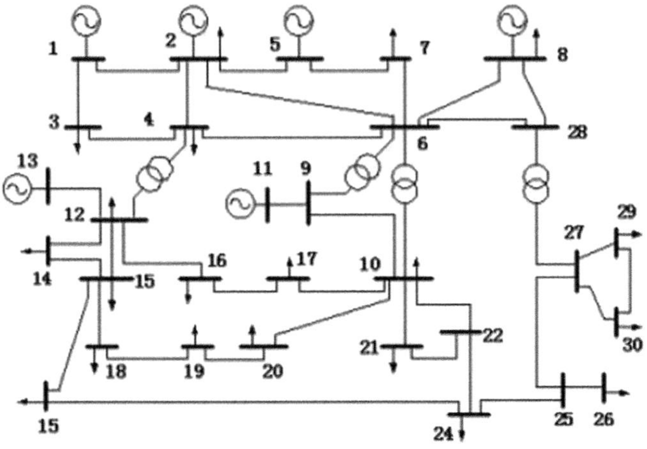

Fig. 2.4 Experimental environment

The simulation adopted a 30-node power system and two 6-node thermal systems, including four thermal power plants and two wind farms, and modified the power supply type and installed capacity of the corresponding nodes. Among them, each thermal power plant includes 2 thermoelectric units and a heat storage unit, the region is heated by thermal power plant 1, region 2 and is heated by thermal power plant 2, and a 50 MW electric boiler is configured at node 4 of region 2 for heat supply. The calculation example is a day-ahead dispatch; the dispatch period is 24 h, with 1 h as the dispatch period, and the electric system load and thermal power forecast power are selected from 9:00 of a certain day to 8:00 of the next day, as shown in Fig. 2.5.

Fig. 2.5 Test result

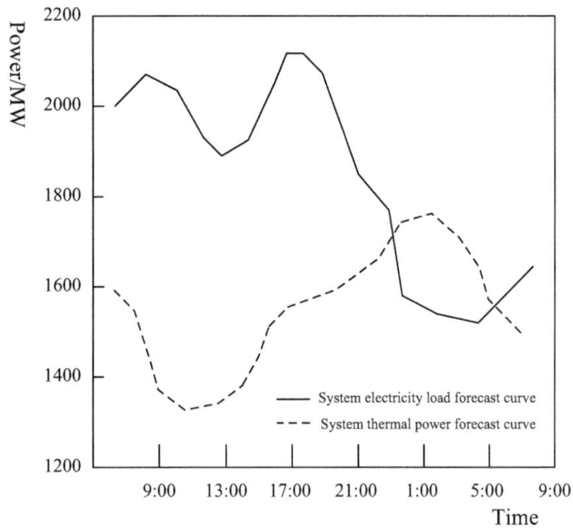

Three scenarios were set up for simulation in this experiment. In Scenario 1, where there are no carbon capture devices, heat storage units, or electric boilers; Scenario 2, which includes a carbon capture unit but lacks heat storage and electric boilers; and Scenario 3, which incorporates a carbon capture device and a heat storage unit but excludes an electric boiler.

In the above scenarios, Ant colony optimization algorithms scheduling method, deep learning scheduling method and particle swarm optimization scheduling algorithm proposed in this paper are respectively used to schedule in three scenarios of Scheme 1~Scheme 3, and the scheduling effect is compared. The ant colony algorithm scheduling method simulates the foraging behavior of ants, finding the shortest path from the origin through multiple specified demand points and returning to the origin. On the other hand, deep learning scheduling employs graph models and deep reinforcement learning to establish an end-to-end framework for solving graph combination optimization problems in various scheduling applications. The paper's objective is to design a general framework for approximate solutions, and it determines the optimal parameter combination through experimentation with the two aforementioned comparative methods.

Experimental Results

The experiments are conducted in the form of comparative analysis by using three dispatching methods to dispatch thermal power plant units under different scenarios and comparing their power, with less power representing a higher dispatching performance of the methods and more proper dispatching of thermal power plant units' heat storage capacity.

It can be seen from the analysis of Figs. 2.6, 2.7 and 2.8 that under the three scenarios, the particle swarm optimization algorithm proposed has low carbon emissions, which are lower than 1900 MW as a whole. Compared with the ant colony algorithm and deep learning, the low-carbon scheduling effect is better. The above experimental results show that the simulated thermal units have different degrees of power mitigation after being dispatched by different dispatching methods, and the numerical comparison shows that the units dispatched by the low-carbon dispatching method of thermal storage capacity of thermal units considering carbon emission rights proposed in this paper have significantly lower power, which indicates that the actual dispatching effect of the method in this paper is more effective. The comparison results of performance parameters of low-carbon scheduling schemes under three scenarios are shown in Table 2.1.

The above experimental results show that the carbon emission cost of the thermal unit storage capacity dispatching method proposed in this paper is significantly lower than that of the two groups of traditional thermal unit dispatching and scheduling methods after the dispatching of the simulated power system, indicating that the dispatching method proposed in this paper is better than the traditional dispatching method in terms of actual dispatching performance.

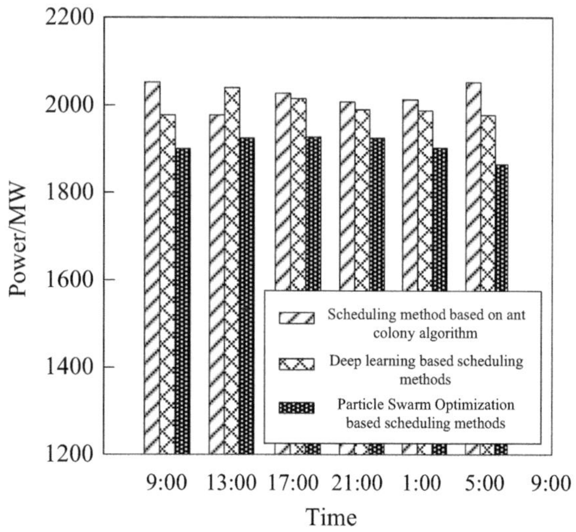

Fig. 2.6 Comparison of low-carbon dispatch of thermal unit storage capacity for thermal units under scenario 1

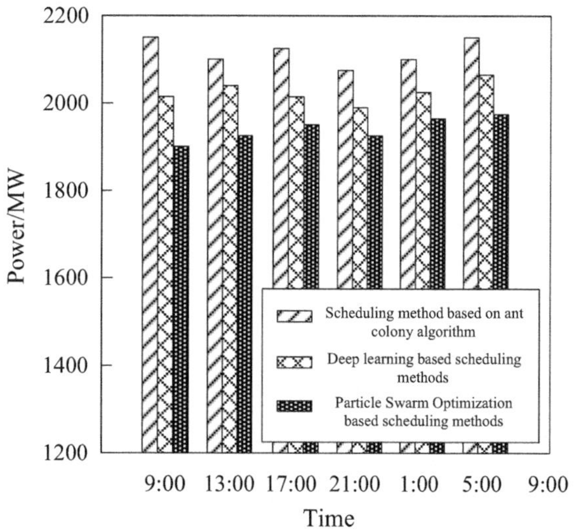

Fig. 2.7 Comparison of low-carbon dispatch of thermal storage capacity of thermal units under scenario 2

2.7 Conclusion

This paper conducts an analysis of the operational characteristics of thermal power units and proposes an allocation of carbon emission allowances based on factors such as baseline division by pressure parameters, capacity level, and fuel type of the generating units. The developed dispatching model efficiently manages the thermal storage capacity of these units, resulting in lower power output while still meeting the normal working demand with reduced consumption. To further enhance the

Fig. 2.8 Comparison of low-carbon dispatch of thermal unit storage capacity for thermal units under scenario 3

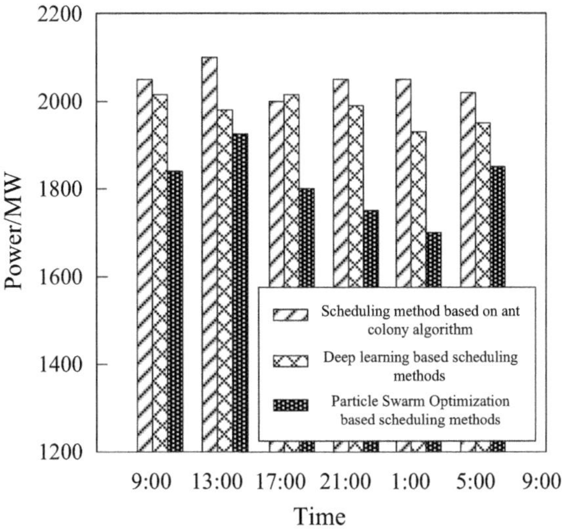

Table 2.1 Comparison of carbon emissions under different time periods

Time period	Particle swarm optimization	Ant colony	Deep learning
1	56.45	60.26	68.49
2	54.26	62.25	68.98
3	54.62	62.37	68.46
4	53.21	63.48	68.75
5	52.16	63.59	69.31
6	54.16	63.51	69.46
7	53.49	63.28	69.35
8	53.96	64.95	69.15
9	53.91	64.26	69.17
10	53.76	64.23	69.32

proposed dispatching scheme, future research should consider additional factors associated with investment, operation, and maintenance, aiming to establish a more comprehensive and rational approach.

References

1. Wang, Z., et al.: A low carbon footprint method for converting low-rank coals to oxygen-containing chemicals. Fuel **315**, 123277 (2022)

2. Bian, B., Du, Z., Zhou, K., Huang, T., Lv, F.: Optimal planning method of integrated energy system considering carbon cost from the perspective of the whole life cycle. IOP Conf. Ser.: Earth Environ. Sci. **897**(1), 012021 (2021)
3. Changsong, M., Bo, H., Tiantong, Y.: Low-carbon manufacturing decisions considering carbon emission trading and green technology input. Environ. Eng. Manag. J., **19** (9), (2020)
4. Somakumar, R., et al.: Optimization of emission cost and economic analysis for microgrid by considering a metaheuristic algorithm assisted dispatch model. Int. J. Numer. Model.: Electron. Netw., Devices Fields **35**(4), e2993 (2022)
5. Qu, Y., et al.: An improved Stöber method towards iron and nitrogen co-doped porous carbon spheres for oxygen reduction reaction in alkaline media. Int. J. Hydrog. Energy **47**(6), 3771–3780 (2022)
6. Cormos, C.C., Cormos, A.M., Petrescu, L., Dragan, S.: Techno-economic assessment of decarbonized biogas catalytic reforming for flexible hydrogen and power production. Appl. Therm. Eng. **207**, 118218 (2022)
7. Iwama, T., Inoue, K.Y., Shiku, H.: Fabrication of high-density vertical closed bipolar electrode arrays by carbon paste filling method for two-dimensional chemical imaging. Anal. Chem. **94**(25), 8857–8866 (2022)
8. Gezović, A., et al.: High Al-ion storage of vine shoots-derived activated carbon: New concept for affordable and sustainable supercapacitors. J. Power Sources **538**, 231561 (2022)
9. Guo, H., et al.: One-pot synthesis of a carbon dots@ zeolitic imidazolate framework-8 composite for enhanced Cu 2+ sensing. Anal. Methods **12**(32), 4058–4063 (2020)
10. Wang, M., Wang, Y., Zhang, C., Yu, F.: Cascade phase change based on hydrate salt/carbon hybrid aerogel for intelligent battery thermal management. J. Energy Storage **45**, 103771 (2022)
11. Yu, T., et al.: Thermal self-protection behavior of energy storage devices using a thermally responsive smart polymer electrolyte. Chem. Sel. **7**(7), e202104499 (2022)
12. Pandey, A., Jayaraman Sridharan, S., Naik, A., Ahmad, A., Sastry, B.S.: A new perspective of wet kata cooling power as a predictor for thermal stress in underground mines. Min., Metall. Explor. **38**(6), 2433–2441 (2021)
13. Ouyang, T., Wu, W., Wang, J., Xie, S.: Multi-energy flow cooperative dispatch for supply-demand balance of distributed power grid with liquid air energy storage system. J. Clean. Prod. **354**, 131710 (2022)
14. Serrano, A., Duran, M., Dauvergne, J.L., Doppiu, S., Del Barrio, E.P.: Tailored transition temperature plastic crystals with enhanced thermal energy storage capacity. Sol. Energy Mater. Sol. Cells **220**, 110848 (2021)
15. Wang, C., et al.: Coordinating thermal energy storage capacity planning and multi-channels energy dispatch in wind-concentrating solar power energy system. J. Clean. Prod. **350**, 131405 (2022)

Chapter 3
Research on "AI Level" in Art Appreciation

Qianqian Kong and Yong Qiao

Abstract In order to improve the intelligent effect of art appreciation, this paper studies the "AI" level of art appreciation, applies artificial intelligence to art appreciation, and proposes a digital art image generation algorithm based on handwritten character information modulation, which uses handwritten character as a benchmark generator to create art images that can adapt to the density and structure information of the original image. Moreover, a color gradient algorithm is designed for line extraction, which first uses the normalization calculation of RGB three-channel plane gradient to highlight edge information in smooth internal area, and uses the maximum variance threshold to segment the threshold to obtain smooth and clear line contour map. This paper realizes the integration of the structure of non-medium floating projection gesture interaction system, and intelligently processes art works. Combined with use analysis, it can be seen that the AI system of art appreciation proposed in this paper has good art appreciation effect and has a good promotion to improve people's art works analysis level.

3.1 Introduction

AI is the abbreviation of Artificial Intelligence, which is called Artificial Intelligence in Chinese. It usually endows the developed wisdom of human beings with machines in the form of creation, manufacturing and processing, and replaces the functions of human speculation, memory, activity and decision-making with machines. Adaptability is to create value for human beings to adapt to the surrounding environment according to the transformation of time and space and the evolution of living environment.

Q. Kong (✉)
College of Fine Arts, Xinjiang Normal University, Urumqi, Xinjiang 830054, China
e-mail: 107622018070019@xjnu.edu.cn

Y. Qiao
Beijing DiDi Infinity Technology and Development Company Ltd, Beijin 100085, China

R. Kountchev (Deceased) et al. (eds.), *AI Methods and Applications in 3D Technologies*, Smart Innovation, Systems and Technologies 388,
https://doi.org/10.1007/978-981-97-2144-3_3

AI focuses on designing products such as smartphones, smart anti-theft doors, smart security alarms, smart switches, and smart furniture in human living environments, which will be adapted according to human habits [1].

Adaptive art appreciation art environment design is the most prominent feature of the current environment, and also the starting point of all the signs of other art appreciation. The feature details or colors or styles of pictures are automatically selected by the algorithm, and all the captured feature details are exaggerated in an amazing and interesting way [2]. Applying artificial intelligence algorithms to the research project of artificial intelligence neural networks, in order to find new solutions in the fields of image recognition classification and speech recognition, public art images are some experimental results in the research process, not intentional professional creative behavior [3].

Using AI as a search keyword on the network, you can find a large number of code to try out the works uploaded by users. From the perspective of artistic creation, AI works are new distribution of styles according to digital algorithms, which seems to be close to the works of an artist who insists on creative style. The point painting series of the Impressionism painter Shura is also similar to the unified style, But the creations of human artists do not have algorithmic average random precision filters, but rather integrate memory and overall body perception in each new creation to achieve a unified style of work [4]. The creation process and result of AI are both collage creation after digital modeling, which is different from the selection of collage materials from early Pop art. Dadaism and Pop art have developed the trend of ready-made art, and AI has made the characteristics of style, form and color become ready-made products [5]

The informationization development of art education activities must be centered around the concept of student-centered development, comprehensively improving the cognitive level and ability of connoisseurs around their growth goals, and cultivating their innovation and creative awareness [6].

Clarify the development direction of art education classroom around the goal, ensure the comprehensiveness and effectiveness of art appreciation activities, truly achieve "people-oriented", highlight the needs of the connoisseur at all times, design everything around the needs, and take the connoisseur's needs as the driving force. Information technology has unlimited creativity, as the diversity of online participants determines the diversity of artistic creativity. Creativity itself represents creativity, and all new things come out of nowhere. They require deep thinking, and then subdivided into information that can be received by the viewer. Only in this way can the correct direction be formed [7].

By creating a data-driven environment to subtly influence and enable connoisseurs to improve and think in specific environments. Thought is the precursor of behavior, and only by optimizing the evaluator's thinking mode can they choose the correct path of growth. They always believe that the growth of thinking is the true growth, and the only thing that can affect the evaluator's life is "habit". Such as thinking habits, social habits, behavioral habits, etc. [8]. Always put effort into guiding thinking and shaping the behavior of the evaluator. The integration of information technology and art appreciation content can have unexpected effects and provide positive guidance to the thinking of the connoisseur.

When participating in art appreciation in art classrooms, connoisseurs should fully expand their cognitive depth, allow them to make unlimited associations, and optimize their methods of understanding the world. Imagination is the cornerstone of human progress, and all creation is based on imagination. Human progress always has infinite assumptions and reasoning, strong relational thinking, and can enable connoisseurs to navigate life with ease and better handle interpersonal relationships [9].

Innovation is one of the core elements of art appreciation activities. Art appreciation activities must meet the growth needs of connoisseurs for innovation, creation, and timely increase communication and interaction time, choose viewpoints consistent with the content of art appreciation for interaction, and effectively improve the efficiency of art appreciation [10].

The informatization of art appreciation is the future development trend, and how to integrate information thinking and value with art appreciation is a topic that must be deeply considered in the field of art appreciation. Choose an open art appreciation mode, timely share knowledge with connoisseurs, and equip them with innovative abilities and awareness. The art appreciation method must be matched with the content, and different art appreciation methods should be chosen for different content [11].

There is no right or wrong way to appreciate art itself, but there are differences in meeting the definition of "innovation". Communication is not a process of asking questions, but rather a communication between both parties under equal conditions. Anyone can become a participant and propose important viewpoints [12]. The main way to appreciate thematic art is through interaction, where the connoisseur obtains comprehensive improvement through communication, participation, interaction, and experience. Therefore, the effectiveness of interaction directly affects the quality of the connoisseur's thematic art appreciation.

In the creative art theme art appreciation activities, the teacher must take himself as the background and give proper voice guidance. The most effective way of creative art is to "moisten things silently". By adopting diversified means of art appreciation, it has a profound impact on the behavior and thought of the appreciators, so that they can sublimate and grow unconsciously, and can better complete various art appreciation activities [13].

This paper studies the "AI" level in art appreciation, and applies artificial intelligence to art appreciation to improve the further application of information technology in art appreciation.

3.2 Technology in Art Appreciation

The interest in art is a necessary factor for learning art knowledge and conducting artistic creation. If a student does not have an interest in painting and design, they will not actively pay attention to this subject, let alone learn the basic knowledge and innovative practice of this subject.

When intelligent technology intervenes in interactive art appreciation, it brings about changes for creators and users. As a creator, I not only formulate rules for intelligent interaction, but also engage in design and production. At present, most interactive works are selected by the participants themselves. Interactive works are changes in content and form, or changes in other aspects. However, with the continuous development and progress of intelligent technology, the behavior, posture, voice, expression, temperature Remote control and response to climate change. Intelligent interactive art works themselves will also have dominance, arbitrariness, and selectivity. People often raise the importance of interactivity and entertainment in works, but overlook their technical intelligence. The enhancement of technological progress is the key to the interactivity of works. Or in this way of understanding, contemporary interactive art works mainly require the common progress and interweaving of concepts and technology, and none of them are indispensable. Artists and designers can create in the form of intelligent technology, and new ideas and concepts in their creations can also be explored through technology. Before the application of AI technology, the creators of interactive art appreciation have experienced three important stages: the design era of electronic images, the design era of the Internet, and the era of new media intelligent Interaction design. In terms of design and creation, the way of expression of creation has changed from a single hand-painted way to a variety of forms of Computer-aided design. Nowadays, design creation focuses more on comprehensive forms such as digitization, precision, intelligence, and virtualization. The selection and application of technology and materials not only affect the style of design, but also determine the success or failure of works to some extent. At present, the research focus of interactive art appreciation is gradually shifting towards the fields of new media interaction, intelligent interaction, and virtual reality, mainly focusing on human-centered interaction research. The high content of intelligent interaction in current technology is interfaceless interaction (human body and speech drive machines), immersive interaction, brainwave interaction, and multimedia performance interaction. Although these intelligent interactive forms provide artists, designers, and participants with greater freedom and space, they still have rules and limitations for interactive art creators, such as the transformation of works, feedback, autonomy, material and space limitations, and the degree of audience participation and interaction. The emergence of non biological intelligence means that the trend of human design will shift towards technological intelligence, and it will also change the innovation of future art appreciation structure, content, form, and its development trend. For designers and artists, if they can effectively utilize technological intelligence to extract new strategies and reconstruct their consciousness, guiding the direction and form of art appreciation, it will bring them an important transformation and create better works. With the continuous impact of intelligent technology, art appreciation has also changed, and such changes will also reconstruct the perspective of creators.

In this study, the steps of line extraction and segmentation of art images based on color gradient are as follows. Step 1: A color original art image with R, G and B channels is input. Step 2: A continuous and clear line outline is obtained by color gradient algorithm. Step 3: The algorithm takes the extracted contour art image as the mask art image, and subtracts the gray art image of the input art image to segment the art image and obtain the background art image. Among them, the specific process of color gradient algorithm is as follows. Firstly, the algorithm inputs the color art image img, and outputs the contour art image counter Img. After that, the algorithm extracts three channels of img, img R, img G and img B, and creates a nonlinear filter f. Finally, the algorithm filters the x and y directions of img R, img G and img B with f, and obtains img RX, img RY, img GX, img GY, img BX and img BY. According to Eqs. (3.1)–(3.4), the angle A of radian is calculated [14]:

$$xy = imgRX \times imgRY + imgGX \times imgGY + imgBX \times imgBY \tag{3.1}$$

$$xx = img\ RX \times img\ RX + img\ GX \times img\ GX + img\ BX \times img\ BX \tag{3.2}$$

$$yy = img\ RY \times img\ RY + img\ GY \times img\ GY + img\ BY \times img\ BY \tag{3.3}$$

$$A = 0.5 \times \left(\arctan\left(\frac{2 \times xy}{xx - yy}\right)\right) \tag{3.4}$$

The normalized matrix Q is obtained according to expressions (3.5)-(3.7):

$$Q1 = 0.5 \times ((xx + yy) + (xx - yy) \times cos(2 \times A) + 2 \times xy \times sin(2 \times A)) \tag{3.5}$$

$$Q2 = 0.5 \times \begin{pmatrix} (xx + yy) + (xx - yy) \times cos(2 \times (A + \pi/2)) \\ +2 \times xy \times sin(2 \times (A + \pi/2)) \end{pmatrix} \tag{3.6}$$

$$Q = max(Q1,\ Q2) \tag{3.7}$$

According to the edge information, the contour art image is obtained by thresholding the maximum variance.

The weighted centroid Voronoi algorithm firstly identifies a set of generator points, and divides the whole plane by the perpendicular bisector between each point and its adjacent points to obtain sub-regions (Voronoi elements). It then moves the position of the set of generator points to the position of the center of mass of the Voronoi cell. The above two steps are repeated until the end of iteration, and the weighted Voronoi diagram of centroid is obtained. The calculation process is as shown in Eq. (3.8).

$$V_{\mathrm{i}} = \left\{ \begin{array}{l} p \in P|L(\mathrm{p,p_i}) < L(\mathrm{p,p_j}), \\ i, j = \{0, 1, ..., n\}, i \neq j \end{array} \right\} \tag{3.8}$$

As an artistic technique that can present rich tones with only one color, the key of stippling is that the distribution and density change of dots can highlight important information, filter unimportant information and show subtle differences in tones. Therefore, when performing the iterative step of centroid movement, the distribution of points should not only be based on the area division of the plane, but also reflect the gray effect of the art image. Furthermore, the centroid shift of points introduces the idea of density function into Voronoi diagram, and weights the calculation of centroid. If the density function $\rho(\theta) \geq 0$ for each position θ in the region Ω is known, the weighted centroid of the Voronoi element is defined as Eq. (3.9):

$$C_{\mathrm{i}} = \frac{\int_{\Omega} \theta\rho(\theta)d\Omega}{\int_{\Omega} \rho(\theta)d\Omega} \tag{3.9}$$

As far as the distribution of handwritten characters is concerned, high-quality stippling is as close to the original tone as possible in aesthetic distribution, and at the same time, it can ensure the uniform distribution and perfect interval of points, which is an important guarantee for the quality of digital art images generated based on handwritten characters in this study. Therefore, the weighted centroid Voronoi algorithm is introduced to generate stippling art images, and the point set data is obtained as the constraint of handwritten character distribution.

The generation steps of stippling art image are as follows. Step 1: The algorithm rasters the background art image. Step 2: The algorithm takes the average gray value of the area corresponding to the original art image as the reference parameter, and completes the placement of the number of generated points in each grid. Step 3: The algorithm discards the bondage of the grid, and each point takes itself as a site to change the position of the point again, so that the point distribution of the background characteristic art image satisfies Voronoi distribution, the distance between points satisfies Voronoi unit property, and the continuous tone art image is quickly converted into stippling art image. Step 4: The algorithm calculates the centroid of Voronoi unit by fitting density function ρ (x, y) with gray scale adjustment as weight coefficient, repeats the iteration for 50 times, puts the position of points in the centroid, divides the operation of Voronoi unit, and constructs the weighted centroid Voronoi diagram. Because the Voronoi element is a convex polygon, the centroid coordinates are quickly calculated by using the optimized triangulation algorithm, and the calculation formulas are shown in Eqs. (3.10) and (3.11) [15].

$$x = \frac{\sum_{k=1}^{n-2} S_{\mathrm{k}} \times (x_{k1} + x_{k2} + x_{k3})}{3\sum_{k=1}^{n-2} S_{\mathrm{k}}} \tag{3.10}$$

$$y = \frac{\sum_{k=1}^{n-2} S_k \times (y_{k1} + y_{k2} + y_{k3})}{3\sum_{k=1}^{n-2} S_k} \tag{3.11}$$

Among them, S_k is the area of each triangle cut by an n-sided polygon, and (x_{k1}, y_{k1}), (x_{k2}, y_{k2}) and (x_{k3}, y_{k3}) are the coordinates of the three vertices of the corresponding triangle.

3.3 AI Model of Art Appreciation

Although technology is constantly advancing, its ultimate target audience is still people. Therefore, a greater emphasis on humanized design will also be a mainstream trend in interactive art design. The intervention of intelligent technology is to create better lifestyles for people, and also brings about a shift in user perspectives. How to make design works or products more humanized needs to be analyzed from the user's perspective. By analyzing users' feelings when using products and facing artwork, the level of interaction itself can be improved.

Interactive art design belongs to a constantly changing field, and on the one hand, the change is in response to the continuous innovation of technology, and on the other hand, it is to meet the needs of users and bring good interactive experience effects to them. The intervention of intelligent technology has turned human–computer interaction from graphical user interface to Natural user interface, and creating more humanized interactive works is the key point of the next development.

Interactive art design spans multiple fields such as physics, biology, and digital, and is a concentrated display of art. At that time, it will no longer be simply a cold tool in life, ranging from nano robots to Earth as interactive art. The development of human–computer art not only discusses the inherent attributes of artificial intelligence as a practical machine, but also requires independent insights and theoretical thinking through unrestricted and self expressive thinking. With the continuous development of artificial intelligence, I believe there will be two main trends in the future. One trend is that artificial intelligence may become our partner and friend in the future; Another trend is that artificial intelligence has become our enemy. What I hope to see more is that in the future, we will gradually expand from the "Unity of Heaven and humanity" relationship between man and nature to the "human–machine in one" relationship, achieve the subconscious integration of "you have me, I have you", and then realize the parallel or interactive development of the two relationships to achieve the ultimate spiritual concept of human–machine integration and human–machine coexistence, which can also be extended to the future interactive art design.

This article is based on the above ideas for the intelligent design of an art appreciation system.

The running calculation of the system needs to choose the appropriate server. This paper chooses the matching server skeleton according to the node scale to ensure that the whole 3D animation process has higher efficiency. The file server sometimes gets stuck because the node is too big, so the file server in this paper uses RAD (Rapid Application Development) technology to improve memory performance and fault tolerance, and adopts multi-port server network card to ensure that it can provide a relatively stable and reliable network channel for animation nodes. This paper also needs to combine VR (Virtual Reality) equipment with three-dimensional animation production, and design a three-dimensional animation virtual system, so as to combine VR equipment, three-dimensional animation production software, technology and network technology, and provide a good interactive environment for the production and development of three-dimensional animation. The hardware structure of the system is shown in Fig. 3.1.

Placing multiple pixel light sources behind a lens width can form viewpoint pictures with different angles in multiple positions in space. Viewers can feel depth information as long as they see two of these viewpoints at the same time in different viewing positions. This technology is called multi-viewpoint technology. The multi-view schematic diagram is shown in Fig. 3.2.

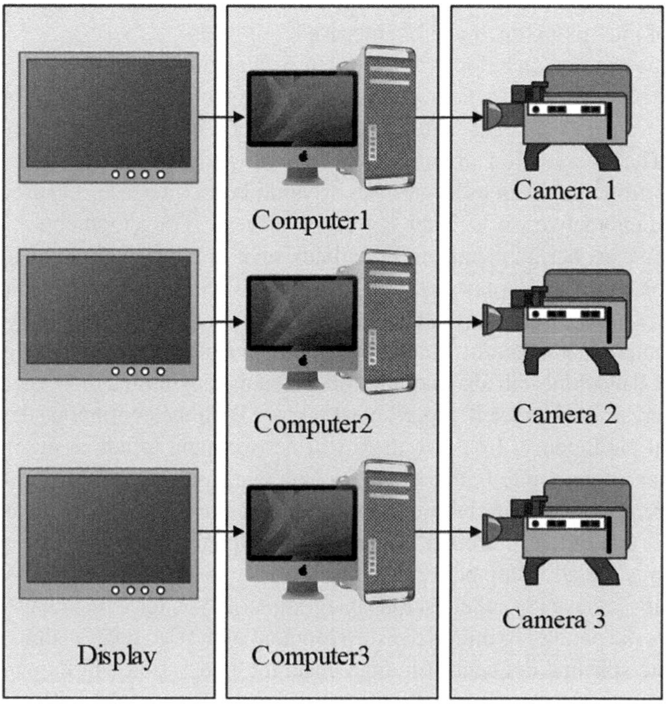

Fig. 3.1 Hardware structure of 3D art appreciation system based on virtual reality technology

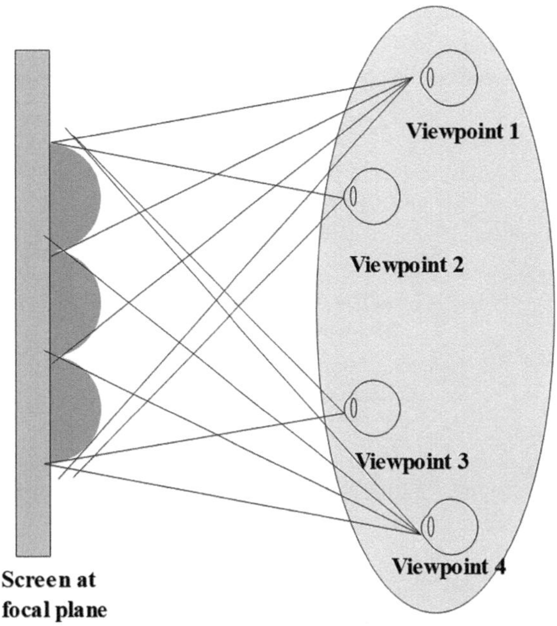

Fig. 3.2 Lens grating imaging

To alleviate the problem of resolution imbalance, the lenticular lens can be placed obliquely, and the reduced resolution is divided into horizontal and vertical directions, so that the resolution received by human eyes is not too low. Using somatosensory interaction technology and naked-eye 3D display technology, natural and immersive interactive input and output are constructed, which improves the interactive experience. The system design framework is shown in Fig. 3.3.

The real mirror image is formed by using micro-lens array in the air, and these micro-lenses are composed of dihedral corner reflectors. As shown in Fig. 3.4, the dihedral corner reflector is composed of two vertical micro-mirrors. In the figure, a dihedral corner reflector facing X axis is placed at the origin of coordinates, and the micro-lens is parallel to Z axis. Point A is the point light source coordinate (X, Y,-Z). The light diverging from the point light source is reflected twice by the dihedral corner reflector and converges to point B (X, Y, Z) to form a real image point. Point A and point B are symmetrical about XOY plane.

As shown in the imaging schematic diagram of Fig. 3.5, the optical micro-lens structure is a new nano-scale optical material-microchannel matrix optical waveguide plate, which is a piece of glass when viewed from the outside. In fact, the inside is a group of micro-lens arrays, and a single array is composed of dihedral corner reflectors. The incident light is reflected on the side of the reflector, and only a virtual image can be formed by one reflection. The light emitted from the light source is reflected twice on the two orthogonal sides of each reflector to form a stable and completely positioned aerial real image.

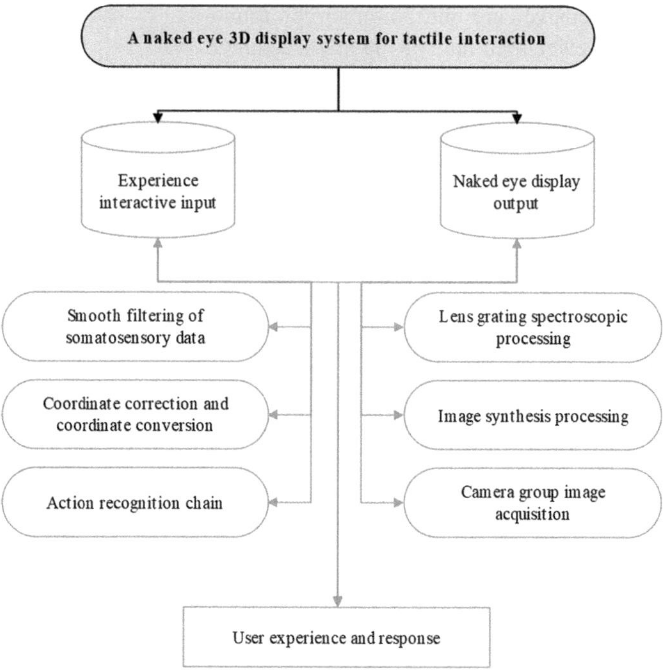

Fig. 3.3 System framework

Fig. 3.4 Schematic diagram of dihedral corner reflector

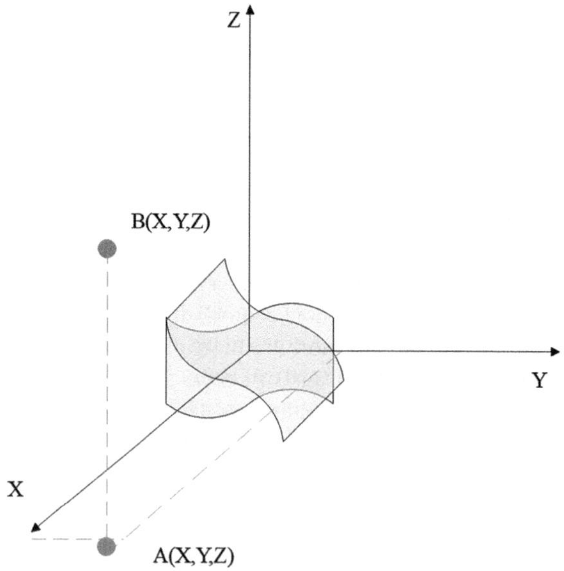

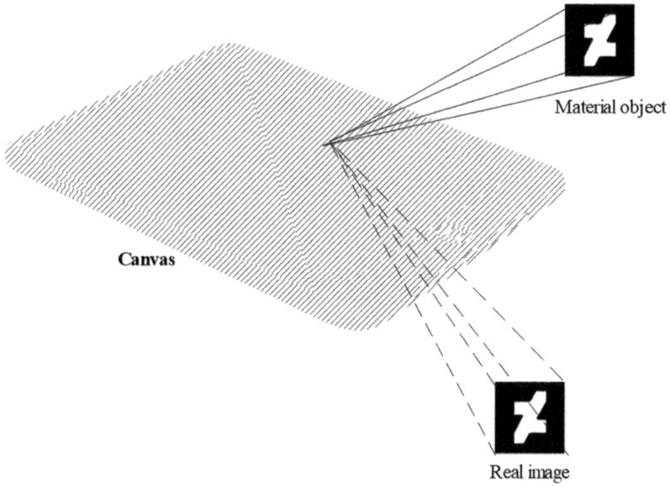

Fig. 3.5 Imaging schematic diagram

In this paper, the structure integration of the dielectric-free floating projection gesture interaction system is realized. As shown in Fig. 3.6, the main components of the dielectric-free floating projection interaction system include the main board, the optical micro-mirror structure board, the infrared array touch sensor and the highlight display. The display screen is placed at a certain inclined angle in the vertical direction, the optical micro-mirror structure plate is placed horizontally, and the infrared array touch sensor is placed parallel to the floating projection area to ensure that the touch area and the imaging area form a mapping relationship.

USB (Universal Serial Bus) system uses the master controller to control the data transmission between the host and USB device. As shown in Fig. 3.7, the simple connection between the slave device and the host requires multiple levels of the host to interact with the device. The function layer provides application functions to the host through proper matching with the client software layer. USB device layer is a USB system software presentation layer that performs general USB operations on devices. Moreover, USB bus interface layer provides physical, signal and data packet connection between host and device. However, in fact, all levels of interaction are completed by USB bus interface layer and host data transmission. The interface between the USB system and the host controller depends on the hardware definition of the host controller.

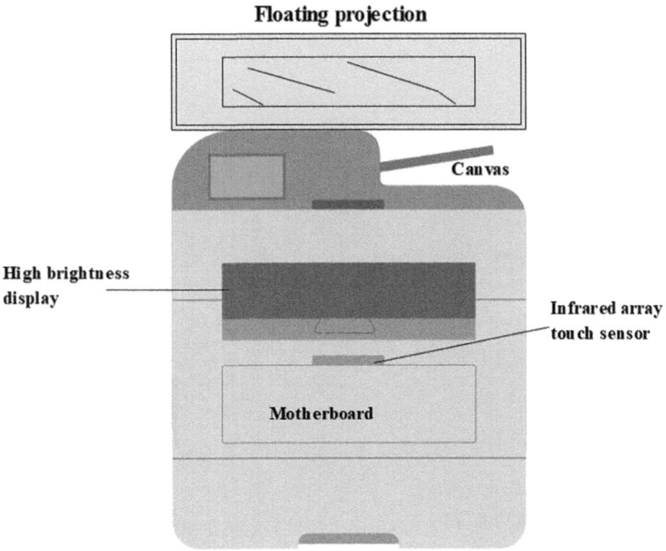

Fig. 3.6 Structure diagram of floating projection interactive system without medium

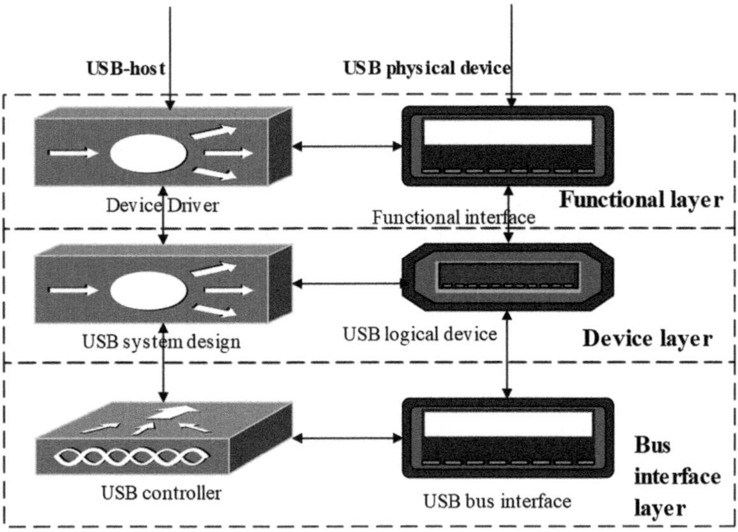

Fig. 3.7 Data flow diagram of USB host and USB device

3.4 Conclusion

In the context of rapid development of AI space, artificial intelligence (AI) affects the development tendency of the whole social environment, while the development of adaptability fixes the development state of the whole social environment, and a harmonious and stable living environment cannot be separated from the maintenance of adaptability close to life. The arrival of AI era injects new vitality into people's lives. The emergence of computer informationization, virtual reality and VR technology can predict the development direction of things in advance and avoid unnecessary detours. This paper studies the "AI" level of art appreciation, and applies artificial intelligence to art appreciation to improve the further application of information technology in art appreciation. Through experimental research, we can see that the art appreciation AI system proposed in this paper has a good appreciation effect of art works, and has a good promotion of art works to enhance people's art works analysis level.

References

1. Liu, H., Zhang, X.T., Fu, X.M., Dong, Z.C., Liu, L.: Computational peeling art design. ACM Trans. Graph. (TOG) **38**(4), 1–12 (2019)
2. Yi, G.: Design research on the network multimedia courseware for art-design teaching. Eurasia J. Math., Sci. Technol. Educ. **13**(12), 7885–7892 (2017)
3. Wan, Q., Song, S.S., Li, X.H., Zhang, Q., Yang, X., Zhang, Y.C., Design, A.: The visual perception of the cardboard product using eye-tracking technology. Wood Res., **63**(1), 165–178
4. McCartney, N., Tynan, J.: Fashioning contemporary art: a new interdisciplinary aesthetics in art-design collaborations. J. Vis. Art Pract. **20**(1–2), 143–162 (2021)
5. Lockheart, J.: The importance of writing as a material practice for art and design students: A contemporary rereading of the Coldstream Reports. Art Des. Commun. High. Educ. **17**(2), 151–175 (2018)
6. Sachdev, G.: Engaging with plants in an urban environment through street art and design. Plants, People, Planet **1**(3), 271–289 (2019)
7. Andreeva, Y.M., Luong, V.C., Lutoshina, D.S., Medvedev, O.S., Mikhailovskii, V.Y., Moskvin, M.K., Veiko, V.P.: Laser coloration of metals in visual art and design. Opt. Mater. Express, **9**(3), 1310–1319 (2019)
8. Liu, C., Chen, S., Sheng, C., Ding, P., Qian, Z., Ren, L.: The art of a hydraulic joint in a spider's leg: modelling, computational fluid dynamics (CFD) simulation, and bio-inspired design. J. Comp. Physiol. A. **205**(4), 491–504 (2019)
9. Luo, Z., Dai, J.: Synthetic genomics: the art of design and synthesis. Sheng wu gong cheng xue bao= Chin. J. Biotechnol., **33**(3), 331–342 (2017)
10. Knight, E., Daymond, J., Paroutis, S.: Design-led strategy: how to bring design thinking into the art of strategic management. Calif. Manage. Rev. **62**(2), 30–52 (2020)
11. Jordan, D., O'Donoghue, H.: Histories of change in art and design education in Ireland: towards reform: the evolving trajectory of art education. Int. J. Art & Des. Educ. **37**(4), 574–586 (2018)
12. Bastogne, T.: Quality-by-design of nanopharmaceuticals–a state of the art. Nanomedicine: Nanotechnol., Biol. Med., **13**(7), 2151–2157 (2017)
13. Maras, K.: A realist account of critical agency in art criticism in art and design education. Int. J. Art & Des. Educ. **37**(4), 599–610 (2018)

14. Ravelomanantsoa, M.S., Ducq, Y., Vallespir, B.: A state of the art and comparison of approaches for performance measurement systems definition and design. Int. J. Prod. Res. **57**(15–16), 5026–5046 (2019)
15. Tsai, K.C.: Teacher-student relationships, satisfaction, and achievement among art and design college students in Macau. J. Educ. Pract. **8**(6), 12–16 (2017)

Chapter 4
Research on Fresh Squid Body Recognition Based on Improved YOLO v5 Model

Chengzhe Xie and Xiaozhen Hu

Abstract In order to improve the intelligence level and efficiency of squid fishing, a new target detection and recognition model, YOLO v5-SQ based on improved YOLO v5, is proposed based on the operation process of squid fishing industry and the sorting characteristics of the squid fishing industry. Based on the stiffer cartilage of squid, which makes its body relatively straight, the length of the body is innovatively measured and graded to synchronize the measurement and grading of fresh squid during the sorting process, so as to improve work efficiency. The improved model YOLO v5-SQ uses image processing to increase sample size and improve learning effect. To better capture the details of squid neck, ConvNeXt is used to preserve the features of small details. To solve the problem of squid posture diversity and large changes in neck characteristics, BiFPN(Bi-directional Feature Pyramid Network) structure is used in the part of network feature fusion, which can enhance the network's ability to fuse multi-scale target features. During squid transport, if no posture correction is made, the body may not be parallel to the final label frame, resulting in too large difference between the converted size and the actual size. Therefore, a mechanical structure is used to modify the squid posture at the front of recognition, so that the middle line of the body is as parallel as possible to the horizontal line of the shooting interface and to reduce errors. The results show that the accuracy of YOLO v5-SQ detection can reach 98.7%, the average error of body length is 1.57%, and the speed of time detection when calling the camera is 47.83 f/s. This method can achieve the expected effect and has high practical value. It also provides some improvement ideas for the intelligent research of other seafood operations.

C. Xie (✉) · X. Hu
School of Ocean Engineering Equipment, Zhejiang Ocean University, Zhoushan, China
e-mail: chzchn@foxmail.com

R. Kountchev (Deceased) et al. (eds.), *AI Methods and Applications in 3D Technologies*, Smart Innovation, Systems and Technologies 388,
https://doi.org/10.1007/978-981-97-2144-3_4

4.1 Introduction

Cephalopoda, with more than 700 extant species covering Tropical, temperate, and cold sea domain species, is distributed throughout the Atlantic, Indian, and Pacific oceans. Cephalopods are important in marine ecology and are an important food source for large fish species such as tunas, marine mammals such as whales, and a few seabirds. It has a short life cycle and rapid growth rate, and a proportion of small and medium-sized species can reach the body length close to their parents after approximately half a year of growth [1]. Two subclasses, Nautiloidea and coleoidea, are classified down according to characteristics such as the number of gills and carpal legs. The vast majority of human species used are concentrated under the Coleoptera, for example: cuttlefish, squid, octopus, etc., which are very important marine fishery resources because they have a relatively high proportion of edible parts, and they have a wide market.

Squid belongs to Cephalopod Coleoptera, which is divided into Loliginidae and Ommastrephidae etc. Because of the genus Cephalopoda, it has a relatively short life cycle (usually one year), a relatively fast growth rate, and strong reproductive capacity, so resources are very rich. The squid has the advantages of high protein, low fat, low heat, and is rich in calcium, phosphorus, and iron, as well as trace elements such as calcium, selenium, iodine, and manganese, but also rich in highly unsaturated fatty acids such as DHA (Docosahexaenoic Acid) and EPA (Eicosapentaenoic Acid), and high in taurine, which is highly nutritious. In terms of worldwide consumption of squid, products from fresh or frozen squid account for more than 70% of the total production. In the daily small to medium squid consumption, again most common in the form of whole squid, visible complete small to medium-sized squid morphology is an important component of the squid product in market circulation.

In the process of squid fishing, the proportion of labor is relatively large, and there is still a large development space for mechanization and intelligence. Especially in offshore squid fishing, the sources of catches are concentrated, the space of processing, storage and transportation sites is limited, and the speed of manual treatment is limited, which makes it difficult to guarantee the freshness of squid before freezing and the operating efficiency can not meet the increasing market demand. In recent years, although the mechanization of fishing vessels has been gradually improved, there is still a large gap in sorting and classification. The space conditions of fishing boats are different from those on land. If common weighing method is used, the first is difficult to meet the requirements of rapid handling on fishing boats. Secondly, due to erosion of high humidity and corrosive air at sea and natural shaking of fishing boats, it is easy to cause unstable weighing and frequent damage to parts. Therefore, visual scheme can effectively solve these problems. Although the mapping environment can be set relatively simple on the ship, the ordinary image processing method is easily disturbed by ambient light, and the processing effect is poor at low brightness, because of the characteristics of squid's own soft, all of these cause the difficulty of squid image detection and make it difficult to have a more standard measurement method. In order to effectively solve the influence of soft body characteristics of squid and provide a standard measurement standard, this paper makes use of the inner shell and relatively rigid characteristics of squid body to make its body relatively straight and classify it according to the measurement characteristics of body length. In order to meet the task requirement of fast recognition, a target detection and recognition model based on one-stage algorithm and YOLO v5 network is proposed, which is applicable to ship scene and is classified according to the length of squid body [2–4]. Compared with RCNN (Region based Convolutional Neural Network), the improved YOLO v5 network greatly speeds up the detection speed. After training, the detection accuracy can reach 98.7%, the average error is 1.57%, and the detection speed is 47.83 f/s when the camera is called. The improved YOLO v5 network has better detection accuracy and speed.

4.2 Establishment of Squid Dataset

Squid is a general term, it can still be subdivided down, usually expressed as Loliginidae and Ommastrephidae. They are similar in morphology and slightly different. This paper takes squid as a discussion, does not study each species individually anymore, reduces the difficulty of collecting samples, increases the sample types, and the final model can also improve a certain degree of generalization.

Based on the variety of squid species and the softer physical characteristics, images of squid in different species, angles and environments were selected during the image collection process. Images are taken from live photos or uploaded by others on the Internet, as shown in Fig. 4.1.

Fig. 4.1 Images of squid

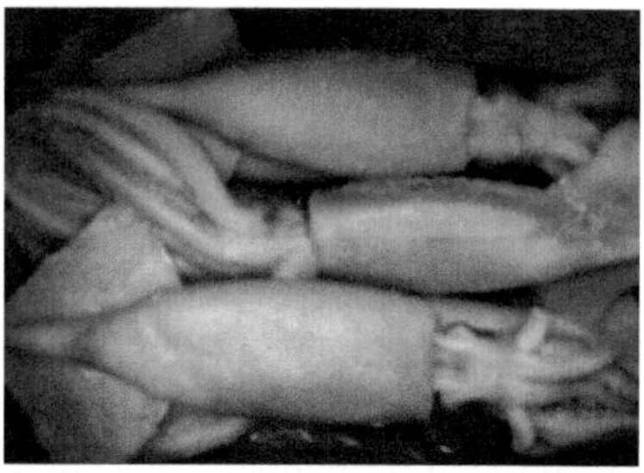

Fig. 4.2 Image of squids

In the collection process, 3527 squid images of different types and different shapes were obtained initially. In the later stage, the collected images were manually checked to remove the images with obscure shooting and serious occlusion features. At the same time, some images containing complex squid and slightly obscured are retained, as shown in Fig. 4.2.

This kind of image can enhance the learning ability of the neural network. The total number of images was 1813. The data set has a slightly smaller number of samples and needs to be preprocessed for the squid image data set.

4.3 Image Preprocessing

In order to expand the dataset and improve the accuracy of recognition, an image enhancement algorithm is used to add the number of times a picture is used, and a mosaic data enhancement method is used to enhance the richness of disguised expanded squid image samples to improve the generalization ability of the recognition network.

Data Set Expansion

Pictures obtained through various channels are filtered layer by layer to get the final image. The captured video is read frame intercepted to get the image. By processing these pictures in different ways, the variety of pictures and the amount of training data are increased, the over-fitting phenomenon is reduced effectively, and the robustness of the final model is improved.

In this paper, the image data is enhanced by rotation, mirror flipping, upside-down flipping, clipping, add noise and other methods, and used individually or in combination. Some of the effect diagrams are shown in Fig. 4.3.

In addition to the above, Mosaic algorithm can be added to the network to process the dataset, which can increase the number of samples selected for each training and reduce the computational overhead.

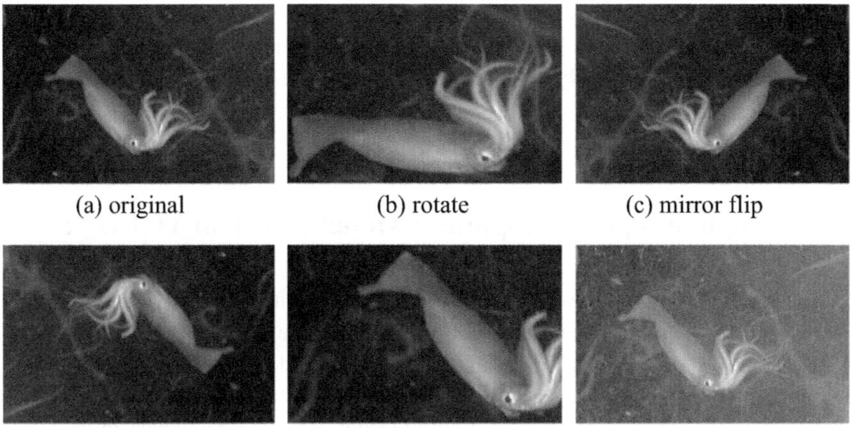

(a) original (b) rotate (c) mirror flip

(d) upside down (e) clipping (f) add noiset

Fig. 4.3 Image enhancement

Data Annotations

The initial model was not trained for the squid dataset, so in order to enable the network model to learn the squid characteristics and location information and improve the accuracy of the output of the model after training, the squid position in the image can be marked, and the locusts in the image can be rectangular marked by the labeling tool LabelImg. The labeling tool is installed by inputting the command pip install labelimg –i. After installation, enter the command labelimg to open, load the picture you want to label, label the rectangular box, save the label file in VOC format, store it as xml suffix file, label interface and xml file content as shown in Fig. 4.4.

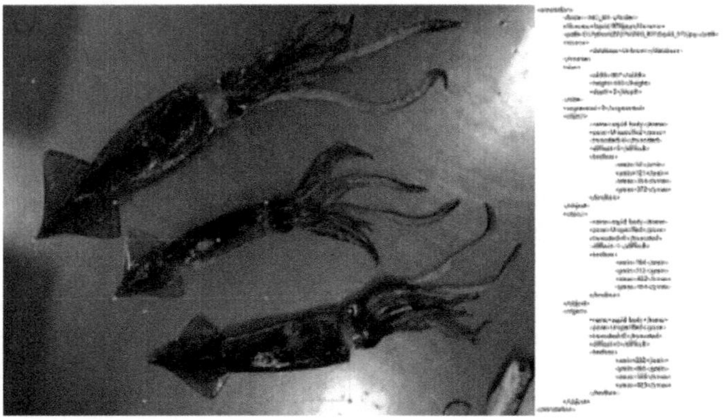

Fig. 4.4 Image labeling and its label file information

The labeling of pictures needs to be as accurate as possible and minimize the bounding box to reduce the impact of the background and make the network model work well. After labeling, the label files and images are used as training sets for the model.

4.4 Design of Squid Recognition Model Based on YOLO v5

YOLO series belongs to one-stage target detection algorithm, which extracts features directly from convolution neural network to predict target classification and location. It has the advantages of fast detection, fast training and learning the generalization characteristics of objects, but its detection accuracy is slightly lower than the two-stage algorithm, and it is not good for small target detection. However, in the application environment in this paper, can make the environment relatively simple, reduce the interference of complex background, make the target size suitable and not easy to be omitted, and apply the target detection algorithm of YOLO v5 because there is a certain requirement for the speed of detection. In this paper, the original backbone network Darknet of YOLO v5 is improved to ConvNeXt to solve the problem that the original backbone network Darknet has insufficient feature extraction ability for small target, that is, squid head and body segment (neck) in practice. At the same time, the feature fusion part Neck is improved to a two-way feature fusion network BiFPN to solve the multiscale feature fusion problem. These improvements improve the ability of the network model to recognize squid body with different postures and improve the accuracy of rectangle selection. The improved network model is named YOLO v5-SQ.

YOLO v5 Structure Model

YOLO v5 has been improved on the basis of previous generation YOLO series. It has faster operation speed and higher recognition accuracy and is widely used. YOLO v5 consists of Backbone, Neck, and Head. The structure model is shown in Fig. 4.5.

Some term abbreviations consist of basic term abbreviations, as shown in Fig. 4.5. The explanation of other basic terminology abbreviations is as follows: Conv: Convolution; BN: Batch Normalization; SiLU: A activation function; CBS: A combination of Conv, BN and SiLU; K: Convolutional kernel; SPPF: Spatial Pyramid Pooling Fast.

The original YOLO v5 network has good performance in detecting large and medium-sized and obvious targets, but the detection and recognition ability of small target, small detail and small sample images without improvement of YOLO v5 is not ideal, so it needs to be improved in feature information extraction and feature fusion for specific targets to adapt to the detection and recognition of only body on intact squid.

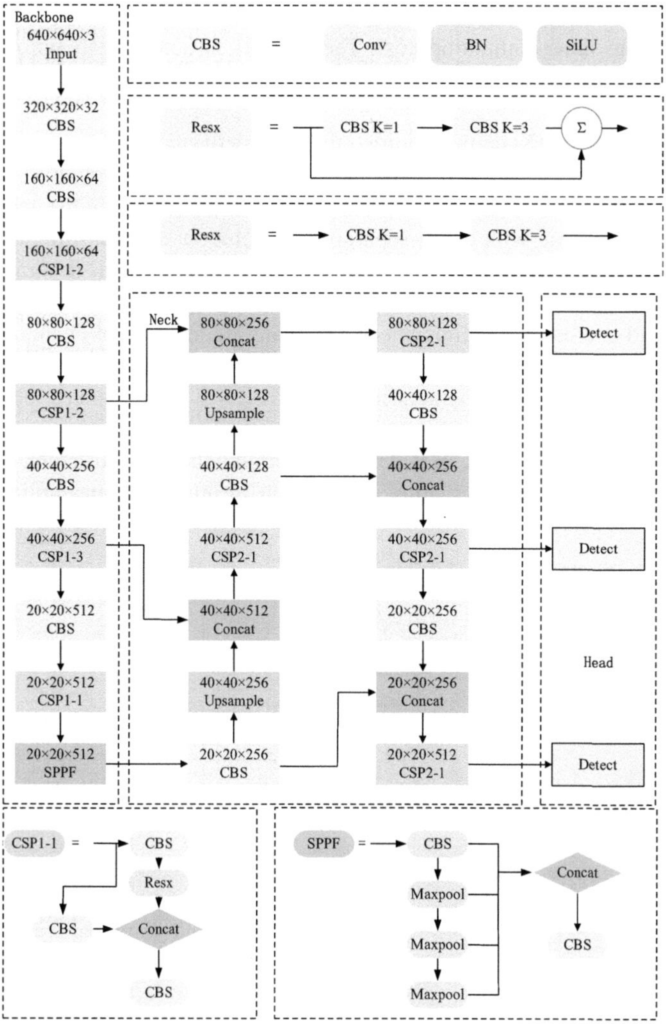

Fig. 4.5 YOLO v5 structure

Feature Extraction

In the squid dataset, it is difficult to grasp the neck, which is a small target detail feature. This small target feature is easily lost in the adoption of YOLO v5. In order to effectively solve the problem of inaccurate neck recognition and insufficient feature extraction, in addition to labeling as accurately as possible, the YOLO v5 model needs to be improved. YOLO v5-SQ uses ConvNeXt network with better performance in feature extraction.

ConvNeXt is a pure convolution network with faster inference speed and higher accuracy at the same computing cost. The network structure is shown in Fig. 4.6. Where *h, w, dim* represent the height, width and number of layers of the feature map, respectively.

In Fig. 4.6, ConvNeXt network adjusts the number of stacked blocks of the original ResNet module from (3, 4, 6, 3) to (3, 3, 9, 3), downsampling through a convolution layer with a convolution core size equal to the step size, balancing the accuracy and computation overhead with depth-separable convolution, and adjusting the number of channels to 96.ConvNeXt Block uses an inverted bottleneck design with 3 × 3 Depth detachable convolution is used to extract features by 1 × 1 Convolution increases and decreases the dimension, increasing the number of channels from 96 to 384 and then returning to 96 to reduce the loss of high-dimensional information. After four stages of feature extraction, the size is 7 × 7 × 768, after global average pooling and LN layer, finally output through linear classifier [5–7].

ConvNeXt model, because of its special inverted bottleneck structure, effectively avoids the loss of small target feature information during the downsampling process,

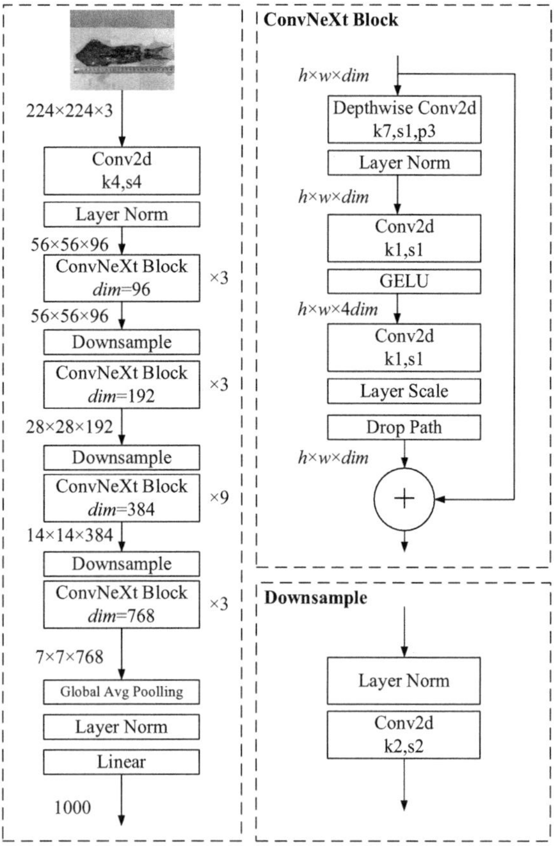

Fig. 4.6 ConNeXt structure

and can extract the required detail features. ConvNeXt adds an LN layer before and after each downsampling and after global mean pooling. This normalization strategy makes the model more stable and reduces gradient oscillations. In addition, the ConvNeXt model not only improves the accuracy of squid neck grasp, but also reduces the number of floating-point operations in the network because it uses the Swin Transformer structure to reduce the use of activation functions and BN layers [8].

Feature Fusion

In the network training process, as the number of layers of the network increases, the more advanced the feature information is obtained, but corresponding, the location information of the target becomes weaker, and for small targets, the location loss will be greater. Therefore, after extracting the characteristics of squid neck, Neck is needed to fuse the features to improve the accuracy of Head's prediction of neck location.

YOLO v5 uses different scales to predict and fuse features using a feature Pyramid Network (FPN) combined with a Pixel Aggregation Network (PAN) [9, 10]. Because the size of squid individuals varies, different types of color and posture also vary, there are many possibilities for posture detection. Abdomen up, back up and side lying will have a large change in the target size of the neck. The original PANet structure of YOLO v5 has different judgments on the neck of squid under different conditions and can not achieve good learning results. To solve the problem of multi-pose multi-scale detection of targets, the improved model YOLO v5-SQ uses BiFPN structure in the feature fusion section, and introduces weights to solve the problem that feature maps of different scales contribute to feature fusion differently. The PANet structure is shown in Fig. 4.7, and the BiFPN structure is shown in Fig. 4.8.

The BiFPN structure enables the network to learn how important these feature maps are to feature fusion in feature maps of different resolutions, and to perform multi-scale feature fusion over and over again [11].

The fusion formula of BiFPN weighted features is as follows:

$$O = \sum\nolimits_i \frac{w_i}{\epsilon + \sum\nolimits_j w_j} \cdot I_i \tag{4.1}$$

In the formula: w_i, w_j—— Weight, which measures the contribution of features of different resolutions to feature fusion (Where $w_i \geq 0$ is ensured by applying a Relu after each w_i).

ϵ——Let ϵ=0:0001 is a small value to avoid numerical instability

I_i——Input characteristics

O——Output, value of weighted feature fusion

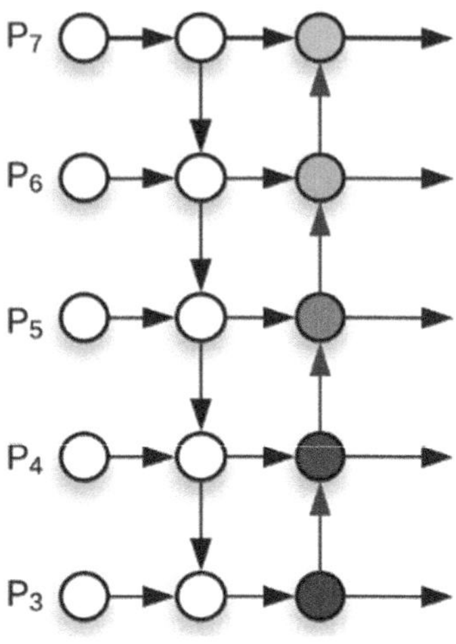

Fig. 4.7 PANet structure

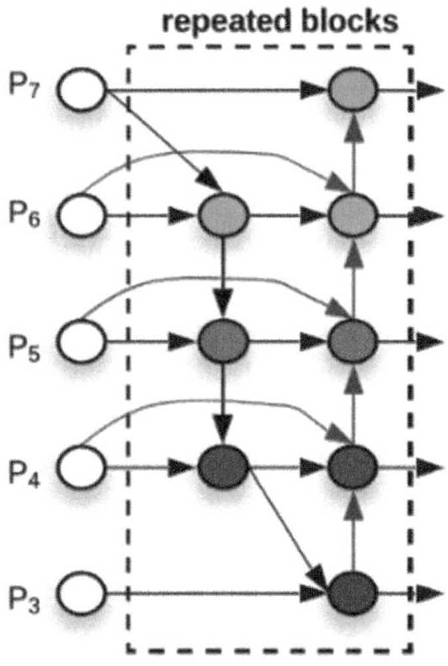

Fig. 4.8 BiFPN structure

BiFPN integrates both the bidirectional crossscale connections and the fast normalized fusion. The two fused features at level 6 for BiFPN shown in down:

$$P_6^{td} = Conv(\frac{w_1 \cdot P_6^{in} + w_1 \cdot Resize(P_7^{in})}{w_1 + w_2 + \epsilon}) \tag{4.2}$$

$$P_6^{out} = Conv\left(\frac{w'_1 \cdot P_6^{in} + w'_2 \cdot P_6^{td} + w'_3 \cdot Resize(P_5^{out})}{w'_1 + w'_2 + w'_3 + \in}\right) \tag{4.3}$$

In the formula: P_6^{td}——The intermediate feature on the topdown pathway.
P_6^{out}——The output feature on the bottom-up pathway.

Fig. 4.9 Mechanical structure diagram

P^{in}—— Input sampling down.
$Resize$——Upsampling function.
w—— Learned parameters.
P^{out}——Output characteristics of fusion.
$Conv$——Convolution.

Because of BiFPN have integrates both the bidirectional crossscale connections and the fast normalized fusion, can fuse more features efficiently when fusing feature maps of different scales. At the same time, using the weighted feature fusion design, the contribution of input features at different scales to output features is obtained by learning the weights of input features at different layers, which solves the problem that different squid postures and different squid species affect the large change of image scale [12].

4.5 Results and Analysis of Image Recognition of Squid Body

Preparation in Advance

The rectangular frame obtained by YOLO target detection algorithm is the largest outer rectangle of squid body. Therefore, the closer the angle between the body center line and the horizontal line is to 45°, the greater the difference between the actual length and the rectangular frame length. So, the orientation device needs to be designed so that the attitude of squid can be adjusted to the ideal state before image detection, so that the body center line is as parallel to the horizontal line of the field of view as possible. The device structure and direction are shown in Fig. 4.9.

Experimental Environment

The model is trained on the server, configured as Inter Core CPU i7-8700, GeForce RTX3060 12GB GPU, environment Python 3.7, framework Pytorch1.8.1.

Fig. 4.10 Label map and data

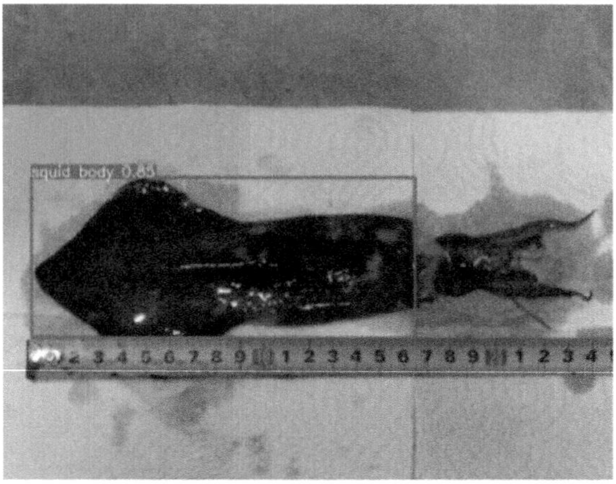

0 0.359375 0.533333 0.628125 0.337500

Identification Results

The improved model YOLO v5-SQ was used to identify and detect the whole body part of squid. The results are shown in Fig. 4.10. It can be seen that YOLO v5-SQ can accurately locate and identify squid body in image. YOLO v5 label file format is.txt, including five items:class, x, y, w, h. x & y are the center point and w&h are the width and height of the rectangle. It is a normalized value, so the actual size needs to be converted if needed.

Size Conversion

X and Y denote the center point, and their values lie between 0 and 1, indicating that the point is located in the percentage of the width and height of the graph. Its tag box and tag file are shown in Fig. 4.10.

For example in Fig. 4.10, body corresponds to w with a value of 0.628125. The whole picture is 1280 pixels wide, corresponding to the true length of about 25.8 cm. The body length of squid is 16.21 cm and true length of squid is 16.02 cm respectively. The absolute error is 0.19 cm and the error rate is 1.19%.

Test Result

The shooting should be taken vertically to reduce errors caused by lens distortion or large errors caused by distortion can be avoided by early adjustment. Take some squid images and measure the data. Make a chart as shown in Table 4.1.

Table 4.1 Dimension measurements

Number	Conversion value	True value	Error rate (%)
1	16.21	16.02	1.19
2	16.96	17.36	–2.30
3	16.01	16.13	–0.74
4	17.12	16.59	3.19
5	29.89	31.13	–3.98
6	18.39	19.21	–4.27
7	19.73	20.26	–2.62
8	18.13	17.57	3.19
9	15.38	16.18	–4.94
10	21.19	22.14	–4.29
11	19.86	20.25	–1.93
12	24.87	24.13	3.07
13	13.84	14.48	–4.42
14	17.84	18.11	–1.49
15	15.76	16.27	–3.13
Average error/%			–1.57

Model Comparison

From Table 4.1, it can be seen that the improved model YOLO v5-SQ can meet the accuracy requirements. Meanwhile, to demonstrate its performance, this is compared with some other models. As shown in Table 4.2.

Among them: FPS (Frame Per Second) represents the detection speed; P is Precision, higher value, fewer false positives; R is Recall, higher value, fewer false negative; AP is Average Precision, it combines Precision and Recall, which can represent the overall performance of the model. From Table 4.2, it is found that the improved model sacrifices some detection speed but achieves higher overall performance. Usually, real-time detection can be achieved when FPS > 30, so the improved model can meet the real-time detection requirements.

Table 4.2 Performance comparison of different models

Model	FPS	AP(%)	P(%)	R(%)
Faster R-CNN	6.41	89.1	72.8	92.1
YOLO v5	53.69	92.9	97.4	93.6
YOLO v5-SQ	47.83	93.7	98.3	94.9

4.6 Conclusion

(a) In view of the special environment on the ship, a visual recognition scheme is adopted to measure squid. At the same time, in order to solve the problem that ordinary image processing is difficult to distinguish head from body of squid, a target detection and recognition model YOLO v5-SQ based on classification of body length of squid is put forward by utilizing the characteristics of inner cartilage of squid body to make its body straight. Detailed features of squid neck are extracted by ConvNeXt and feature fusion is carried out by combining BiFPN structure. A detection model for identifying body parts of complete squid is constructed and tested in practice.

(b) When squid body detection and identification is carried out for this part of squid, the improved YOLO v5-SQ model in this paper has a recognition accuracy of 98.7%. Compared with the true value, the maximum error is 4.94% and the average error is 1.57%. The squid body length can be used as the standard for squid classification.

(c) Combining with the actual application scenario, call the camera to test, the detection speed can reach 47.83 f/s, which can meet the real-time requirement of a certain speed, improve the hardware performance can get better results.In this method, the detection speed is fast.Just adjust the squid posture through mechanical structure before detection so that the squid body center line is parallel to the horizontal line of the field of view as far as possible, which can achieve a good detection effect. The detection process needs no contact, the result is reliable and the labor cost is saved. This method has very good application and promotion value.

References

1. Chen, X.J.: World cephalopod economic resources and their development and utilization (in chinese). World Science **3**, 25–26 (2005)
2. Qiu, X., Sun, X., Chen, Y., Wang, X.: Pedestrian detection and counting method based on YOLOv5+DeepSORT. In: 4th International symposium on power electronics and control engineering (ISPECE 2021), SPIE (2021)
3. Li, R., Wu, Y.P.: Improved YOLO v5 wheat ear detection algorithm based on attention mechanism. Electronics **11**(11), 1673–1673 (2022)
4. Zhang, Y., et al.: Real-time vehicle detection based on improved YOLO v5. Sustainability **14**(19), 12274–12274 (2022)
5. Liu, Z., Mao, H., Wu, C. Y., Feichtenhofer, C., Darrell, T., Xie, S.: A ConvNet for the 2020s. arXivpreprint arXiv:2201.03545, (2022)
6. Liu, Z., Lin, Y., Cao, Y., Hu, H., Wei, Y., Zhang, Z., et al.: Swin transformer: Hierarchical vision transformer using shifted windows. In: Proceedings of the IEEE/CVF international conference on computer vision, pp. 10012–10022 (2021)
7. Xie, S., Girshick, R., P Dollár, Tu, Z., He, K..., et al.: Aggregated residual transformations for deep neural networks. In: Proceedings of the IEEE conference on computer vision and pattern recognition, pp. 1492–1500 (2017)

8. Liu, Z., Lin, Y. T., Cao, Y., et al.: Swin transformer: hierarchical vision transformer using shifted windows. In: Proceedings of the IEEE/CVF international conference on computer vision, pp. 10012–10022 (2021)
9. Lin, T. Y., Dollar, P., Girshick, R., He, K., Hariharan, B., Belongie, S.: Feature pyramid networks for object detection. IEEE Comput. Soc., (2017)
10. Tan M., Pang R., Le Q.V.: EfficientDet: Scalable and efficient object detection. In: 2020 IEEE/CVF conference on computer vision and pattern recognition (CVPR), IEEE (2020)
11. Hendrycks, D., Gimpel, K.: Gaussian error linear units (gelus). arXiv preprint arXiv:1606.08415, (2016)
12. Howard, A., Sandler, M., Chu, G., Chen, L. C., Chen, B., Tan, M., et al.: Searching for MobileNetV3. In: 2019 IEEE/CVF international conference on computer vision (ICCV), IEEE (2019)

Chapter 5
Applications of Explainable Artificial Intelligent Algorithms to Medicine: A Bibliographic Study

Mini Han Wang, Lumin Xing, Xiangrong Yu, Chenyu Yan, Ruoyu Zhou, Kelvin K. L. Chong, Fengling Wang, Xiaoshu Zhou, Guoqiang Chen, Qing Wu, Zhiyuan Lin, Peijin Zeng, and Qide Xiao

Abstract In recent years, Explainable Artificial Intelligent (XAI) algorithms have been widely applied and deeply studied in the field of medicine. Based on the bibliometric method, this paper summarizes and analyses 315 non-repetitive related literature retrieved from the literature databases of Scopus and Web of Science

---

M. H. Wang (✉) · L. Xing · R. Zhou · K. K. L. Chong (✉) · F. Wang · Q. Xiao
The Faculty of Data Science, City University of Macau, Macau, China
e-mail: D20092100037@cityu.mo

K. K. L. Chong
e-mail: chongkamlung@cuhk.edu.hk

M. H. Wang · K. K. L. Chong
The Faculty of Medicine, The Chinese University of Hong Kong, Hong Kong, China

M. H. Wang · X. Yu
Zhuhai People's Hospital (Zhuhai Hospital Affiliated With Jinan University), Guanghui Hou, Zhuhai, China

M. H. Wang · Z. Lin · P. Zeng
The Department of AI and Big Data Application, Zhuhai Institute of Advanced, Zhuhai, China

L. Xing
The First Affiliated Hospital of Shandong First Medical University & Shandong Provincial, Qianfoshan Hospital, Jina, China

C. Yan (✉)
Suzhou Institute of Biomedical Engineering and Technology, Chinese Academy of Sciences, Suzhou, China
e-mail: 542916022@qq.com

F. Wang
Hezhou University, Hezhou, China

X. Zhou
Gree Polytechnic, Zhuhai, China

G. Chen · Z. Lin · P. Zeng
Beijing Institute of Technology, Zhuhai, China

R. Kountchev (Deceased) et al. (eds.), *AI Methods and Applications in 3D Technologies*, Smart Innovation, Systems and Technologies 388,
https://doi.org/10.1007/978-981-97-2144-3_5

(WOS). Developing trends, key authors, contributed countries, journals, references, and keywords are explored. Based on the method of text mining, representative articles, and their core opinions are qualitatively analyzed. Methods are clustered, applications, strengths, and drawbacks are discussed, and issues and future directions are exhibited. Findings show that applications of the interpretable algorithm to the medical field have vital potential clinical value. Medical diagnosis based on computer vision with an explainable mechanism is aimed at breaking the trust between artificial intelligence and medicine disciplines. In addition, cooperative research and talent training in this interdisciplinary area are also worthy of attention in the future. Thus, a reference significance for XAI applied to the medical field is contributed by this study.

5.1 Introduction

Artificial Intelligence (AI) is an advanced technology that can help doctors improve efficiency and patient care and reduce medical costs through early detection and perfect treatment algorithm [1]. For instance, based on 129,450 clinical images, Esteva et al. [2] indicated that the deep learning method of CNN has a competitive advantage over dermatologists for skin lesion detection tasks [2]. However, it is difficult for AI to explain in the clinic. There is also widespread concern and distrust among doctors, where advanced technologies, such as artificial intelligence, may replace doctors, or at least replace some tasks that clinicians are used to performing. Moreover, most doctors also trust their own experience and medical skills and do not trust the results of AI diagnose. Furthermore, many patients do not trust computer-aided diagnosis and prefer face-to-face diagnosis and treatment. In addition, there is a big difference between the data calculation direction and the result diagnosis direction. AI technology is mainly based on the judgment of digital orientation, while clinical practice focuses on clinical manifestations. The logic of Artificial intelligence diagnosis is to have the diagnosis results first and then explain the reasons, which is prone to errors and omissions in focus identification. As far as clinical diagnosis is concerned, it is based on pathological characteristics and emphasizes priority. Due to the problems of black-box and the robustness of methods based on artificial intelligence, they cannot be fully trusted in the medical field. The methods based on artificial intelligence are highly dependent on data quality, and there are significant differences based on a small amount of clinical data in the experimental process.

Thus, based on the theory of Koch's Postulates and evidence-based medicine (EBM), the research on interpretable algorithms in the field of medicine is of great significance to improve the transparency, interpretability, and translatability of the assisted diagnosis algorithm model, which plays a vital role to break the cross-domain understanding and trust barrier between doctors and artificial intelligence.

Q. Wu
Zhuhai Science and Technology Development Promotion Association, Zhuhai, China

5.2 Methodology

This paper delivers a bibliographic analysis based on Scopus and Web of Science (WOS) literature databases. The retrieval statement is set as ST = "Explainable Artificial Intelligen*" AND "Medic*". The retrieval time is unlimited, and the retrieval date is February 27, 2022. CiteSpace V.6.1, R6 (64 bit) basic [1, 3] is applied for a bibliographic analysis of the retrieved literature. An exploration of contributing countries, authors, journals, and references is implemented. Based on a text mining of the representative literature. Classification and discussion are implemented on XAI-based methods. Advantages and disadvantages of methods, and related studies are explored.

5.3 Results and Discussion

Bibliographic Analysis Results

Through the retrieval of Scopus and WOS, a total of 315 nonduplicate data were obtained, 46 valid data were obtained by WOS, and 300 valid data were obtained by Scopus. The developing trend, distribution of countries, journals and authors are identified. Opinions of core references and topics are explored.

As shown in Fig. 5.1, the earliest relevant literature can be traced back to 1987 [4], Swartout and Smoliar proposed that clinical practice must explain the results of medical diagnosis. The study of causality of diagnosis is very meaningful for patients' understanding and clinical operation. They put forward three hypothetical frameworks to realize the derivation of diagnostic interpretability. After that, the study did not attract people's attention for a long time. It was not until 2017 that this research field reappeared in front of people, attracted people's attention, and developed rapidly. Therefore, this paper predicts that this topic will increase in the next few years.

As shown on Fig. 5.2 and Table 5.1, based on the number of publications, the top-6 contributed countries are listed as the USA, Germany, Italy, The United Kingdom, South Korea, and China. However, with the number of nodes = 112 and Link = 104 (Density = 0.0167), the cooperation between countries is still needed to be concentrated.

According to the number of published papers, the top 5 journals are ranked in this paper, and the ranking results are shown in Table 5.2. The most contributed journals are recognized as Nature, IEEE Access, Journal of the American Medical Association (JAMA), and Journal of Natural Medicines, and Advances in Neural Information Processing Systems. Journals are all highly qualified with high impact factors, which are listed in Table 5.2, the retrieval time is 29th march, 2022. Thereby, this article concluded that research on XAI in medicine has a high potential academic value in the next few years.

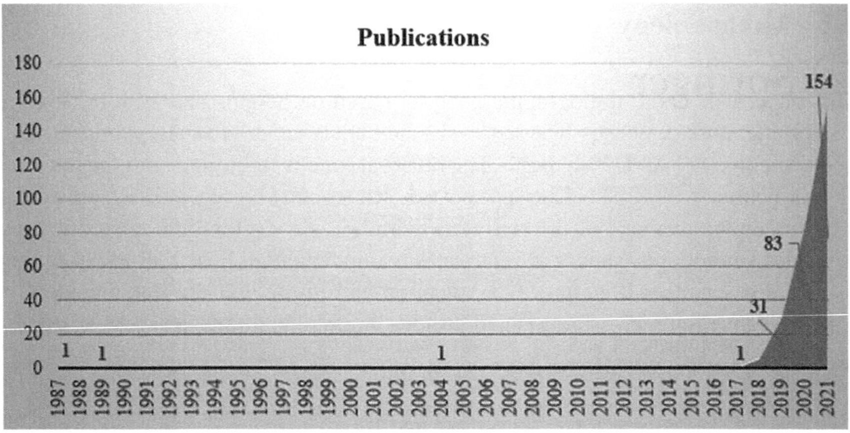

Fig. 5.1 The developing trend of XAI

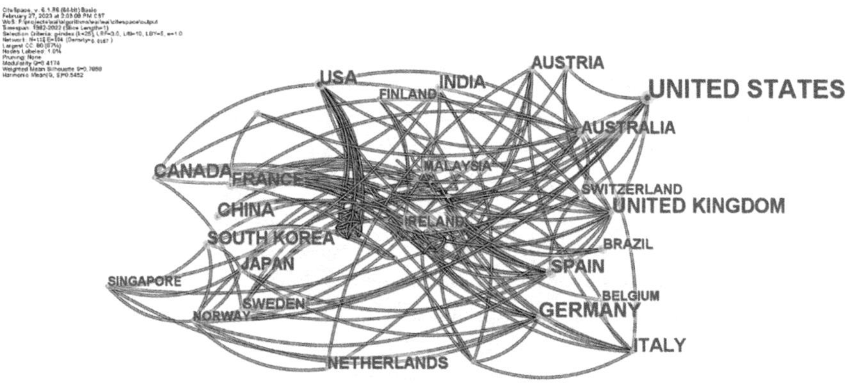

Fig. 5.2 The distribution of countries

Table 5.1 Top 6 countries according to the number of publications

Ranking	Country	Count
1	USA	76
2	Germany	21
3	Italy	17
4	The United Kingdom	16
5	South Korea	12
6	China	12

Table 5.2 The top 5 co-cited journals according to the centrality

Ranking	Journals	Count	Impact factor (2020)
1	Nature	75	49.96
2	IEEE Access	56	3.367
3	Journal of the American Medical Association (JAMA)	46	56.27
4	Journal of Natural Medicines	44	2.343
5	Advances in Neural Information Processing Systems	43	16.54

Based on the number of citations, the top 5 journals are shown in Table 5.3. The most contributed article is [5], which is cited 1888 times according to Google scholar (retrieving time is 29th March 2022). It highlights the value of XAI and encourages the scholars to pay more attention to the development and challenges of this method. The second paper is identified from Ribeiro et al. [6], which is cited 8293 times according to Google scholar (retrieving time is 29th March 2022). This article proposed a method of LIME, which has been applied in multiple situations by the following studies [7]. Thereby, this study concluded, due to the drawbacks of black-box AI models (model explicability, outcome interpretability, model inspection, model design transparency) [8], it is necessary to consider the XAI as a solution instead of black-box models when it comes to high-stakes decision supporting tasks. Especially for the fields of criminal justice, healthcare and medical diagnosis, and computer vision, XAI presented great advantages for avoiding making incorrect, incomprehensible, and incorrect decisions for loan decisions, safety guidance, and health suggestions [5, 9]. XAI could be classified by the capability of specific explanation problem solving, the type of explanator adoption, the level of model transparency, and input data type [8]. It could be measured by the System Causability Scale (SCS), which enables users to adapt the measuring scale based on the particularity of disciplines [10]. As far as the field of medicine is concerned, XAI is regarded as the sine qua non for facilitating transparency and trust between AI and Medicine domains. Data sources of images, omics data, and text are the major types of datasets for XAI applied to Medicine fields. High accuracy is required in the medical field. To enhance the performances, the hybrid distributional models and AI models with optical computing should be considered [11]. Besides, it is essential for workflow optimization and medical error detection for the system, it is vital of accessing personal data and analyzing results for patients. However, challenges are identified as transparency, privacy, bias, and security [12]. The typical XAI algorithms are LIME, which has a great performance for classification tasks [6]. Knowledge-based XAI has a great advantage for medical rule mining results [13]. Another two methods are proved to be effective. One is to enhance the expandability of models by measuring the sensitivity of prediction results on the change of input data. The other is that decomposing results meaningfully concerning input features [14].

However, the most challenge of XAI is that the expandability is decided by engineers' subjectivity, the possibility is not as essential as causal links [15].

Table 5.3 Top 5 cited references according to the number of being cited

Ranking	Count	Year	Cited Reference	Conclusion
1	28	2018	Peeking inside the black-box: a survey on explainable artificial intelligence (XAI)	To solve the issue of the black box of AI, XAI is a new solution and a new opportunity for multiple applications. It should be paid more attention to in future research [5]
2	19	2016	"Why should I trust you?" Explaining the predictions of any classifier	A novel method of LIME is proposed, which is identified as an efficient algorithm for explainable prediction methods by considering local features [6]
3	17	2019	Explanation in artificial intelligence: Insights from the social sciences	The most challenge of XAI is that the expandability is decided by engineers' subjectivity, the possibility is not as essential as causal links [15]
4	12	2017	What do we need to build explainable AI systems for the medical domain?	XAI is regarded as the sine qua non for facilitating transparency and trust between AI and Medicine domains. Data sources of images, omics data, and text are the major types of datasets for XAI applied to Medicine fields. High accuracy is required in the medical field. To enhance the performances, the hybrid distributional models and AI models with optical computing should be considered [11]
5	12	2019	High-performance medicine: the convergence of human and artificial intelligence	The most important of XAI for clinicians is the efficiency and accuracy of image interpretation. It is essential for workflow optimization and medical error detection for the health system, it is vital of accessing personal data and analyzing results for patients. The challenges are identified as transparency, privacy, bias, and security [12]
6	11	2017	Explainable Artificial Intelligence: Understanding, Visualizing, and Interpreting Deep Learning Models	It is encouraged for XAI applied in medical fields, especially in the aspects of understanding, visualizing, and interpreting deep learning models. Two methods are regarded as efficient approaches to classification tasks. One way is that enhancing the expandability of models by measuring the sensitivity of prediction results on the change of input data. The other one is that decomposing results meaningfully concerning input features [14]

(continued)

Table 5.3 (continued)

Ranking	Count	Year	Cited Reference	Conclusion
7	10	2020	Measuring the Quality of Explanations: The System Causability Scale (SCS): Comparing Human and Machine Explanations	An XAI measuring system named System Causability Scale is proposed, which enables users to adapt the measuring scale due to the particularity of different medical disciplines [10]
8	10	2020	eXplainable Artificial Intelligence (XAI) for the identification of biologically relevant gene expression patterns in longitudinal human studies, insights from obesity research	They proposed a rule-based XAI model and applied it to 57 subjects for the task of biological correlation between human gene expression and obesity performances. Results show that knowledge-based XAI has a great advantage for medical rule mining results [13]
9	9	2019	Stop explaining black-box machine learning models for high stakes decisions and use interpretable models instead	For high-stakes decision supporting tasks, it is necessary to consider XAI as a priority over black-box models when it comes to criminal justice, healthcare and medical diagnosis, and computer vision. The reason is that it will cause a huge influence if the incorrect, incomprehensible, and incorrect decisions for loan decisions, safety guidance, and health suggestions [9]
10	7	2018	A survey of methods for explaining black box models	The drawbacks of black-box AI models are classified as issues of model explicability, outcome interpretability, model inspection, and model design transparency. The classification of XAI is decided by the capability of specific explanation problem solving, the type of explanator adoption, the level of model transparency, and input data type [8]

The top 25 keywords of the strongest burst keywords are summarized, which are learning system, medical computing, decision making supporting, process computing, aided instruction, AI, decision supporting system, user interface, medical application, computer-aided diagnosis, deep learning and machine learning algorithms with digital pathological datasets, natural language processing, visualization, procedure, intensive, care unit, behavioral research, medical diagnostics, feature extraction, random forest, retrospective study, clinical practice, sensitivity and specificity burst during the latest five years. Thus, this study indicated that AI-based methods play a significant role in medical applications, especially for computer-aided diagnosis. The techniques of deep learning and machine learning are included, which are utilized for user interface developing, natural language processing, results automatic visualizing, behavioral researching, feature extraction, retrospective study, clinical practicing, and digital pathology research. The parameters of sensitivity and specificity are the major measurement indexes for AI models.

Classification and Discussion on XAI

At present, the most widely used pattern recognition algorithm in computer vision is a deep convolution neural network. Understand which part of the image DL network make decisions on is the key for AI-based model interpretable. The intuitive interpretation of the deep learning network model will not only help researchers debug the code and find the obvious mistakes made by the "black box" neural network in making a decision, but also help to improve the model. Knowledge discovery can also be realized by finding the key factors in the final decision-making in the input image [16]. Applying deep learning to image-based diagnose often requires end-to-end training of a very deep neural network with a large number of parameters. End-to-end deep learning algorithms usually take advantage of a high-dimensional hidden layer, which is abstract for the human brain. Therefore, people have raised various questions, such as whether the output of the artificial intelligence model can be explained and questioned; Whether differential diagnosis and confidence estimation are provided. In one case, despite strong diagnostic validation, there are still differences between clinicians' diagnoses and artificial intelligence algorithms. When rationalizing treatment decisions and managing risks, you must have the ability to explain and question models [17, 18]. To better interpretability and clinician acceptance, a lot of research work has been done to visualize and explain the function of features generated by DL model evaluation, which is the role of the interpretable algorithm [19, 20]. To alleviate this problem, several visualization methods have been proposed, such as occlusion, deconvolution and saliency map [21], Interpretive mask [22], and class startup diagram [23]. Key prediction features are visualized for the medical figures. The methods of XAI have been summarized in 7 clusters.

XAI-based method through effective dimensionality reduction. Automatic recognition through important feature extraction is the most traditional XAI-based method. The advantage of using this method is that the model volume is smaller, it is easier to understand, and it is not easy to fit; The disadvantage is that there may be omissions in the extraction of key features, so the performance of specific classification and recognition is not high [24]. Principal Component Analysis (PCA), Independent Component Analysis (ICA), Laplacian Eigenmaps, K-nearest neighbor algorithm, genetic algorithm, and decision tree algorithm based on the multi-label are effective XAI-based methods for dimensionality reduction. In addition, Zhao et al. [25] realized the classification task of a large number of lung cancer patients in the same period through supervised learning of representative cases [25].

Feature importance extraction based XAI. Based on the importance of features and the relationship between features, the characteristics, and importance of extracted features are explained to realize XAI. Through the extraction of feature importance, XAI can not only explain the essential features but also reflect their relative importance to clinical interpretation; However, the importance of numerical weight is not easy to understand or may be misunderstood [24]. For example, Shrikumar et al. Used the back-propagation mechanism of a deep learning network to measure the output gradient of input and output with the importance of features [26].

Attention-based XAI. Attention is the most important part of the translation task, which is to extract and deal with the most important information in the natural language. The attention mechanisms can evaluate the importance of words for classification purposes and generate meaning representation vectors. There are various types of attention mechanisms, including using all words to construct the global attention of the context, relying only on the local attention of a subset of words, or realizing the self-attention of several attention mechanisms at the same time, trying to find each relationship between word pairs [27]. This method is also commonly used in the interpretable algorithm of medical images [28]. The advantage of this method is that it can analyze the AI algorithm through global or partial attention features, to improve the trust and tolerance of clinical users to the model; The disadvantage is that excessive or wrong important attention information extraction will cause trouble for users' judgment and understanding, and make users tend to ignore or key non highlighted areas. Kwon et al. [29] proposed an intuitive and interpretable heart failure and cataract risk prediction model based on RETAINVIS, which improves the interpretability of the deep learning model by highlighting the image sequence features in the sequence. In these positions, these input features can affect the prediction results [29].

Class activation mapping (CAM)-based XAI, which has been used in digital medicine, especially in the field of medical image analysis [30]. Through visual interpretability, it is recommended to use these cam salient maps, so that clinicians and users can have an intuitive understanding of the principle of the AI model and can make a visual judgment and understanding of the principle. The advantage of this method is that it can effectively display important feature areas on the picture; The disadvantage is that it has no obvious effect on other forms of data. Lee et al. [31 based on cam, proposed an XAI algorithm for detecting acute intracranial hemorrhage from a small data set, which is one of the most famous studies using cam [32].

XAI through knowledge extraction and rule extraction. Knowledge refining is a form of model-specific XAI. It summarizes knowledge from complex model to simplified model, so that it can train the "student model" with the "teacher model" which is difficult to explain. Yes, the student model is interpretable. Rule extraction is another widely used XAI-based method, which is closely related to knowledge extraction and can be directly applied to digital medicine [33]. XAI-based method relies on knowledge extraction and rule extraction. In theory, it is a more stable model. The summary representation of complex clinical data can intuitively provide interpretable results for clinical end-users. However, if the interpretation of these XAI results cannot be intuitively understood by non-professional clinical end-users, these expressions may make it more difficult for end-users to understand. Xiao et al. (2018)

established a deep learning model to describe the general situation of patients by transforming EMR events into embedded clinical principles, to realize the dynamic correlation between readmission and possible risk factors for patients.

XAI through presenting alternatives. For the XAI medical identification model of health factors or pathological characteristics, the Local Interpretable Model-Agnostic Explanation (lime) method can be used. For the effective application of this method in the medical field of representing XAI through agents, the local interpretable model agnostic interpretation (lime) method is used to identify individual health-related factors leading to disease prediction. This method establishes a supplementary model, Based on the real data, the neighborhood data is generated, and the interpretation is generated by learning the linear model in the neighborhood [34]. This method has strong generalization and is suitable for most scenes; However, the model is only a vague interpretation of the AI model, and if it is too complex or too abstract, it may affect the understanding of clinicians. Meldo et al. [34] used lime to select important features from the special feature representation of segmented lung suspicious objects [34].

Generative Adversarial Networks (GAN) based XAI. A generative adversarial network was first proposed in 2014, using the idea of a "zero-sum" game to generate realistic images. Gan-based image-to-image conversion is one of the innovative methods to solve the small number of medical images. Conan attempts to use additional information to make Gan controllable [35]. Infogan and Clustergan both use the introduction of external auxiliary networks to realize conditional image generation [36, 37]. Pix2pix Gan uses u-net combined with countermeasure training to realize pixel-based countermeasure training [38]. CycleGAN uses two sets of GAN and increases the cycle loss to realize the style conversion of unpaired data [39]. StyleGan improves the style of the original image and realizes image enhancement based on style migration [40]. These methods have been applied to many medical image processing tasks, such as PET-CT conversion, MR motion artifact correction, and PET image denoising. Cross-modal translation is helpful to deepen the mutual understanding between medical texts and images; The disadvantage is that the deviation generated in the process of image enhancement is used to produce the deviation of the result. Wang et al. [41] proposed a Gan image synthesis and generation structure (path-GAN) based on the pathological description and realized the task of Dr image enhancement and recognition [42] (Table 5.4).

5.4 Contributions

This study applied a bibliographic analysis and text mining method to deliver a review of XAI. The key authors, countries, and journals are identified. Opinions of core articles and burst topics are explored. The classification, advantages, disadvantages, and examples are discussed. This research indicates that XAI is a vital technique for medical diagnosis, it exhibits a great strength over black-box models. All kinds of digital medical and pathological datasets could be considered, for instance images,

Table 5.4 Analysis of XAI-based method applied to medical fields

Index	Methods	Advantages	Disadvantages	Cases
1	XAI-based method through effective dimensionality reductions	The volume of the model is smaller, easier to understand, and it is not easy to fit	There may be omissions in the extraction of key features, so the performance of specific classification and recognition is not high	Zhao et al. [25] realized the classification task of a large number of lung cancer patients in the same period through supervised learning of representative cases [25]
2	Feature importance extraction based XAI	Through the extraction of feature importance, XAI can not only explain the essential features but also reflect their relative importance to clinical interpretation	Is numerical weight important? It is not easy to understand or maybe misunderstood	Shrikumar et al. Used the back-propagation mechanism of a deep learning network to measure the output gradient of input and output with the importance of features [26]
3	Attention-based XAI	The AI algorithm can be analyzed through global or partial attention features, to improve the trust and tolerance of clinical users to the model	Excessive or wrong important attention to information extraction will cause trouble for users' judgment and understanding, and make users tend to ignore or key non highlighted areas	Kwon et al. [43] proposed an intuitive and interpretable heart failure and cataract risk prediction model based on retains, which improves the interpretability of the deep learning model by highlighting the image sequence features in the sequence. In these positions, these input features can affect the prediction results [43]
4	CAM-based XAI	It can effectively display important feature areas on the picture	For other forms of data, the effect is not obvious	Lee et al. [31] based on cam, proposed an XAI algorithm for detecting acute intracranial hemorrhage from a small data set, which is one of the most famous studies using cam [31]

(continued)

Table 5.4 (continued)

Index	Methods	Advantages	Disadvantages	Cases
5	XAI through knowledge extraction and rule	XAI-based method relies on knowledge extraction and rule extraction. In theory, it is a more stable model. The summary representation of complex clinical data can intuitively provide interpretable results for clinical end-users	If the interpretation of these XAI results cannot be intuitively understood by non-professional clinical end-users, these expressions may make it more difficult for end-users to understand	Xiao et al. [44] established a deep learning model to describe the general situation of patients by transforming EMR events into embedded clinical principles, to realize the dynamic correlation between readmission and possible risk factors for patients [44]
6	XAI through presenting alternatives	This method has strong generalization and is suitable for most scenes	The model is only a vague interpretation of the AI model, and if it is too complex or too abstract, it may affect the understanding of clinicians	Meldo et al. [34] used lime to select important features from the special feature representation of segmented lung suspicious objects [34]
7	GAN-based XAI	Cross-modal translation helps to deepen the mutual understanding between medical texts and images	This method uses the deviation generated in the process of image enhancement to produce the deviation of the result	Wang et al. [41] proposed a GAN image synthesis and generation structure (path GAN) based on the pathological description and realized the task of Dr image enhancement and recognition [42]

omics data, and text. The hybrid distributional models and AI models with optical computing could be considered to be apply to XAI in the future. Workflow optimization and health system medical error detection, XAI algorithm development, and XAI measurement systems are also potential directions for future researchers. Challenges of data transparency, privacy, bias, and security should also be paid attention to. However, this article is implemented on the limited literature databases in English, where more databases in more than one language should be considered.

Acknowledgements This research was funded by the 2020 Key Technology R & D Program of GuangDong Province, grant number of 2020B1111540001, Zhuhai Technology and Research Foundation, grant number of ZH01110405180056PWC, Zhuhai Technology and Research Foundation, grant number of ZH22036201210034PWC and 2220004002412, Zhuhai Basic and Application Research Project, grant number of ZH22017003200011PWC.

References

1. Wang, H., Li, Z.: The application of machine learning and deep learning to Ophthalmology: A bibliometric study (2000–2021). In: Tareq, A., Taiar, R., (eds.), Human interaction & emerging technologies (IHIET-AI 2022): Artificial intelligence & future applications, vol. 23, (2022)
2. Esteva, A., et al.: Dermatologist-level classification of skin cancer with deep neural networks. Nature, **542**(7639), 115–118 (2017)
3. Wang, H.: A bibliographic study and quantitative analysis of age-related macular degeneration and fundus images. Ann. Ophthalmol. Vis. Sci., **5**(1027), 1–8 (2022)
4. Swartout, W.R., Smoliar, S.W.: Explaining the link between causal reasoning and expert behavior. In: Proceedings—The eleventh annual symposium on computer applications in medical care, pp. 37–42. IEEE, New York, USA (1987)
5. Adadi, A., Berrada, M.: Peeking inside the black-box: a survey on explainable artificial intelligence (XAI). IEEE Access, **6**, 52138–52160 (2018)
6. Ribeiro, M. T., Singh, S., Guestrin, C.: Why should i trust you? Explaining the predictions of any classifier. In: Proceedings of the 22nd ACM SIGKDD international conference on knowledge discovery and data mining, pp. 1135–1144 (2016)
7. Sarzynska-Wawer, J., et al.: Detecting formal thought disorder by deep contextualized word representations. Psychiatry Res. **304**, 114135 (2021)
8. Guidotti, R., Monreale, A., Ruggieri, S., Turini, F., Giannotti, F., Pedreschi, D.: A survey of methods for explaining black box models. ACM Comput. Surv., **51**(5), 1–42 (2018)
9. Rudin, C.: Stop explaining black box machine learning models for high stakes decisions and use interpretable models instead. Nat. Mach. Intell., **1**(5), 206–215 (2019)
10. Holzinger, A.A.-O., Carrington, A., Müller, H.: Measuring the quality of explanations: the system causability scale (scs): comparing human and machine explanations. Kunstliche Intell (Oldenbourg), **34**(1610–1987) (Electronic), 193–198 (2020)
11. Holzinger, A., Biemann, C., Pattichis, C.S., Kell, D.B.: What do we need to build explainable AI systems for the medical domain? ed: arXiv (2017)
12. Topol, E.A.-O.: High-performance medicine: the convergence of human and artificial intelligence. (in eng), Nat Med, **25**(1546–170X) (Electronic), 44–56 (2019)
13. Anguitaruiz, A., et al.: EXplainable Artificial Intelligence (XAI) for the identification of biologically relevant gene expression patterns in longitudinal human studies, insights from obesity research. PLoS Comput. Biol. **16**, e1007792 (2020)
14. Samek, W., Wiegand, T., Müller, KR.: Explainable artificial intelligence: understanding, visualizing and interpreting deep learning models. arXiv (2017)
15. Miller, T.: Explanation in artificial intelligence: Insights from the social sciences. Artif. Intell., **267**, 1–38 (2019)
16. Livia, F., et al.: A clinician's guide to artificial intelligence: how to critically appraise machine learning studies. Transl. Vis. Sci. & Technol., **9**(2), (2020)
17. Vollmer, S., et al.: Machine learning and AI research for patient benefit: 20 critical questions on transparency, replicability, ethics and effectiveness. arXiv preprint arXiv:1812.10404 (2018)
18. Verghese, A., Shah, N.H., Harrington, R.A.: What this computer needs is a physician: humanism and artificial intelligence. JAMA, **319**(1), (2018)
19. Stuart, K., et al.: Visualizing deep learning models for the detection of referable diabetic retinopathy and glaucoma. JAMA Ophthalmol., **137**(3), (2019)
20. Felix, G., et al.: A deep learning algorithm for prediction of age-related eye disease study severity scale for age-related macular degeneration from color fundus photography. Ophthalmology, **125**(9), (2018)
21. Simonyan, K., Vedaldi, A., Zisserman, A.: Deep inside convolutional networks: Visualising image classification models and saliency maps. arXiv (2013), arXiv:1312.6034. (2013)
22. Phillips, L., Goh, G., Hodas, N.: Explanatory masks for neural network interpretability. arXiv(2019), arXiv:1911.0687. (2019)

23. Holzinger, A., Langs, G., Denk, H., Zatloukal, K., Muller, H.: Causability and explainability of artificial intelligence in medicine. Wiley Interdiscip Rev Data Min Knowl Discov **9**(4), e1312 (2019)
24. Yang, G., Ye, Q., Xia, J.: Unbox the black-box for the medical explainable AI via multi-modal and multi-centre data fusion: A mini-review, two showcases and beyond. Inf. Fusion, **77**, 29–52 (2022)
25. Zhao, L.P., Bolouri, H.: Object-oriented regression for building predictive models with high dimensional omics data from translational studies. J. Biomed. Inform., **60**, 431–445 (2016)
26. Shrikumar, A., Greenside, P., Kundaje, A.: Learning important features through propagating activation differences. In: International conference on machine learning, pp. 3145–3153 (2017)
27. Putelli, L., Gerevini, A.E., Lavelli, A., Serina, I.: Applying self-interaction attention for extracting drug-drug interactions. In: International conference of the Italian association for artificial intelligence, pp. 445–460 (2019)
28. Chou, Y. L., Moreira, C., Bruza, P., Ouyang, C., Jorge, J.: Counterfactuals and causability in explainable artificial intelligence: Theory, algorithms, and applications. Inf. Fusion, Article **81**, 59–83 (2022)
29. Kwon, B.C., et al.: RetainVis: Visual analytics with interpretable and interactive recurrent neural networks on electronic medical records. IEEE Trans. Visual Comput. Graphics **25**(1), 299–309 (2019)
30. Zhou, B., Khosla, A., Lapedriza, A., Oliva, A., Torralba, A.: Learning deep features for discriminative localization. In: Proceedings of the IEEE conference on computer vision and pattern recognition, pp. 2921–2929 (2016)
31. Lee, H., Kim, S.T., Ro, Y.M.: Generation of multimodal justification using visual word constraint model for explainable computer-aided diagnosis. In: The 2nd international workshop on interpretability of machine intelligence in medical image computing, iMIMIC 2019, and the 9th international workshop on multimodal learning for clinical decision support, ML-CDS 2019, held in conjunction with the 22nd international conference on medical imaging and computer-assisted intervention, MICCAI 2019, vol. 11797, pp. 21–29 (2019)
32. Lee, H., et al.: An explainable deep-learning algorithm for the detection of acute intracranial haemorrhage from small datasets. Nat. Biomed. Eng., **3**(3), 173–182 (2019)
33. Salahuddin, Z., Woodruff, H.C., Chatterjee, A., Lambin, P.: Transparency of deep neural networks for medical image analysis: A review of interpretability methods. Comput. Biol. Med., **140**, (2022)
34. Meldo, A., Utkin, L., Kovalev, M., Kasimov, E.: The natural language explanation algorithms for the lung cancer computer-aided diagnosis system. Artif. Intell. Med. **108**, 101952 (2020)
35. Ma, Y., Chen, X., Zhu, W., Cheng, X., Xiang, D., Shi, F.: Speckle noise reduction in optical coherence tomography images based on edge-sensitive cGAN. Biomed. Opt. Express, **9**(11), 5129–5146 (2018)
36. Chen, X., Duan, Y., Houthooft, R., Schulman, J., Sutskever, I., Abbeel, P.: Infogan: Interpretable representation learning by information maximizing generative adversarial nets. Adv. Neural Inf. Process. Syst., **29**, (2016)
37. Mukherjee, S., Asnani, H., Lin, E., Kannan, S.: Clustergan: Latent space clustering in generative adversarial networks. In: Proceedings of the AAAI conference on artificial intelligence, vol. 33, no. 01, pp. 4610–4617 (2019)
38. Qu, Y., Chen, Y., Huang, J., Xie, Y.: Enhanced pix2pix dehazing network. In: Proceedings of the IEEE/CVF conference on computer vision and pattern recognition, pp. 8160–8168 (2019)
39. Almahairi, A., Rajeshwar, S., Sordoni, A., Bachman, P., Courville, A.: Augmented cyclegan: Learning many-to-many mappings from unpaired data. In: International conference on machine learning, pp. 195–204 (2018)
40. Karras, T., Laine, S., Aittala, M., Hellsten, J., Lehtinen, J., Aila, T.: Analyzing and improving the image quality of stylegan. In: Proceedings of the IEEE/CVF conference on computer vision and pattern recognition, pp. 8110–8119 (2020)
41. Wang, M.H., Chong, K.K.-l., Lin, Z., Yu, X., Pan, Y.: An explainable artificial intelligence-based robustness optimization approach for age-related macular degeneration detection based

on medical IOT systems. Electronics, **12**, 2697. https://doi.org/10.3390/electronics12122697 (2023)
42. Niu, Y., Gu, L., Zhao, Y., Lu, F.: Explainable diabetic retinopathy detection and retinal image generation. IEEE J. Biomed. Health Inform., **26**(1), 44–55 (2022)
43. Kwon, J., Choi, J., Shin, J. W., Lee, J., Kook, M.S.: Glaucoma diagnostic capabilities of foveal avascular zone parameters using optical coherence tomography angiography according to visual field defect location. J. Glaucoma, **26**(12), 1120–1129 (2017)
44. Xiao, C., Ma, T., Dieng, A. B., Blei, D. M., Wang, F.: Readmission prediction via deep contextual embedding of clinical concepts. PloS one, **13**(4), e0195024 (2018).

Chapter 6
Online Task Admission and Resource Allocation in Edge Computing-Based Cloud Gaming

Zhongyang Wang, Gang Sun, and Du Xu

Abstract Cloud gaming refers to an online gaming paradigm that runs game program on remote server and streams the game video back to client device. It allows players to access high-quality games without the need of powerful hardware. The streaming of cloud gaming requires a high-bandwidth, low-latency and stable Internet connection. Conventional cloud gaming is based on cloud computing, but it is gradually being replaced by edge computing. Edge computing is an emerging network computing pattern that deploys edge servers at the edge of the network, close to users. In this way, edge computing is able to provide low-latency and high-bandwidth network computation service, which fits the requirement of cloud gaming perfectly. An edge server needs to serve multiple gaming tasks simultaneously. The CPU and GPU resource requirements of each task varies over time and cannot be predicted accurately. Meanwhile, the arrival and the departure time of a task is unknown until it happens. These online characteristics makes the decision of task admission and resource allocation problem hard to solve. In this paper, the inherited online optimization problem is formulated into a Mixed Integer Nonlinear Programming problem (MINLP). We proposed an online resource reservation and parameter estimation-based greedy (REG) algorithm. According to the result of simulation tests, the performance of the proposed algorithm is 92.12% of the optimal solution.

6.1 Introduction

Cloud gaming is an innovative form of online gaming that utilizes remote servers to operate video games and streams the game video back to players' devices. In contrast to the conventional video gaming pattern, where games operate locally on a PC or console, cloud gaming has the potential to transform the gaming industry by

Z. Wang · G. Sun (✉) · D. Xu
School of Information and Communication Engineering, University of Electronic Science and Technology of China, Chengdu 611731, P. R. China
e-mail: gangsun@uestc.edu.cn

R. Kountchev (Deceased) et al. (eds.), *AI Methods and Applications in 3D Technologies*, Smart Innovation, Systems and Technologies 388,
https://doi.org/10.1007/978-981-97-2144-3_6

providing users with access to high-quality game services on thin clients like smartphones and set-top boxes. However, cloud gaming imposes a strict requirement on network computation and communication. The user experience in cloud gaming is largely influenced by user-perceived latency, which refers to the delay between a user's input and the corresponding game video displayed on their client's screen. Conventionally, cloud gaming is based on cloud computing, which fails to provide a low-latency and stable network connection. Fortunately, the emerging network computing paradigm of edge computing offers a promising alternative to cloud computing in cloud gaming. By situating the computing and caching resources at the edge of the internet, edge computing can provide a low-latency, high-bandwidth, and less-fluctuating network computation service.

Live streaming and cloud gaming are often compared due to their common requirement for low-latency video transmission. However, several differences exist between the two technologies. In cloud gaming, latency encompasses not only the latency of video streaming but also the latency of uploading the user's input stream and the time required to execute the game program. Additionally, cloud gaming systems demand strict response latency requirements, often requiring a latency of less than 100 ms [1]. As a result, research on live streaming is not entirely applicable to cloud gaming problems.

In edge computing, an area is typically divided into several small areas, each of which receives computing services from an edge server. However, the computation resource of an edge server is limited, and each game program requires different CPU and GPU resources at various time points that cannot be precisely predicted. Therefore, task admission decisions and computing resource allocation on edge servers are crucial in achieving high system efficiency. Additionally, for a game running on an edge server, its video resolution can be dynamically adjusted, which affects both the clarity of the video and the delay of video encoding [2]. To jointly solve these optimization problems, we proposed an online resource reservation and parameter estimation-based greedy (REG) algorithm, whose performance has been evaluated through simulation tests. In the context of evaluating the performance of an online algorithm, it is customary to compare it to the offline optimal solution, which is assumed to possess complete knowledge not only of the past but also of the future. The results indicate that the proposed algorithm's solution is close to the optimum. The contributions of this work are as follows:

- The task admission and resource allocation problems in cloud gaming systems based on edge computing have been identified and formulated as an online integer-mixed non-linear optimization problem.
- To address the challenges of task arrival randomness and unknown parameters at the current time, a resource reservation and parameter estimation-based online greedy algorithm (REG) has been designed.
- Through simulation experiments, the performance of the proposed REG algorithm in this paper is shown to be close to the offline optimal solution.

6.2 Related Work

In this section, we will introduce the existing works concerning resource allocation in edge computing and cloud gaming. In [3], the impact of delay on cloud gaming QoE was studied by identifying game characteristics influencing the delay perception of the users. Foveated video encoding (FVE) reduces bandwidth requirements by exploiting the non-uniform acuity of the human visual system and understanding where the user is looking. Work [4] proposed a prototype of FVE cloud gaming platform and the corresponding FVE parameter settings. Zadtootaghaj et al. [5] provides a game video quality data set. This data set uses the H.264 standard hardware acceleration engine for video compression. Xu et al. [6] provided the first measurement research on the public edge cloud gaming platform which had been densely deployed in China. This work studied the experience comparison between cloud-based and edge-based cloud gaming from user's perspective, as well as the workload difference of cloud-based and edge-based cloud gaming from service provider's perspective. In [7], a cloud gaming platform combining edge computing and software-defined networks (SDN) is invented. A suboptimal dynamic server and path selection method is proposed to achieve a fair resource allocation among a group of players. Paper [8] studied the optimization problem for QoE-aware resource allocation for multiple cloud gaming users sharing a bottleneck link. The objective jointly considered both quality and QoE fairness. In edge computing, in order to efficiently model requests in service function chains, [9] proposes a service request-like workflow. An efficient scheduling algorithm for online service function chaining requests is proposed in [10]. To better deploy service function chains, [11] invented a breadth-first search-based online algorithm. In [12], a meta-heuristic algorithm is proposed to determine the framerate of game video for each user under the CPU and GPU resource constraint.

6.3 System Model and Problem Formulation

The system model under consideration in this study comprises multiple user clients and a single edge server. Figure 6.1 depicts the flow diagram of the cloud gaming process. Initially, a user client acquires input from the user and transmits it to the edge server. The edge server processes the input, runs the game program, generates the original game frames, encodes the frames into game video, and sends it back to the client. Subsequently, the client decodes and displays the game video on the screen. To optimize system performance, the controller collects system information and adjusts the resolution of the original game frames dynamically. The video encoder encodes the game video without modifying the resolution.

A discrete time model consisting of $|T|$ time slots is considered. Each time slot is labeled with a serial number $t \in T$, where T is the set of all time slots. The duration of each time slot is T^{slot} and $|T|$ input segments are assumed to be generated by

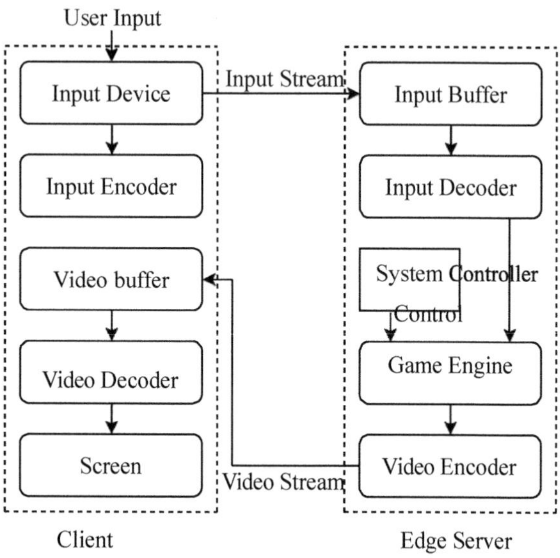

Fig. 6.1 Flow diagram of cloud gaming

the client, each of length T^{slot}, for convenience. The t-th input is received at the beginning of time slot t, and the delay between the user generating the k-th input and its arrival at the edge server is denoted as T^{input} t. The length of the t-th video segment is also T^{slot} as this model does not consider changing the playback speed of the video. Each video segment is encoded into a Group of Picture (GoP) and transmitted to the client.

In total, there are $|I|$ game tasks. Assuming task arrival in order, task i is marked as the i-th arriving task. At the beginning of some random time slot, each task arrives. Each task i has a start time (i.e., arriving time) and end time, during which uninterrupted cloud gaming service should be provided. The duration of task i is $T_i \in T$. The decision variable $y_i \in \{0, 1\}$ indicates whether game task i is served or not. The indirect variable $x_{it} \in \{0, 1\}$ indicates whether task i is served at time t. Constraint (6.1) guarantees that task i is served for the duration.

$$x_{it} = y_i, \forall i \in I, t \in T_i \tag{6.1}$$

In running the game, the CPU is responsible for executing the game logic, while the GPU is responsible for rendering the game graphics and encoding the game video. The CPU resource requirement c_{it} for each game task varies over time and is difficult to predict with accuracy. However, its distribution can be analyzed by examining historical records. The CPU demand c_{it} for each game task at each time slot can be treated as independent and identically distributed. C and G denote the total amount of CPU and GPU resource that can be provided by the edge server in one time slot, respectively.

$$\sum_{i \in I} c_{it} x_{it} \leq C, \forall t \in T \tag{6.2}$$

$$\sum_{i \in I} g_{it} x_{it} \leq G, \forall t \in T \tag{6.3}$$

The original game graphics are produced by the GPU. The resolution r_{it} of the original game graphics can be dynamically adjusted. However, the time required by the GPU to generate the original game graphics, denoted as T^{GPU}, remains constant regardless of the resolution. The value of the resolution is determined by the product of the number of horizontal pixels and the number of vertical pixels. The scaling factor β_{it} represents the computing power required by the GPU to render a pixel. The GPU resource allocated to task i on time slot t is denoted as g_{it}. The difficulty of rendering the game graphics for task i on time slot t is reflected by the value of β_{it}. The greater the difficulty, the more GPU computing power is required. It should be noted that β_{it} varies with time and cannot be accurately predicted. The resolution of the game graphics is typically set to 360p, 480p, 720p, 1080p, or 2K. The minimum resolution is set to 360p, which corresponds to a resolution of 640×360, or 230,400 pixels. The maximum resolution for each task depends on the service level selected by the user.

$$g_{it} = \beta_{it} r_{it}, \forall i \in I, t \in T \tag{6.4}$$

$$r^{\min} \leq r_{it} \leq r^{\max}, \forall i \in I, t \in T \tag{6.5}$$

The H.264 format is used by the video encoder to convert the original game graphics into a video suitable for network transmission. The bitrate of the video is determined by three parameters, namely resolution, frame rate, and bits per pixel (BPP), whose product yields the bitrate. As this work focuses on enhancing the Quality of Service (QoS) by optimizing GPU resource allocation, we assume fixed values for frame rate and BPP. Specifically, the frame rate is set to $f = 60$ frames per second (fps), while the BPP is $\kappa = 8$ bits per frame. Consequently, the quality of a video segment is represented by its bitrate b_{it}. The length of each video segment is fixed at T^{slot}, such that the bitrate determines the size of the video segment file. The bitrate at minimum resolution r^{min} is denoted as b^{base}.

$$b_{it} = f \kappa r_{it}, \forall i \in I, t \in T \tag{6.6}$$

The focus of the system model is on server-side task scheduling and its impact on quality of service; hence, the relationship between bitrate and network transmission packet loss is not taken into account. The model does not consider the variation of transmission latency caused by packet loss and congestion. The quality of video segment i on time slot t is defined as $Q_{it}^{bitrate}$ according to [13]:

$$Q_{it}^{bitrate} = \ln\left(\frac{b_{it}}{b^{base}}\right), \forall i \in I, t \in T \tag{6.7}$$

It is noteworthy that the relationship $b^{base} = f\kappa r^{min}$ implies that $Q_{it}^{bitrate} = \ln\left(\frac{r_{it}}{r^{\min}}\right)$ can be inferred. The encoding time for a given video segment i during time slot t is represented by T_{it}^{encode}. Based on the empirical findings presented in [14], when encoding quality is held constant, a regression model of encoding delay as a function of resolution r_{it} is fitted to Eq. (6.8), as demonstrated in Fig. 6.2.

$$T_{it}^{encode} = a \cdot r_{it} + e, \forall i \in I, t \in T \tag{6.8}$$

where $a = 9.169 \times 10^{-6}$, $e = 10.82$. For instance, the pixel count for a 1440p video is computed as $2560 \times 1440 = 3{,}686{,}400$. The actual encoding delay for a 1440p video is 44.05 ms, while the encoding delay obtained from Eq. (6.8) is 44.62 ms. The discrepancy between the estimated and actual values is below 1.30%. Consequently, the fitted function's error is negligible, and it can be employed in problem formulation.

Equation (6.9) presents the four components that constitute the user perceived latency, $T_{it}^{perceived}$. The collection and transmission latency of user input is denoted by T^{uplink}, whereas the latency associated with running game programs at the edge server and encoding video are represented by $T^{program}$ and T_{it}^{encode}, respectively. Furthermore, $T_{it}^{downlink}$ signifies the latency associated with video transmission and playback, which is a stochastic process. Given the low bandwidth of the user input stream and its minimal susceptibility to network fluctuations, T^{uplink} is considered a constant. In addition, the delay associated with game program execution and the generation of original game frames, $T^{program}$, remains stable. The encoding delay, T_{it}^{encode} is determined by the video resolution, as presented in Eq. (6.8). To simplify the notation, the sum of T^{uplink}, $T^{program}$, and $T_{it}^{downlink}$ is denoted as T_{it}^{other}, as illustrated in Eq. (6.10).

Fig. 6.2 Encoding time versus resolution (pixel number)

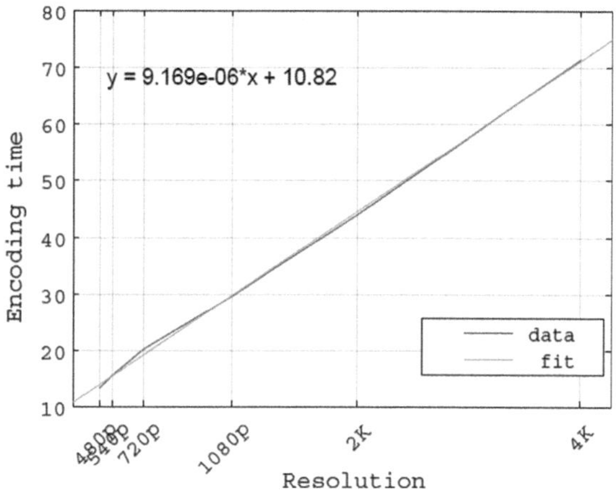

$$T_{it}^{perceived} = T^{upllink} + T^{program} + T_{it}^{encode} + T_{it}^{downlink} \tag{6.9}$$

$$T_{it}^{other} = T^{upllink} + T^{program} + T_{it}^{downlink} \tag{6.10}$$

$$Q_{it}^{latency} = T_{it}^{threashold} - (T_{it}^{encode} + T_{it}^{other}), \forall i \in I, t \in T \tag{6.11}$$

The weight factors, $\gamma^{bitrate}$ and $\gamma^{latency}$, are utilized to respectively adjust the proportion of $Q_{it}^{bitrate}$ $Q_{it}^{latency}$ in the overall quality of service (QoS). The objective of the formulated optimization problem is to maximize the overall QoS.

$$\begin{aligned}
&\max_{y_i, r_{it}} \sum_{t\in T}\sum_{i\in I} x_{it}(\gamma^{bitrate} Q_{it}^{bitrate} + \gamma^{latency} Q_{it}^{latency}) \\
&\text{s.t.} Q_{it}^{bitrate} = \ln(\frac{r_{it}}{r^{\min}}), \forall i \in I, t \in T \\
&Q_{it}^{latency} = T^{threshold} - (a \cdot r_{it} + e + T_{it}^{other}), \forall i \in I, t \in T \\
&\sum_{i\in I} c_{it} x_{it} \le C, \forall t \in T \\
&\sum_{i\in I} \beta_{it} r_{it} x_{it} \le G, \forall t \in T \\
&r^{\min} \le r_{it} \le r^{\max}, \forall i \in I, t \in T \\
&y_i, x_{it} \in \{0, 1\}, \forall i \in I, t \in T
\end{aligned} \tag{6.12}$$

6.4 Algorithm Design

The proposed online optimization problem is characterized by its decision variables, which include y_i and r_{it}. The former represents the decision to serve task i, while the latter denotes the resolution setting for task i on each time slot t. The problem addresses two primary concerns: whether to serve a particular task, and the optimal resolution setting for that task on each time slot. When multiple game tasks compete for the CPU and GPU resources of the edge server on a given time slot t, rejecting some tasks due to high system load results in a decline in the overall QoS. Hence, to maximize the objective function, the edge server should serve as many tasks as possible. The optimal resolution setting for each task i on time slot t is a complex trade-off between the increase in $Q_{it}^{bitrate}$ with higher resolution r_{it}, and the decrease in $Q_{it}^{latency}$ due to increased encoding delay. These two metrics are inherently contradictory, and the algorithm must find an optimal balance between them.

We propose an online algorithm named the Reservation and Estimation-based Greedy algorithm (REG). The main idea of the REG algorithm is to decompose the online problem into a series of single-time-slot sub-problems and solve them individually. To account for the inelasticity of a task's CPU resource demand on a time slot, the algorithm takes into account the expectations of each game task's CPU requirement when decomposing the problem. In the subproblems, the downlink latency $T_{it}^{downlink}$ is estimated using a sliding window method. The subproblems are then solved optimally as they are convex optimization problems. It is worth noting that there are two manually adjustable parameters, p and d, for the REG algorithm. The role of p is to adjust the virtual upper limit of the CPU resources of the edge server. d is the length of the sliding window used to estimate $T_{it}^{downlink}$. The details of the REG algorithm are as follows.

The online optimization problem under consideration poses a challenge in obtaining accurate future information due to its dynamic nature. In particular, the values of c_{it}, β_{it} and $T_{it}^{downlink}$ cannot be precisely predicted, but they can be estimated. The historical data can be used to obtain the expectation $c_i^{\exp}$ and the standard deviation c_i^{sd} of c_{it}. The real-time value of β_{it} can be estimated by using the previous time slot value, that is, $\overset{\wedge}{\beta_{it}} = \beta_{i(t-1)}$. A sliding window method is employed to estimate $T_{it}^{downlink}$, where the length of the window is d. The estimated value of $T_{it}^{downlink}$ is denoted by $\hat{T}_{it}^{downlink}$. Different values of d have varying sensitivities to fluctuations in T values. When d is large, the estimated value $\hat{T}_{it}^{downlink}$ is closer to the long-term downlink latency. It is worth noting that the impact of network fluctuations can be alleviated by the video cache on the client. Therefore, it is advisable to avoid selecting a value of d that is too small.

To ensure that the tasks being served are not interrupted due to insufficient CPU resources, this algorithm proposes a virtual CPU resource upper limit, denoted as $p \cdot C$, where p is a number between 0 and 1. At each time slot t, the algorithm sums up the expected CPU requirement of the currently running tasks, $c_i^{\exp}$ and compares it with the virtual CPU upper limit, $p \cdot C$. The choice of p is nontrivial and depends on the variability in c_i^{sd}. In practice, p can be adjusted to achieve optimal performance. By utilizing both the virtual CPU upper limit and the expected CPU requirement, the original problem (6.12) is decomposed into a series of subproblems for each time slot t. This approach ensures that the CPU resources are allocated efficiently, and the tasks are served with minimum interruption.

$$
\begin{aligned}
&\max_{r_{it}} \sum_{i \in I} x_{it}(\gamma^{bitrate} Q_{it}^{bitrate} + \gamma^{latency} Q_{it}^{latency}) \\
&\text{s.t.} Q_{it}^{bitrate} = \ln(\frac{r_{it}}{r^{\min}}), \forall i \in I, t \in T \\
&Q_{it}^{latency} = T^{threshold} - (a \cdot r_{it} + e + T_{it}^{other}), \forall i \in I \\
&\sum_{i \in I} \beta_{it} r_{it} x_{it} \le G \\
&r^{\min} \le r_{it} \le r^{\max}, \forall i \in I
\end{aligned}
\tag{6.13}
$$

In the subproblem on time slot t, denoted by (6.13), the aim is to maximize a combination of concave functions, namely the logarithmic functions, and affine functions, namely the linear functions. Thus, the objective function can be expressed as minimizing a convex function. The constraints in this subproblem are linear combinations of variable r_{it}. Both the objective function and the constraints of subproblem (6.13) adhere to the canonical form of a convex function. Therefore, (6.13) is a convex problem, which can be solved efficiently using solver software.

Algorithm 1 : Reservation and Estimation-based Greedy (REG) algorithm

Data: system status on each time slot $t \in T$
Input: algorithm parameters d, p
Result: $y_{it}, r_{it}, \forall i \in I, t \in T$
Initialize: $y_i, r_{it}, x_{it}, \forall i \in I, t \in T$
for $t \in T$
    Denote the set of tasks arriving on time slot t as $I_t^{arrival}$;
    $I_t^{serving} = \{i | x_{it} = 1, \forall i \in I\}$;
    if $I_t^{arrival} \neq 0$ then
        Sort all elements in set $I_t^{arrival}$ in ascending order based on the value of c_i^{exp}, then put the sorted elements into an array M;
        $C^{available} = p \cdot C - \sum_{i \in I_t^{serving}} c_i^{exp}$;
        for $\{i | i \in M\}$ do
            if $C^{available} - c_i^{exp} \geq 0$ then
                $C^{available} = C^{available} - c_i^{exp}$;
                $y_i, x_{it} = 1, \forall T \in T_i$;
                Add element i into set $I_t^{serving}$;
            end
        end
    end
    for $\{i | i \in I_t^{serving}, \forall i \in I\}$ do
        If $t - d \geq 0$ then
            $\hat{T}_{it}^{downlink} = \frac{1}{d} \sum_{t \in T}^{t-d \leq T \leq t} T_{it}^{downlink}$;
        else
            $\hat{T}_{it}^{downlink} = \frac{1}{d} \sum_{t \in T}^{1 \leq T \leq t} T_{it}^{downlink}$
        end
        $\hat{T}_{it}^{other} = T^{upllink} + T^{program} + \hat{T}_{it}^{downlink}$;
        Solve subproblem (13) using a solver, and set values for r_{it};
    end
end

6.5 Discussion

The discussion is based on the assumption that the distribution of *cit* for a certain task i on time each time slot is normal distribution. The distribution of the sum of independent normal distribution variables is also a normal distribution. In this case, the distribution of $\sum_{i \in I} c_{it} x_{it}$ is a normal distribution with the expectation of $\mu = \sum_{i \in I} c_i^{\exp} x_{it}$ and the variation of $\sigma^2 = \sum_{i \in I} (c_i^{sd})^2 x_{it}$. In the empirical sciences, the so-called three-sigma rule of thumb (or 3σ rule) expresses a conventional heuristic that nearly all values are taken to lie within three standard deviations of the mean, and thus it is empirically useful to treat 99.7% probability as near certainty. By utilizing the 3σ rule, decisions about which new tasks to accept can be made easily without using the virtual CPU resource upper limit method. Specifically, as long as the calculation is performed for each newly arrived task in turn to ensure that the following inequality (6.14) can be established, one can determine which tasks can be accepted, while ensuring the CPU computing resource constraint will not be violated with a high probability. However, the underlying assumption of c_{it}'s normal distribution might be too strong, and we didn't find enough evidence to support this assumption, thus the idea of this discussion is not used in this paper.

$$\mu + 3\sigma = \sum_{i \in I} x_{it} c_i^{\exp} + 3\sqrt{\sum_{i \in I} x_{it} (c_i^{sd})^2} \leq C, \forall t \in T \tag{6.14}$$

6.6 Evaluation

The proposed algorithm is evaluated through simulation tests, where the parameter settings are provided in Table 6.1. The performance of the GPU is typically measured in floating-point operations per second (FLOPS, flops, or flop/s). For this test, we assume that the GPU of the edge server has a capacity of 120×10^{12} FLOPS. Therefore, in one time slot of 100 ms, the GPU can perform up to 12×10^{12} FLOPS. The GPU requirement for computing a single pixel of game video during a time slot is denoted by β_{it}. The values of $\gamma^{bitrate}$ and $\gamma^{latency}$ are determined jointly from the objective function (6.13) and constraints. Denote the objective function as $f(r_{it})$. The stationary points of $f(r_{it})$ are $r_{it} = \frac{\gamma^{bitrate}}{a \cdot \gamma^{latency}}, \forall i \in I$. However, the stationary points should lie within the feasible region; otherwise, the solution will always be r_{min} or r_{max}. Therefore, we have $r_{\min} \leq \frac{\gamma^{bitrate}}{a \cdot \gamma^{latency}} \leq r_{\max}$, which yields $2.11 \leq \frac{\gamma^{bitrate}}{\gamma^{latency}} \leq 33.8$. In our simulations, we set the value of $\gamma^{bitrate}$ to 10 and $\gamma^{latency}$ to 1.

Two benchmark algorithms are proposed to evaluate the performance of REG algorithm. The first one is a algorithm named No Reservation (NR) that doesn't reserve CPU resource for potential fluctuation of each task's CPU requirement. The

Table 6.1 Simulation test parameter setting

|T|	500
|I|	30
T^{slot}	100 ms
C	3.2 billion cycles
G	12×10^{12} FLOPS
Duration of tasks	U [200, 400] time slots
c_{it}	$N\ (c_i^{\exp},\ (c_i^{sd})^2)$
$c_i^{\exp}$	U [0.2, 0.4] billion cycles
c_i^{sd}	U [0.05, 0.1]
β_{it}	$N\ (\beta_i^{\exp},, (\beta_i^{sd})^2)$ FLOPS
$\beta_i^{\exp}$	$U\ [2 \times 10^5, 4 \times 10^5]$ FLOPS
β_i^{sd}	3×10^4
r_{min}	$640 \times 360 = 230{,}400$
r_{max}	$2560 \times 1440 = 3{,}686{,}400$
$T^{threshold}$	200 ms
T^{other}	$N\ (100, 40^2)$ ms
$\gamma^{bitrate}$	10
$\gamma^{latency}$	1

second benchmark algorithm is called Random Rejection (RR) algorithm, which is introduced in [15]. For each incoming task, the probability of rejecting it equals to the CPU usage on current time slot. The principle is that the higher the system load, the less it should receive new tasks.

Figures 6.3, 6.4, and 6.5 present the results of the conducted simulation tests. The starting times of tasks are within the range of [0, 300] ms, hence the parts with high CPU usage are concentrated in the first 300 slots. In the case of high CPU load, late arriving tasks are rejected. Figure 6.3 depicts the change in CPU usage over time of the proposed algorithm REG and benchmark algorithms NR, RR. It can be inferred that the CPU usage of NR and RR often exceeds the upper limit of resources, which is not desirable in real scenarios. When the value of p is set to 0.8, the CPU load of the system is low, which results in a waste of computing resources. On the other hand, the CPU usage of REG under p = 0.9 is more appropriate and able to handle more tasks. At the same time, the number of times the CPU usage exceeds the upper limit of resources is rare. During low system load, i.e., for $t \in [250, 500]$, $p = 0.9$ exhibits higher CPU usage than NR and RR. This phenomenon can be attributed to the fact that the system load is too high when using NR and RR algorithm, leading to the rejection of tasks that could have been accepted.

Figure 6.4 depicts the GPU occupancy ratio over time for different algorithms. Due to the elasticity of GPU resource allocation, the GPU usage does not surpass the GPU resource upper bound. When the system load is high, the GPU usage for $p = 0.9$ and $p = 0.8$ are similar. As seen with the CPU usage under low system load,

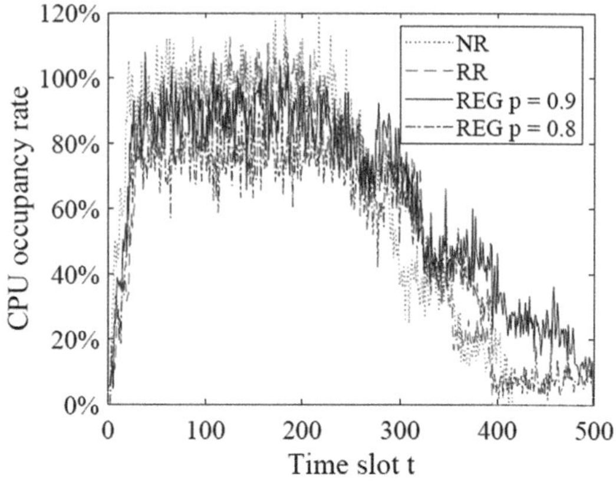

Fig. 6.3 CPU usage ratio under different p versus t

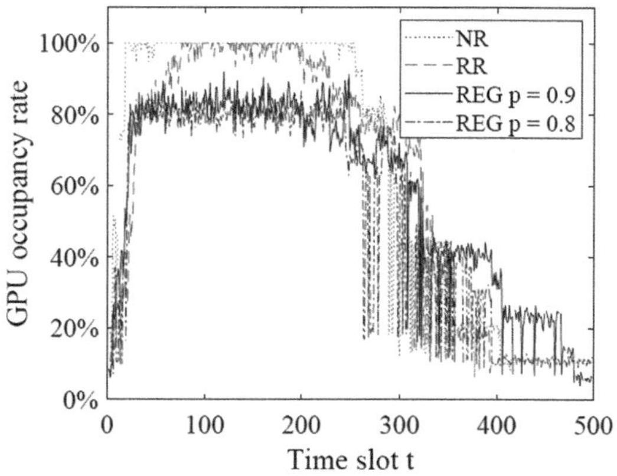

Fig. 6.4 GPU usage under different p versus t

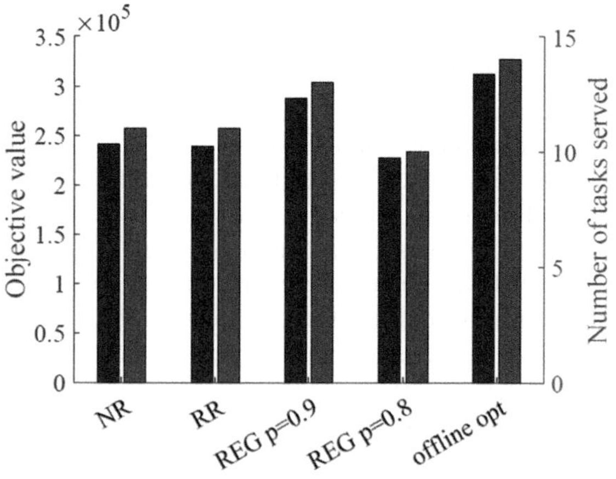

Fig. 6.5 Objective values and served task number of different p and offline optimal solution

the GPU usage for p = 0.9 is higher than that for $p = 0.8$, due to the acceptance of more tasks. Figure 6.5 demonstrates the objective values and served task numbers achieved by different algorithms and the offline optimal solution. The optimal solution is calculated by branching and trimming the decision of task admission. The optimal solution for each task can be calculated in short time since the subproblems are convex optimization problems. The results indicate that our proposed algorithm, REG, performs well under p = 0.9, as its performance closely approaches the offline optimal solution. Specifically, the objective value and the number of served tasks by the REG algorithm reached 92.12% and 92.86% of the optimum solution, respectively. In conclusion, by manually adjusting the value of the parameter p, our proposed algorithm, REG, can achieve system performance that is close to optimal.

6.7 Conclusion

This study addresses the optimization of Quality of Service (QoS) in edge computing based cloud gaming systems by studying the game task admission and resource allocation problem. Specifically, the problem is formulated as an online MINLP problem, and a resource reservation and parameter estimation-based greedy (REG) algorithm is proposed and evaluated through simulation tests. The results demonstrate that the performance of the proposed REG algorithm is close to the offline optimal solution, achieving 92.12% of the optimal value. For future work, a promising direction is the joint optimization of edge server and client, which could further improve the QoS of the system.

References

1. Claypool, M., Claypool, K.: Latency and player actions in online games. Commun. ACM **49**(11), 40–45 (2006)
2. ©2004–2022, Epic Games, I.: dynamic resolution. https://docs.unrealengine.com/5.1/en-US/dynamic-resolution-in-unreal-engine/. Accessed 29 Sept 2022
3. Sabet, S.S., Schmidt, S., Zadtootaghaj, S., Griwodz, C., et al.: Delay sensitivity classification of cloud gaming content. In: Proceedings of the 12th ACM International Workshop on Immersive Mixed and Virtual Environment Systems, pp. 25–30 (2020)
4. Illahi, G.K., Gemert, T.V., Siekkinen, M., Masala, E., Oulasvirta, A., et al.: Cloud gaming with foveated video encoding. ACM Trans. Multimed. Comput., Commun., Appl. (TOMM) **16**(1), 1–24 (2020)
5. Zadtootaghaj, S., Schmidt, S., Sabet, S.S., et al.: Quality estimation models for gaming video streaming services using perceptual video quality dimensions. In: Proceedings of the 11th ACM Multimedia Systems Conference, pp. 213–224 (2020). https://doi.org/10.1145/3339825.3391872
6. Xu, M., Fu, Z., Ma, X., Zhang, L., Li, Y., Qian, F., Wang, S., Li, K., Yang, J., Liu, X.: From cloud to edge: a first look at public edge platforms. In: Proceedings of the 21st ACM Internet Measurement Conference, pp. 37–53 (2021)

7. Parastar, P., Pakdaman, F., Hashemi, M.R.: Frame-SDN: a fair resource allocation for multi-player edge-based cloud gaming in SDN. In: Proceedings of the 25th ACM Workshop on Packet Video, pp. 21–27 (2020). https://doi.org/10.1145/3386292.3397120
8. Slivar, I., Skorin-Kapov, L., Suznjevic, M.: QOE-aware resource allocation for multiple cloud gaming users sharing a bottleneck link. In: 2019 22nd Conference on Innovation in Clouds, Internet and Networks and Workshops (ICIN), pp. 118–123. IEEE (2019)
9. Sun, G., Li, Y., Li, Y., Liao, D., Chang, V.: Low-latency orchestration for workflow-oriented service function chain in edge computing. Futur. Gener. Comput. Syst. **85**, 116–128 (2018)
10. Sun, G., Chen, Z., Yu, H., Du, X., Guizani, M.: Online parallelized service function chain orchestration in data center networks. IEEE Access **7**, 100147–100161 (2019)
11. Sun, G., Xu, Z., Yu, H., Chen, X., Chang, V., Vasilakos, A.V.: Low-latency and resource-efficient service function chaining orchestration in net
12. Seyed Aboutorabi, S.J., Rezvani, M.H.: An optimized meta-heuristic bees algorithm for players' frame rate allocation problem in cloud gaming environments. Comput. Games J. **9**, 281–304 (2020)
13. Zhang, X., Chen, H., Zhao, Y., Ma, Z., Xu, Y., Huang, H., Yin, H., Wu, D.O.: Improving cloud gaming experience through mobile edge computing. IEEE Wirel. Commun. **26**(4), 178–183 (2019). https://doi.org/10.1109/MWC.2019.1800440
14. Team, H.: Handbrake documentation-performance. https://handbrake.fr/docs/en/1.3.0/technical/performance.html. Accessed 21 Feb 2023
15. Buttazzo, G., Lipari, G., Abeni, L., Caccamo, M.: Stochastic scheduling. In: Soft Real-Time Systems: Predictability vs. Efficiency, pp. 235–258 (2005)

Chapter 7
Studying Artificial Intelligence Neural Network Based TCM Diagnosis and Treatment Information Management System

Haixia Huang and Yonghua Mo

Abstract Traditional Chinese medicine is a priceless legacy of Chinese medicine that has developed over thousands of generations of practical accumulation into a whole medical system with a distinctive theoretical foundation and extensive clinical experience. However, because of its heavy reliance on diagnostic experience, it is unsuitable for management and advancement. The idea of digital medicine is therefore progressively becoming more broadly recognised, particularly in the area of medical diagnostics. The effectiveness and precision of medical diagnosis directly affect people's lives and health, making it a critical problem that cannot be disregarded. In order to accomplish the goal of Traditional Chinese medicine (TCM) diagnosis, an expert system is constructed using the continuously expanding case base as the knowledge basis for reasoning. The purpose of this study is to use artificial neural networks (ANNs), which have great learning, adaptability, and massively parallel processing capabilities, to deal with tough, unstructured, nonlinear problems in management information systems. Following an experimental investigation, the total error of the modified system is decreased by around 75.3%, and the average forecast precision near 75%. The back propagation (BP) algorithm is upgraded to optimise the inaccuracy of the weight, repair, or energy parameters in order to achieve this.

7.1 Introduction

Traditional database management systems can only run straightforward queries and generate statistics, which is insufficient for today's demands. People aim to process and analyse large databases of data at a higher level in order to determine the data's general properties and forecast future trends. A modelling of the human nervous

H. Huang · Y. Mo (✉)
Guilin Institute of Information Technology, Guangxi, China
e-mail: 1056029876@qq.com

R. Kountchev (Deceased) et al. (eds.), *AI Methods and Applications in 3D Technologies*, Smart Innovation, Systems and Technologies 388,
https://doi.org/10.1007/978-981-97-2144-3_7

system is called an artificial neural network (ANN). Though it is a simplified and abstract model, it captures many fundamental aspects of how the human brain works. However, it does not quite replicate how the human brain nervous system actually functions.

In order to develop useful solutions, it is important to understand the fundamental mechanics of information processing and storage in the human brain. By applying these concepts to real-world engineering problems, artificial neural networks may be studied. Even if the traditional BP paradigm has many important benefits, it also has a lot of issues. In actuality, choosing the right parameters is the secret to neural network modelling. Experiments and experience indicate that the BP model's starting weight should be a tiny random number. In order to fully utilise the potential of artificial neural network technology in TCM diagnosis and to encourage the advancement of TCM diagnosis in the direction of informatization, this research focuses on the use of artificial neural networks in the informatization of TCM diagnosis. It is envisaged that the enhanced BP neural network algorithm would enable the management information system's performance to be optimised.

7.2 Research on Artificial Neural Network

A neuron is the fundamental building block of an artificial neural network. It has three parts: an input portion, an output part, and a calculation function component. The neuron may generate a scalar output by taking a weighted sum of the input feature values and their related weights and running them through a non-linear activation recursive function. In a neural network, the input unit, the connection weight, and the threshold—which may be thought of as a functional relationship between the three—determine the actual output of a neuron. The input signal is frequently provided as a vector in practical applications; this is referred to as an input sample. Typically, three activation functions are used to process these samples [1].

An intelligent programme with an expert level in a certain topic and the ability to tackle related issues is known as an expert system. The evolution of neural networks has made it feasible to create a new, effective way for a medical intelligent diagnosing system. The sample input values are delivered to the input layer nodes and the sample data is utilised to train the artificial neural network in TCM diagnosis [2]. The input values and weights are computed and analysed by the neural network, which then communicates the computed findings via the output layer. The neural network may modify the network weights once the output pattern diverges from the sample pattern [3].

A system with experience is typically thought of as a computer programme in a certain topic, with professional knowledge and expertise in that field, and the ability to replicate the thinking style of experts in that field and resolve issues that are typically only accessible to experts and academics [4]. The knowledge base, knowledge acquisition mechanism, comprehensive database, reasoning system, man–machine interface, and text interpreting system are the five components that make up the expert

Table 7.1 Application of LRD value in system analysis

LRD value	Number of diagnoses	Precision of prediction (%)
LRD > 3.0	22	82
3.0 < LRD < 4.0	44	95
4.0 < LRD < 5.5	67	93
5.5 < LRD < 7.0	75	97
7.0 < LRD	20	97

system. The knowledge base and reasoning engine are the most important, difficult, but also necessary, of these five components [5]. A full expert system should contain all five of the aforementioned components at once, even though certain specialised sectors may not have an explanation portion and knowledge acquisition section.

A knowledge base is a repository for the expertise and knowledge of subject specialists. Since computers cannot comprehend human language, it is required to assemble expert information in accordance with certain rules and utilise the proper values and symbols to translate it into machine language that computers can understand in order to create the knowledge base of the expert system [6]. Whether an intelligent knowledge base is dynamic or static is the key distinction between them. Experts in the field may keep tabs on the expert system's performance while it is in use [7]. They can change, extend, or remove the knowledge base if errors or omissions have been discovered. Meridian theory and visceral differentiation theory make up the majority of the theoretical underpinnings of the TCM diagnostic system [8].

The exterior performance of the human body is intimately correlated with the state of the interior viscera, according to the TCM Zang-fu syndrome differentiation hypothesis [9]. The human body is fundamentally a control system. We can learn more about issues with internal organs by evaluating the body's outside signs, including impedance data (Table 7.1).

7.3 Information Management Model Construction and Algorithm Optimization

It is often important to create a development model based on user demands before creating a software system for a TCM diagnostic and treatment information management system [10]. The preference is for those with lower structural sizes when fitness performance is equivalent. The new individual with enhanced performance would only take the place of the least-performing member of its parent generation for the hidden unit addition if a structural mutation could increase the performance of weight evolution learning, which is represented by structural evolution [11]. For the deletion mutation of hidden units, the replacement of the worst-performing member of the parent generation by the new individual with superior performance, which also reflects structural evolution, will occur only when the structural variation does not impair the weight evolution learning performance [12].

The system can gradually broaden its capabilities and link different diagnostic information until the informatization of diagnosis is achieved when vital diagnostic information is digitised. The information management system for TCM diagnosis and treatment created in this manner provides clear modification features [13]. Due to the inconsistent existing software standards, it is not only unreliable in many medical disciplines but also challenging to switch between several diagnostic kinds in the same field.

As a result, a cross-domain management information system development platform based on ANN is suggested, with the aim of resolving a number of issues with traditional customised development, cutting down on development time, and lowering development costs [14]. The information management system's algorithm will be optimised in the following stage so that it can handle data reliably and quickly when used in practise.

Both of them stages of the learning process in the BP algorithm are the forward propagation of each layer signal and the reverse propagation of the error for weight modification [15]. When the network output differs from the planned output, an error will happen, and its mathematical formulation is as follows:

$$E = \frac{1}{2}(D - Y)^2 \tag{7.1}$$

$$= \frac{1}{2}\sum_{k=1}^{l}(d_k - y_k)^2 \tag{7.2}$$

Following transmission of the previous definition of error to the hidden layer, the following is determined:

$$E = \frac{1}{2}\sum_{k=1}^{l}\left[d_k - f\left(\sum_{j=0}^{m}\omega_{jk}h_j\right)\right]^2 \tag{7.3}$$

The mathematical calculation is then extended to the input layer in the manner described below:

$$E = \frac{1}{2}\sum_{k=1}^{l}\left[d_k - f\left(\sum_{j=0}^{m}\omega_{jk}f\right)\left(\sum_{i=0}^{n}v_{ij}x_i\right)\right] \tag{7.4}$$

It can be seen from the above formula that the network input error is a function of the weight of each layer of W_{jk} and v_{ij}, so adjusting the weight can change the error E. Obviously, the principle of adjusting the weight is to continuously reduce the error [16], so the adjustment of the weight should be proportional to the negative gradient of the error;

$$\Delta\omega_{jk} = -\eta\frac{\vartheta E}{\vartheta\omega_{jk}} \tag{7.5}$$

$$\Delta v_{ij} = -\eta\frac{\vartheta E}{\vartheta v_{ij}} \tag{7.6}$$

$j = 0, 1, 2, \ldots, m; k = 1, 2, \ldots l, i = 0, 1, 2, \ldots, n; j = 1, 2, \ldots m$. The formula's negative sign indicates that the gradient is entirely downward, and the constant denotes a proportionality factor, which reflects the sample's diagnostic rate [17]. Additionally, the knowledge base should be specified as $W = (U, R)$, where U is the domain and R is the relationship of equivalence on U, while it is being designed. therefore,

$$IND(B) = \{X_i, X_j)\epsilon(U \times U)\big|\forall\alpha \in R, f(X_i, \alpha) = f\left(X_j, \alpha\right)\} \tag{7.7}$$

Following that, represents all equivalence connections of *IND*(*B*), represents *IND*(*B*) the indistinguishable relationship of every subset *B*, and represents *U*/*B*. There are now two different categories of approximate definitions. The first is the upper approximate definition, which is typically written as

$$R_{-}(X) = U\left\{Y \in \frac{U}{R|Y \subseteq X}\right\} \tag{7.8}$$

The other is what is roughly defined as follows, which is commonly represented as

$$R^{-}(X) = U\left\{Y \in \frac{U}{R|Y \cap X} \neq \varphi\right\} \tag{7.9}$$

Set $BN_R(X) = R^{-}(X) - R_{-}(X)$ that uses X and R to establish the border domain. Based on this, the improved BP algorithm adds an iterative term $\alpha\Delta W_{ij}(k)$ to the formula to speed up the accurate correction of weights [18]. This is equivalent to including a momentum category in the weights and successfully filtering out the high-frequency variables of the error surface of the weight space, as follows:

$$\Delta W_{ij}(k+1) = \eta\delta_i x_j + \alpha\Delta W_{ij}(k)$$

$$\alpha = \{\begin{matrix} 1.7, \ \delta_i(k+1) < \delta_i(k) \\ 0.7, \ \delta_i(k+1) < \delta_i(k) \end{matrix}$$

where α stands for the relaxation factor, $\Delta W_{ij}(k+1)$ for the weight correction in the following cycle, and $\Delta W_{ij}(k)$ for the weight correction in the cycle before. As a result, the aforementioned equation may be thought of as a first-order difference equation. The weighted total value rises when all the parameters have the same sign.

At this point, the weight's steady adjustment quickens its rate of adjustment [19]. If the sign is the reverse, oscillation is present. The outcome and exponential weighting will be lowered at this point, with a steady impact.

7.4 Result Analysis

The computer system currently uses a hybrid CPU + GPU processing mode, which gives it the capacity to train neural networks. In order to confirm the validity of the system design, this study conducts experimental analysis based on prior research and analysis. It also examines the system's usability and the management of TCM diagnosis and treatment data. The objective is to create a system for managing information on TCM diagnosis and therapy that is practical, scientific, and effective [20]. In order to reach conclusions, this article will analyse and assess crucial indicators including the average error (Fig. 7.1).

The change in the mean square error of the model for a hidden layer with A, B, and C number of nodes after N iterations is depicted in Fig. 7.2 where the horizontal axis denotes the number of BP neural network fitting iterations and the vertical axis the mean square error after each iteration. The training is complete when the total error is less than 0.4 or when the maximum number of iterations has been reached. The model's training is stopped after N iterations because there are too many parameters. The image shows that the mean square error following iterations A, B, and C in the n training procedure is 48.7%. The current cross-validation result is the best, and after optimisation, the system's total error is decreased by around 75.3%. The fitness value clearly increases after n trainings using the aforementioned approach to train the illness diagnostic model based on BP network. The system's fitness is likely to remain steady after numerous searches. When the distribution rate stays close to 65%, the situation is the most stable. Let Q and M represent two representative samples of TCM diagnostic data for experimental research in terms of data precision and node prediction precision. Figures 7.3 and 7.4 display the data analysis charts.

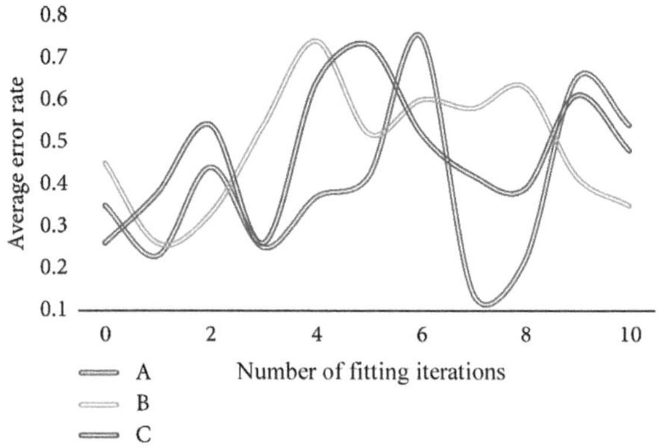

Fig. 7.1 Average error analysis chart

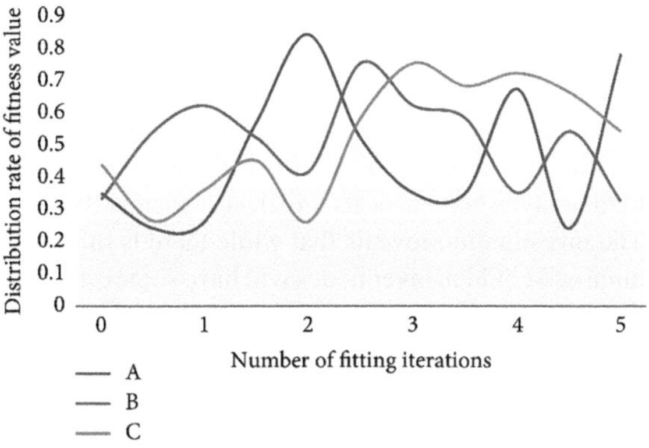

Fig. 7.2 Fitness value distribution rate analysis

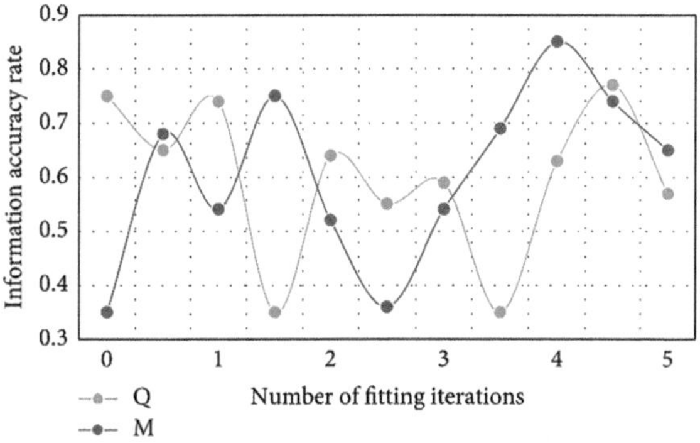

Fig. 7.3 Diagram of an investigation of information correctness

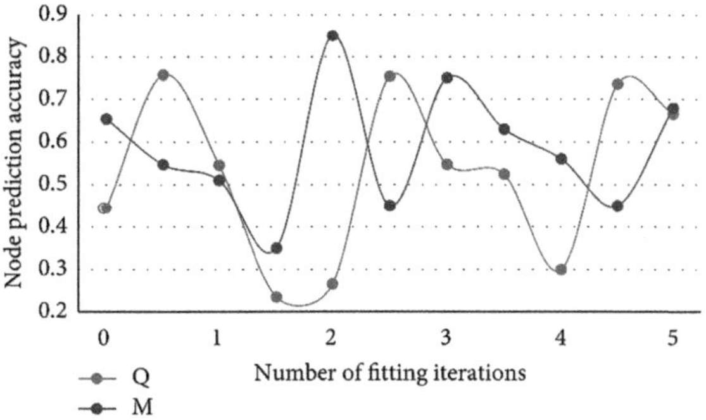

Fig. 7.4 Accuracy of node prediction analysis

Attention should be made to the screening of information since the correctness of the information influences the dependability of the management system. As a result, this component is designed in the experimental section of this study. After entering the data, it is discovered that there is a significant variation between various sample sets as a result of the early entry of diagnostic information not having a full information chain and analysis mode. The stability of the system proposed in this paper is optimised, and it essentially keeps increasing after intervals 3–4, with the growth rate essentially maintaining at the average level of 68.4%. However, as the number of iterations increases, the results will obviously increase or decrease with the same trend. Figure 7.4 shows that the BP neural network's prediction accuracy is at its maximum, or 85%, when the hidden layer node is at position 2. When the hidden layer node is at its smallest, which is 3 s, the training time is at its shortest. The investigation reveals that while there is minimal change in prediction time, the number of hidden layer nodes will have some effect on prediction accuracy. In light of the application environment, it should be taken into account to choose 2 hidden layer nodes that have a high prediction accuracy of 75% and can satisfy the application's performance requirements.

7.5 Conclusion

This study develops a TCM diagnosis and treatment information management optimisation system based on artificial neural networks, which uses the learning, adaptability, and parallel processing capabilities of artificial neural networks to solve complex nonlinear and unstructured problems in the management information system. The intelligence of TCM diagnosis has been further enhanced with the aid of artificial neural networks and different information technologies, which has contributed to the technical support and long-term growth of TCM. In contrast to conventional data processing techniques, this research integrates natural language processing technology with data preparation techniques to mine the semantic content of symptoms and illnesses by training symptom word vectors. This research also builds an optimised neural network algorithm while outlining the steps and traits of the pertinent neural network algorithm. According to experimental study, the optimised system's total error has decreased by around 75.3%, which satisfies the performance requirements of the application. However, there are still some restrictions in this study that need to be addressed and clarified in next research, such as the node prediction accuracy not taking into account various application contexts.

7.6 Discussion

TCM diagnosis and treatment information management system is an important tool that combines TCM theory with modern information technology, which can improve the efficiency and quality of TCM clinical work. However, there are still some problems in the current TCM diagnosis and treatment information management system, such as low data quality, inaccurate algorithms, poor user experience, and imperfect security guarantees. Therefore, it is necessary to carry out optimization research on TCM diagnosis and treatment information management system to further improve the performance and application effect of the system. As a computational model that simulates biological neural networks, artificial neural networks can simulate the learning and reasoning capabilities of the human brain. In TCM diagnosis and treatment, artificial neural network can be applied in disease analysis, disease diagnosis, generation of treatment plan and so on.

Data quality is the foundation of TCM diagnosis and treatment information management system, which has an important impact on the performance and accuracy of the system. To improve data quality, the following strategies can be adopted:

Data collection standardization: develop unified data collection standards and specifications to ensure data accuracy and consistency.

Data cleaning and denoising: Clean and denoise the collected data, remove errors and redundant information, and improve data quality.

Data Integrity Guarantee: Ensure data integrity and prevent data loss or damage, such as using backup and disaster recovery technologies.

Algorithm optimization Algorithm is the core of TCM diagnosis and treatment information management system, which has an important impact on diagnosis and treatment decision-making. In order to improve the accuracy and efficiency of the algorithm, the following aspects can be considered:

Optimize the feature extraction algorithm: Aiming at the complexity and diversity of TCM diagnosis and treatment features, research and improve the feature extraction algorithm to improve the diagnostic accuracy and curative effect prediction ability.

Acknowledgements Guangxi Department of Education Project: Exploration and Practice of Cultivating Applied Talents in Network Engineering Based on Collaboration between Government, School and Enterprise (2022JGA415).

References

1. Zhang, H., Ni, W., Li, J., Zhang, J.: Artificial intelligence–based traditional Chinese medicine assistive diagnostic system: validation study. JMIR Med. Inform. **8**(6), e17608 (2020)
2. Yan, E., Song, J., Liu, C., Luan, J., Hong, W.: Comparison of support vector machine, back propagation neural network and extreme learning machine for syndrome element differentiation. Artif. Intell. Rev. **53**, 2453–2481 (2020)
3. Zhou, W., Yang, K., Zeng, J., Lai, X., Wang, X., Ji, C., Li, S.: FordNet: recommending traditional Chinese medicine formula via deep neural network integrating phenotype and molecule. Pharmacol. Res. **173**, 105752 (2021)
4. Zhou, X., Chen, S., Liu, B., Zhang, R., Wang, Y., Li, P., Yan, X., et al.: Development of traditional Chinese medicine clinical data warehouse for medical knowledge discovery and decision support. Artif. Intell. Med. **48**(2–3), 139–152 (2010)
5. Wang, Y., He, X., Nie, H., Zhou, J., Cao, P., Ou, C.: Application of artificial intelligence to the diagnosis and therapy of colorectal cancer. Am. J. Cancer Res. **10**(11), 3575 (2020)
6. Feng, C., Shao, Y., Wang, B., Qu, Y., Wang, Q., Li, Y., Yang, T.: Development and application of artificial intelligence in auxiliary TCM diagnosis. Evid.-Based Complement. Altern. Med. 1–8 (2021)
7. Wang, J., Duan, L., Li, H., Liu, J., Chen, H.: Construction of an artificial intelligence traditional Chinese medicine diagnosis and treatment model based on syndrome elements and small-sample data. Engineering **8**, 29–32 (2022)

8. Alice, L.Y.L., Binghe, G., Shuang, C., Hoyin, C., Kawai, K., Wenjung, L., Jiangang, S.: Artificial intelligence meets traditional Chinese medicine: a bridge to opening the magic box of sphygmopalpation for pulse pattern recognition. Digit. Chin. Med. **4**(1), 1–8 (2021)
9. Sheikhtaheri, A., Sadoughi, F., Hashemi Dehaghi, Z.: Developing and using expert systems and neural networks in medicine: a review on benefits and challenges. J. Med. Syst. **38**, 1–6 (2014)
10. Wang, X., Liu, J., Wu, C., Liu, J., Li, Q., Chen, Y., Chen, J., et al.: Artificial intelligence in tongue diagnosis: using deep convolutional neural network for recognizing unhealthy tongue with tooth-mark. Comput. Struct. Biotechnol. J. **18**, 973–980 (2020)
11. Li, N., Yu, J., Mao, X., Zhao, Y., Huang, L.: The research and development thinking on the status of artificial intelligence in traditional Chinese medicine. Evid.-Based Complement. Altern. Med. (2022)
12. Cheng, G., Wang, C., Wang, X., Zhang, Q., Cheng, W., et al.: Construction and feasibility analysis of a new artificial intelligence assisted traditional Chinese medicine diagnosis and treatment system. In: International Conference on Biomedical and Intelligent Systems (IC-BIS 2022), December, vol. 12458, pp. 717–723. SPIE (2022)
13. Wang, S., Hou, Y., Li, X., Meng, X., Zhang, Y., Wang, X.: Practical implementation of artificial intelligence-based deep learning and cloud computing on the application of traditional medicine and western medicine in the diagnosis and treatment of rheumatoid arthritis. Front. Pharmacol. **12**, 3844 (2021)
14. Song, Z., Xie, Y., Huang, W., Wang, H.: Classification of traditional Chinese medicine cases based on character-level bert and deep learning. In: 2019 IEEE 8th Joint International Information Technology and Artificial Intelligence Conference (ITAIC), May, pp. 1383–1387. IEEE (2019)
15. Wang, Y., Wang, L., Song, J., Zhang, M.: TCM syndrome differentiation based on artificial bee colony optimization BP neural network algorithm. In: 2021 36th Youth Academic Annual Conference of Chinese Association of Automation (YAC), May, pp. 782–787. IEEE (2021)
16. Tian, Z., Wang, D., Sun, X., Fan, Y., Guan, Y., Zhang, N., Wang, H., et al.: Current status and trends of artificial intelligence research on the four traditional Chinese medicine diagnostic methods: a scientometric study. Ann. Transl. Med. **11**(3) (2023)
17. Miao, Y., Zhou, Y.: The construction of smart Chinese medicine cloud health platform based on deep neural networks. Int. Trans. Electr. Energy Syst. (2022)
18. Huang, Z., Miao, J., Chen, J., Zhong, Y., Yang, S., Ma, Y., Wen, C.: A traditional Chinese medicine syndrome classification model based on cross-feature generation by convolution neural network: model development and validation. JMIR Med. Inform. **10**(4), e29290 (2022)
19. Li, C., Zhang, D., Chen, S.: Research about tongue image of traditional Chinese medicine (TCM) based on artificial intelligence technology. In: 2020 IEEE 5th Information Technology and Mechatronics Engineering Conference (ITOEC), June, pp. 633–636. IEEE (2020)
20. Liu, Z., He, H., Yan, S., Wang, Y., Yang, T., Li, G.Z.: End-to-end models to imitate traditional Chinese medicine syndrome differentiation in lung cancer diagnosis: model development and validation. JMIR Med. Inform. **8**(6), e17821 (2020)

Chapter 8
Design of Intelligent Recommendation Algorithm for Music Website Based on Deep Learning and Artificial Intelligence

Tai Jia, Chunqiu Wang, and Xucheng Geng

Abstract As an important way of communication and expression, music has become a part of people's daily activities. Judging from the growth of the music market, the application of digital multimedia technology is more extensive, and many music industries have begun to pay attention to online music services. The introduction of data mining (DM) in the construction of music websites can effectively solve the problem of how users choose their favorite music in the face of a large number of music provided by websites. In this article, an intelligent recommendation algorithm for music websites based on deep learning is proposed. Based on the traditional collaborative filtering (CF) recommendation, situational information is integrated, and the problem of data sparsity is solved by implicit feedback score conversion. The results show that the recommendation algorithm with situational information is better than the traditional recommendation algorithm, which has certain reference significance for optimizing the music recommendation system. The recommendation engine in the intelligent recommendation system actively recommends the music that users may like, which can change the mode of users actively searching for music, increase the number of visits to music websites and bring commercial benefits to music websites.

8.1 Introduction

With the growth of science and technology and the progress of the times, Internet information and electronic music websites have been growing by leaps and bounds, and music resources have become more and more abundant. However, all kinds of

T. Jia · C. Wang (✉)
College of Music and Dance, Huaihua University, Huaihua, China
e-mail: 554342852@qq.com

X. Geng
Planning and Finance Department, Huaihua University, Huaihua, China

R. Kountchev (Deceased) et al. (eds.), *AI Methods and Applications in 3D Technologies*, Smart Innovation, Systems and Technologies 388,
https://doi.org/10.1007/978-981-97-2144-3_8

music have not been effectively integrated, which leads to people's information overload and inability to find their favorite music in the first place [1]. How to quickly obtain the types of music that users need and recommend the music that users are interested in to the vast amount of music media information on the Internet has become the main way for online media websites to enhance their market share [2]. Judging from the growth of the music market, the application of digital multimedia technology is more extensive, and many music industries have begun to pay attention to online music services [3]. Music recommendation system is an important application in the music field. It mainly digs deeply by recording and analyzing users' personal information and listening habits, so as to provide different users with music that meets their needs [4]. In the process of its application, the recommendation system firstly analyzes the user's previous actions, and then completes the content push based on this to push the needed or interested information for the user [5].

As an important way of communication and expression, music has become a part of people's daily activities [6]. The essence of recommendation system is proved to be an information filtering system to solve the problem of information overload for users by providing the items they need. The recommendation system uses various data related to users and generates recommendation results according to different needs expressed by users, and at the same time users can browse the recommended results [7]. With the rapid growth of music streaming service industry and the rapid progress of portable device technology, thousands of music are within reach, and music appears around people in a more flexible form [8]. Although people can get a lot of digital music, it is difficult and time-consuming for users to choose their favorite music from a complete database [9]. The introduction of DM in the construction of music websites can effectively solve the problem of how users choose their favorite music in the face of a large number of music provided by websites [10]. Under the background of AI development, this article proposes an intelligent recommendation algorithm for music websites based on deep learning, which integrates situational information on the basis of traditional CF recommendation, and solves the problem of data sparseness by using implicit feedback score conversion.

8.2 Methodology

Intelligent Music Recommendation Based on User Interest Preferences

Generally speaking, every website will have an interactive system, which interacts with users through interfaces and displays the content to users by displaying web pages and other means. The log processing system is responsible for collecting all kinds of behavior records generated by users in the interactive system, formatting the records and storing them in the storage system. The storage system includes database, file system and memory cache. It is the core of music recommendation system to obtain user's preference through human–computer interaction or user's behavior habit information. Implicit feedback can effectively solve the problem of data sparsity and data noise caused by explicit feedback, effectively reduce the extra human–computer interaction behavior, complete the collection of feedback information without the user's awareness, and make the results more objective [11]. The extraction of music audio features is the extraction of its essential information from audio signals, which can effectively distinguish music. In music recognition and other fields, the most critical link is to extract audio features. In the music recommendation system, the effective extraction of audio features is inevitable. Audio features mainly include two types, namely, time domain and frequency domain features. A high-quality audio feature representation method can effectively represent audio signals and has an efficient calculation method [12].

The requirements for specific project attributes are relatively low, and it is difficult to obtain information. If there are many attribute information of music projects, it takes a lot of time to collect information during the application of content-based recommendation algorithm. However, when CF algorithm is applied, it mainly focuses on the historical behavior data of users, and then provides guidance for subsequent recommendation, which effectively guarantees the recommendation efficiency and the actual recommendation effect is better. Because everyone's preferences are different, some people like to be recommended for their interests, some people like to be recommended for the latest items, while others like to be recommended for previous items, so music recommendations for individuals vary from person to person. The recommendation engine with this architecture can make different recommendation engine combinations for different users by analyzing the feedback from users in order to get satisfactory results. Through the content recommendation algorithm of deep learning, the feature information extraction process and content recommendation process between users and songs are integrated, which can effectively solve the problem of image data extraction between users and songs and the cold start of new songs.

Intelligent Recommendation Algorithm for Music Websites

The key point of content-based recommendation system is to obtain the most effective basic data characteristics, and deep learning is also a system with good freedom, which may explore some characteristics in the deep data without setting the characteristics. Because of this, in the following recommendation algorithm operation, we use the deep neural network system to capture the characteristics behind the basic data, that is, the characteristics between users and songs. Through this implicit representation, users' preferences can be understood, and songs with high similarity can be pushed to users by matching the similarity between songs and users. These implicit representation information can improve the accuracy after pushing. Usually, many recommendation systems are offline computing, and the system will not collect data when users use them, which also leads to the problem of timeliness. Quality and speed can't be guaranteed at the same time. Pushing songs requires processing a large number of user information, which is another reason for the low speed.

Deep learning is progressive to some extent, which mainly shows that it is an ability to simulate human thinking. Therefore, deep learning is like our brain's information introspection thinking to extract information features. No matter how complicated and massive the information is, it can always extract features with high efficiency and high accuracy. The modeling process of music website resource recommendation system is shown in Fig. 8.1.

Because users don't deliberately rate their favorite songs, but express their love for this song by playing the same song repeatedly, it is inferred that if a song has a high number of plays, it indicates that the song is liked by users and can be given a higher rating. In order to recommend that the system matrix can be understood as user behavior data in essence, and a large proportion of it belongs to sparse matrix, the matrix can be completed if the matrix decomposition method meets the above conditions, so that it can be applied to recommendation. In the recommendation

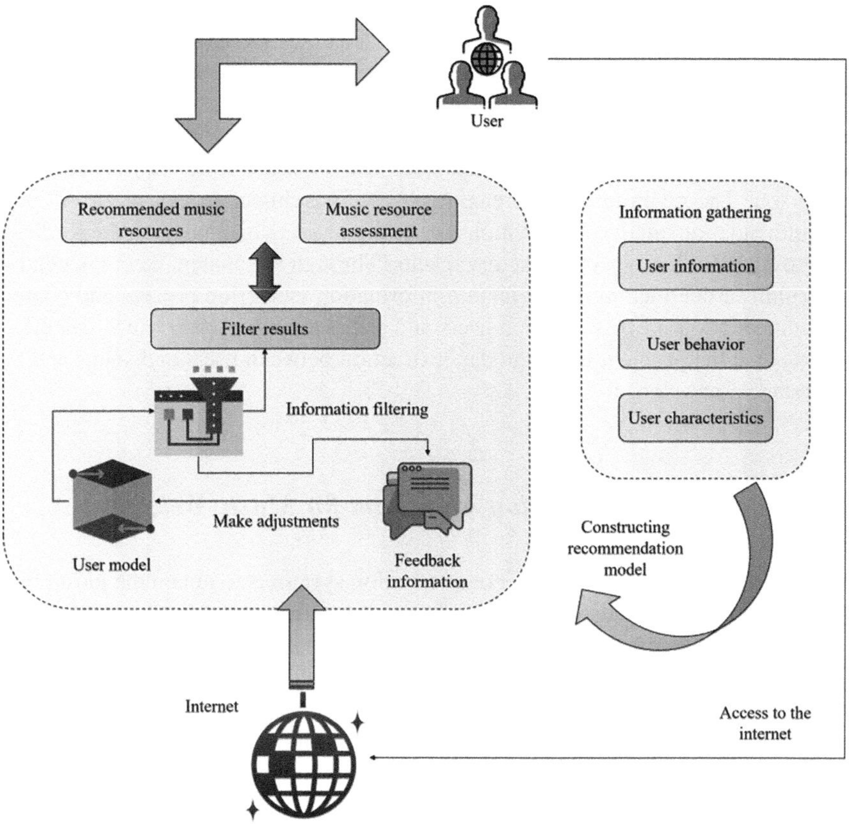

Fig. 8.1 Modeling process of music website resource recommendation system

system, the general idea is to process the data reasonably first, and then calculate the similarity, so as to get the result. First, we need to obtain the data of users or items, and construct Embedding of users or items by isomorphic and reasonable representation functions, and then calculate the similarity between users.

Intelligent recommendation system is a software system that applies various algorithms to recommend information that users may be interested in according to their interest characteristics and access behavior. The main purpose of applying intelligent recommendation system in websites is to provide personalized information acquisition services. In this article, the CF method based on the analysis of users' access behavior is used to design an intelligent recommendation system for music websites. The main method is to derive users' evaluation values of music from the data warehouse, so as to find a set of users with similar interests by clustering calculation, so that the recommendation engine can make directional recommendations to users. When using the user's behavior history to construct preferences, considering that each item has different performance in reflecting the user's interests, an encoder based on self-attention mechanism is designed based on self-encoder. Distinguish users' preferences for items from all aspects, and then integrate the embedding vectors of items with different weights to get accurate user representation vectors and predict users' preferences. Calculate the similarity between music resources by calculating the cosine of the included angle between two vectors;

$$sim(i,j) = \frac{\vec{i} \cdot \vec{j}}{\|i\| \|j\|} \tag{8.1}$$

Based on the similarity calculation of association, calculate the degree of association between two vectors:

$$sim(i,j) = \frac{\sum_{u \in U} \left(R_{u,i} - \overline{R}_i\right)\left(R_{u,j} - \overline{R}_j\right)}{\sqrt{\sum_{u \in U} \left(R_{u,i} - \overline{R}_i\right)^2} \sqrt{\sum_{k \in R_{mn}} \left(R_{u,j} - \overline{R}_j\right)^2}} \tag{8.2}$$

$R_{u,i}$ represents the score of the music resource i by the user u, and $\overline{R}_i$ represents the average score of the i-th music resource. Eliminate the effect of different user scoring habits by subtracting the average of user ratings:

$$sim(i,j) = \frac{\sum_{u \in U} \left(R_{u,i} - \overline{R}_u\right)\left(R_{u,j} - \overline{R}_u\right)}{\sqrt{\sum_{u \in U} \left(R_{u,i} - \overline{R}_u\right)^2} \sqrt{\sum_{k \in R_{mn}} \left(R_{u,j} - \overline{R}_u\right)^2}} \tag{8.3}$$

$\overline{R}_u$ is the average of the user's ratings for all music resources, and $R_{u,i}$ represents the user's u's rating of the music resource i. Assuming that the user's nearest neighbor user set can be represented as U, the predicted score of the target user for the resource item i is:

$$P_{x,i} = \overline{S}_x + \frac{\sum\limits_{y \in U} \left(sim(x, y) \times \left(S_{y,i} - \overline{S}_y \right) \right)}{\sum\limits_{y \in U} \left(sim(x, y) \right)} \tag{8.4}$$

$sim(x, y)$ represents a measure of similarity between user x and user y, $S_{y,i}$ represents the score of user y to resource item i, and $\overline{S}_x$ and $\overline{S}_y$ represent the average resource score of user x and user y, respectively.

On the basis of designing the overall architecture of data acquisition module, offline data processing module and online recommendation module, considering the user's music interest preference and personalized needs, collaborative filtering algorithm is introduced to determine the music recommendation preference by using the number of neighbors, so as to realize accurate music recommendation to users. Based on the current usage of the target user, the situational information of the user, including the location, time and corresponding weather and temperature, is obtained, and the scenario data set that conforms to the user's scenario is found according to the latter three factors. For the scenario where the target user is located, the scenario pre-filtering technology is adopted. Firstly, the scenario is divided into small data sets, and according to the current scenario of the target user, the corresponding data set is found to generate recommendation results by using the previous CF technology. Deep learning can automatically obtain the hidden links in various complex information, map these information into a certain dimension or space, and achieve a unified representation of information. According to this, it can solve some phenomena such as sparse information and cold start in the traditional recommendation system by using systematic recommendation technology.

8.3 Result Analysis and Discussion

In the field of music, explicit scoring data is relatively scarce, even if it can be obtained, it is even rarer than other fields. Therefore, uninterrupted listening behavior is usually regarded as implicit positive feedback. Compared with other fields, music recommendation system relies heavily on content-based recommendation, which means it will face corresponding challenges. All music data records are sorted out and stored in the form of "project-user", and the music data set is divided into two partial training sets and test sets. Here, the music records of the target users are mainly split, and the same "project-user" records cannot appear in the two sets of target users at the same time. Due to the particularity of music recommendation field, we should consider not only the user's own preferences, but also the user's context information and the popularity of music, which requires the recommendation algorithm to be real-time. However, if some recommendation algorithms with strong real-time performance are applied to music recommendation field, the accuracy of recommendation will be reduced due to their strong real-time performance.

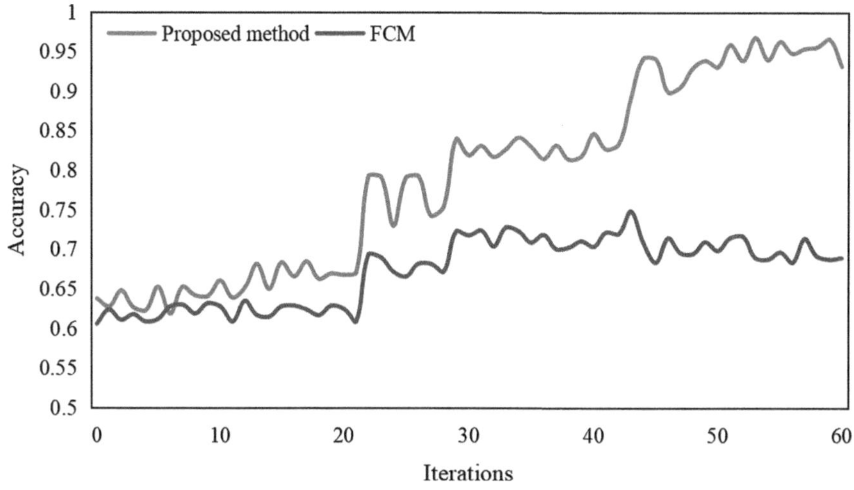

Fig. 8.2 Accuracy comparison of algorithms

The music intelligent recommendation system designed in the text is based on the user's access behavior. By analyzing the user's access behavior, the system scores the music implicitly, calculates the similarity between users, and applies mining algorithm to make intelligent recommendations to users. Therefore, the user's access behavior data is taken as the main body of mining, and it is designed as a fact table, and the user information, music information and other data are designed as a dimension table. For the construction method of music resource recommendation model based on deep learning in this article, a control experiment is set up, and the experimental results are shown in Fig. 8.2.

As can be seen from Fig. 8.2, when the algorithm runs continuously, the accuracy of music recommendation of the proposed algorithm is significantly higher than that of the comparative algorithm, reaching more than 95%, which can provide users with resources that meet their personal music preferences. From the point of view of model training, the basic goal is to strictly control the loss value. Fundamentally speaking, the loss function belongs to the basic category of non-negative real function. If the loss value can be effectively controlled, the performance of the model will be very ideal.

In order to effectively weaken the influence of non-subjective factors, for example, one user will play more songs because he is relatively idle, and another user will play less songs because he is busy. In this study, after collecting the relevant information of users' playing times, we first do some preprocessing, and convert it into the score given by users for music based on a specific way. After the matrix decomposition of the hidden semantic model is implemented, it can be mapped to the potential factor space, and finally presented as a vector, so that the semantic association can be truly established in the subspace, and then the effective description of user preferences can be realized. The hidden semantic model is not simply limited to the

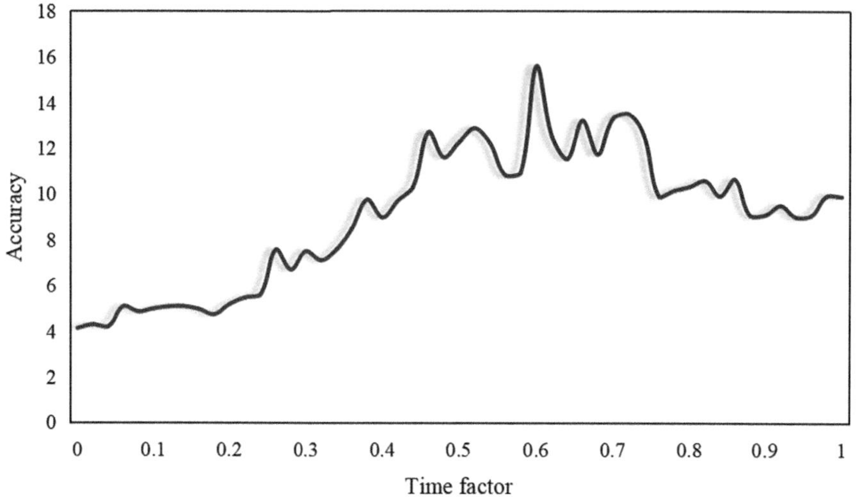

Fig. 8.3 Relationship between time factor and accuracy

single-dimensional analysis of the relevance between users or objects, but focuses on mining the potential semantic relations between them, and then obtains features and preferences. Figure 8.3 shows the change of accuracy when the time factor is between 0 and 1.

It can be seen that the accuracy first increases and then decreases with the increase of time factor. When the time factor is 0.6, the accuracy reaches the peak, so the best time factor is 0.6.

The input of the intelligent recommendation model of music website mainly includes the user's registration information and resource rating data, and the output is the music resource recommendation result calculated by the recommendation model. After data preprocessing and DM, the model of user access behavior is obtained and stored in the knowledge base. When users browse information in the information system, the intelligent recommendation system combines the models in the knowledge base according to the user's behavior. MAE of different music recommendation algorithms is shown in Fig. 8.4. The recalls of different music algorithms are shown in Fig. 8.5.

The results show that the proposed music recommendation model has lower error and higher recommendation accuracy. Compared with the traditional FCM algorithm, the error is reduced by 20.55% and the recall is increased by 24.77%. The research in this article can be used as a supplement to the music recommendation algorithm, which can effectively solve the problems of low recommendation accuracy and insufficient feature analysis in the traditional recommendation system, and then better meet the music recommendation needs. Extracting the features of music audio through digital signals can effectively solve the problem of huge and complex calculation when directly processing the original data, and the obtained features are more distinct, which can be processed and used more conveniently.

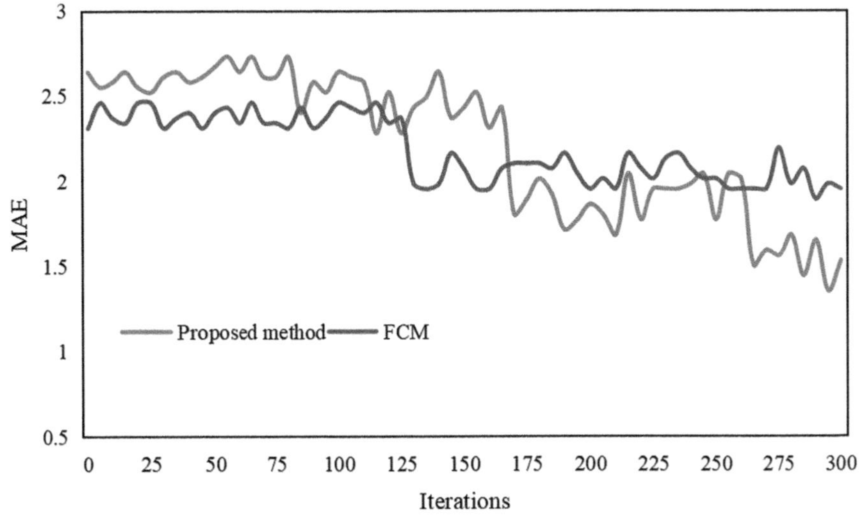

Fig. 8.4 MAE of different algorithms

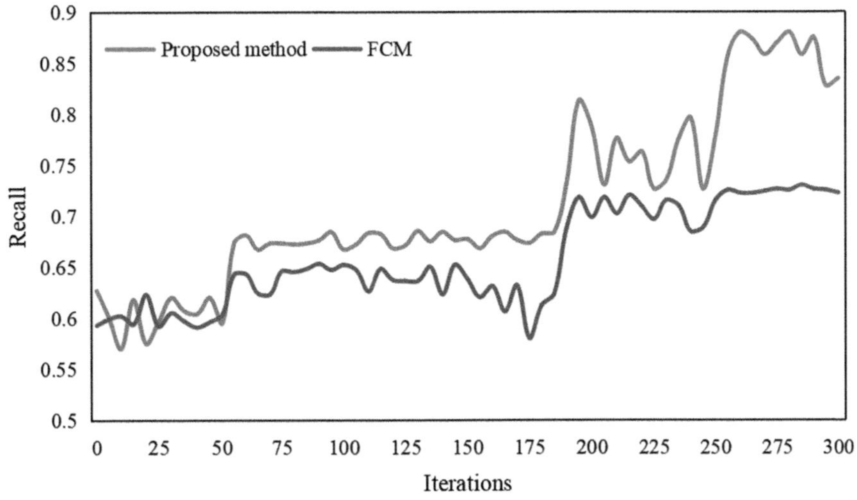

Fig. 8.5 Recall of different algorithms

8.4 Conclusion

Judging from the growth of the music market, the application of digital multimedia technology is more extensive, and many music industries have begun to pay attention to online music services. Music recommendation system is an important application in the music field. It mainly digs deeply by recording and analyzing users' personal information and listening habits, so as to provide different users with music that meets their needs. Under the background of the growth of AI, this article proposes an intelligent recommendation algorithm for music websites based on deep learning, which can change the mode of users actively searching for music, increase the number of visits to music websites and bring commercial benefits to music websites. The results show that the music recommendation model in this article has lower error and higher recommendation accuracy. Compared with the traditional FCM algorithm, the error is reduced by 20.55% and the recall is increased by 24.77%. With the rapid growth of music business, music types are becoming more and more abundant, and users' differences in music preferences are also increasing. Situational information can improve the recommendation accuracy of recommendation algorithm to a certain extent, and it can help users find their favorite music faster and more accurately. In the future research, it will further solve the reality of scene division and the problem of data sparseness, improve the accuracy and intelligence of music recommendation, and provide new ideas and references for recommendation algorithms in the music field.

References

1. Chen, J., Lu, Y., Shang, F., et al.: A novel recommendation scheme with multifactorial weighted matrix decomposition strategies via forgetting rule. Eng. Appl. Artif. Intell. **101**(10), 104191 (2021)
2. Kim, S.T., Oh, J.H.: Music intelligence: granular data and prediction of top ten hit songs. Decis. Support Syst. **145**(2), 113535 (2021)
3. Dm, A., Xi, A., Qd, A., et al.: Humming-query and reinforcement-learning based modeling approach for personalized music recommendation. Procedia Comput. Sci. **176**, 2154–2163 (2020)
4. Wang, C.Y., Wang, Y.C., Chou, S.: A context and emotion aware system for personalized music recommendation. J. Internet Technol. **19**(3), 765–779 (2018)
5. Wei, J., Fei, M., Arunkumar, N.: A personalized authoritative user-based recommendation for social tagging. Futur. Gener. Comput. Syst. **86**(9), 355–361 (2018)
6. Ng, Y.: MovRec: a personalized movie recommendation system for children based on online movie features. Int. J. Web Inf. Syst. **13**(4), 445–470 (2017)
7. Lin, K.P., Chang, Y.W., Shen, C.Y., et al.: Leveraging online word of mouth for personalized app recommendation. IEEE Trans. Comput. Soc. Syst. **5**(4), 1061–1070 (2018)
8. Chen, J., Liu, X.J., Li, B., et al.: Personalized microblogging recommendation based on dynamic interests and social networking of users. Acta Electron. Sin. **45**(4), 898–905 (2017)
9. Huang, Y., Huang, W.J., Xiang, X.L., et al.: An empirical study of personalized advertising recommendation based on DBSCAN clustering of sina weibo user-generated content. Procedia Comput. Sci. **183**(8), 303–310 (2021)
10. Schedl, M., Bauer, C.: An analysis of global and regional mainstreaminess for personalized music recommender systems. J. Mob. Multimed. **14**(1), 95–112 (2018)
11. Huo, Y., Wong, D.F., Ni, L.M., et al.: Knowledge modeling via contextualized representations for LSTM-based personalized exercise recommendation. Inf. Sci. **523**, 266–278 (2020)
12. Hu, Z., Wang, J., Yan, Y., et al.: Neural graph personalized ranking for Top-N Recommendation. Knowl.-Based Syst. **213**(8), 106426 (2020)

Chapter 9
Multi-terrain Motion Control Method for Quadruped Robot Based on Reinforcement Learning

Xiang Luo, Qimin Li, and Ke Su

Abstract Quadruped robot has become a research hotspot because of its good flexibility and adaptability. Most of the current research work has been conducted on flat ground or simple slopes, and it is difficult to achieve robust locomotion in complex environment and multi-terrain. In this paper, a motion control method combining Central Pattern Generator (CPG) and reinforcement learning is proposed. CPG is used to simulate the leg rhythm information of quadruped animals and generate reference motion of robots. By building a multilayer perceptron (MLP) to realize the state detection of the robot, and using reinforcement learning algorithms to optimize the parameters of the network, the robot's attitude adaptive adjustment and motion balance control can be realized. The simulation experiment verifies that the robot using this controller can move smoothly on flat ground, mountainous terrain with strong self-adaptive capability.

9.1 Introduction

Traditional quadruped robots use the combination of dynamic model and forward and inverse kinematics for gait control and trajectory planning [1–5], and the controller parameters based on this approach are predetermined and require targeted settings when the robot is used in different environments. With the dramatic improvements in the computing device performance, learning-based control methods have gradually become a hot research spot.

In contrast to traditional control, methods based on reinforcement learning (RL) do not require robot dynamics modeling and extensive mathematical deduction, which greatly reduces the development difficulty and shortens the development period, and reinforcement learning-based controllers have self-learning capabilities and can learn different control parameters according to various environments. Therefore, the

X. Luo · Q. Li (✉) · K. Su
College of Mechanical and Vehicle Engineering, Chongqing University, Chongqing, China
e-mail: qimin_li@cqu.edu.cn

R. Kountchev (Deceased) et al. (eds.), *AI Methods and Applications in 3D Technologies*,
Smart Innovation, Systems and Technologies 388,
https://doi.org/10.1007/978-981-97-2144-3_9

RL control methods have been widely applied to the motion control of quadruped robots in recent years [6]. Tan et al. [7] implemented trot and gallop gait motions on a quadruped robot Minitaurs using deep reinforcement learning methods and improved the robustness of the control strategy by introducing random disturbances in the simulation environment. In the literature [8], a quadruped robot was realized with smooth multi-step switching. Rudin et al. of ETH Zurich [9] proposed a massively parallel training method which used thousands of robots trained simultaneously in a virtual environment to reduce the training time. Kim et al. [10] proposed a hierarchical controller for a quadruped robot, and the high-level controller implemented a speed command tracking controller using reinforcement learning methods, which learned multiple gaits while tracking speed commands and matched the optimal gaits to different speed ranges to improve energy utilization efficiency. Bellegarda and Quan [11] studied the foot position in Cartesian space by deep reinforcement learning coordinates and used damping control to accomplish the task of running as fast as possible under environmental disturbances, achieving a speed of 4 m/s with no load, 3.5 m/s with a 10 kg load, and its maximum load being able to exceed 100% of its own weight in rugged terrain. Peng et al. [12] used PPO reinforcement learning method for demonstrative learning to enable the robot to learn skills such as martial arts and acrobatics. Peng et al. [13] transferred the movements of quadrupeds to a quadruped robot through imitation learning. In the literature [14], a deep deterministic policy gradient algorithm with a preferential experience replay mechanism was used to speed up the training process.

The above studies mainly take flat ground or slope as the evaluation environment of controller performance, while the diversity and complexity of real-world terrains pose a great challenge to controller performance. In order to improve the walking adaptability of the robot in the field real environment, We have done the following work:

(1) A Central Pattern Generator (CPG) is used to generate biological rhythmic movement information, which makes full use of the advantages of biological evolution in nature and is more in line with objective evaluation criteria in behavioral performance;
(2) A hierarchical control framework is proposed to improve the adaptive capability of the robot. The low-level controller uses CPG to generate the robot motion benchmark, and the high-level controller uses reinforcement learning to adapt the benchmark to the terrain;
(3) The proposed method is verified to be much less expensive than the end-to-end reinforcement learning motion control method in terms of training cost, and the robot based on this control algorithm has multi-terrain adaptation capability.

9.2 Method

The controller is composed of CPG (Central pattern generator) and reinforcement learning. The CPG generates rhythmic periodic signals and uses mapping function to generate robot reference motion. The robot senses its own pose and terrain contact state in real time as input parameters of reinforcement learning. The optimal strategy is selected according to the state parameters, and the robot joint angles are adjusted to achieve attitude adjustment and robust control. The adaptive ability of the quadruped robot and the robustness of the control algorithm are verified by simulation experiments. The overall frame is shown in Fig. 9.1.

The CPG Motion Generator

Since reinforcement learning methods have difficulty in learning periodic rhythmic leg motion information, and the CPG is able to generate rhythmic output, this paper uses the CPG as a reference motion generator. The mathematical model of a single Hopf oscillator is as follows:

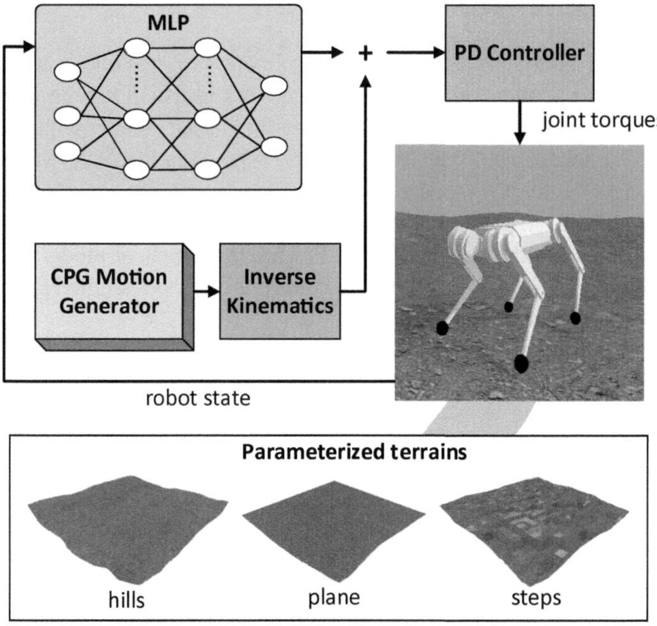

Fig. 9.1 Overview of the presented approach

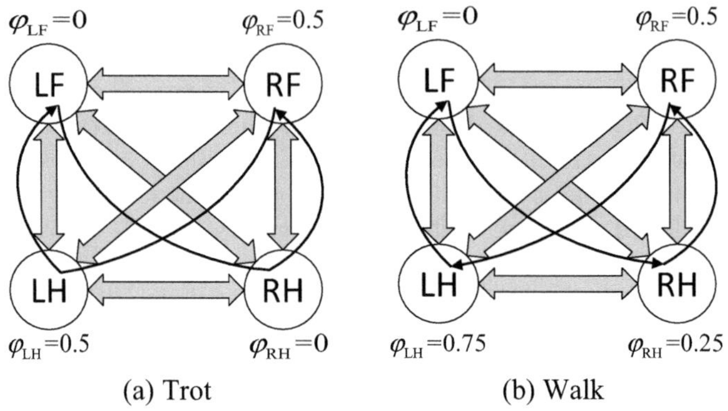

Fig. 9.2 The CPG network structure of typical gait

$$\begin{cases} \dot{x} = \alpha(\mu - r^2)x - \omega z \\ \dot{z} = \alpha(\mu - r^2)z + \omega x \end{cases} \tag{9.1}$$

where x and z are model output parameters; r is the disturbance of the system, $r = \sqrt{x^2 + z^2}$; where α is used to characterize the speed at which the oscillator converges to the limit circle; μ affects the output signal amplitude; ω represents the signal frequency.

A single Hopf oscillator has two output parameters, so it is necessary to use four mutually coupled oscillators to form a CPG network for robot leg motion planning. Oscillators excite or inhibit each other to generate output signals with stable phase differences. The topology of the CPG network is shown in Fig. 9.2, where the black arrow represents the leg lifting sequence.

In Fig. 9.2, The LF leg is defined as the reference leg, $\varphi_{LF} = 0$, and φ_i is the phase of the leg i relative to the reference leg LF. The mathematical model of the above CPG network is generated by adding the coupling term to a single Hopf oscillator and can be expressed as:

$$\begin{cases} \dot{x_i} = \alpha(\mu - {r_i}^2)x_i - \omega_i z_i + \sum_{j=1}^{4}(\cos\varphi_{i,j}x_j - \sin\varphi_{i,j}z_j) \\ \dot{z_i} = \alpha(\mu - {r_i}^2)z_i + \omega_i x_i + \sum_{j=1}^{4}(\sin\varphi_{i,j}x_j + \cos\varphi_{i,j}z_j) \\ \omega_i = \frac{\omega_{st}}{e^{-az_i}+1} + \frac{\omega_{sw}}{e^{az_i}+1} \\ \omega_{st} = \frac{1-\beta}{\beta}\omega_{sw} \end{cases} \tag{9.2}$$

As shown in Eq. (9.2), Legs LF, RF, LH, and RH are represented by the numbers 1, 2, 3, and 4 respectively, $i, j = 1, 2, 3, 4$; $\varphi_{i,j}$ represents the phase difference of i, j legs; ω_i is the frequency of leg i, ω_{sw} and ω_{st} represent the frequency of swing phase and support phase respectively. β is the load factor, which determines the proportional relationship between the support phase and the swing phase.

In this paper, the CPG network output is mapped to generate a foot reference trajectory with phase information. In order to express the position coordinates of the foot, the coordinate system for each leg is established, as shown in Fig. 9.3.

The hip joint motion is not considered in the reference motion trajectory generating process, so the y coordinate of the foot is ignored. The foot coordinates of the LF leg and the phase relationship between each leg when using the CPG network to output the trot gait are shown in Fig. 9.4a, b, respectively.

As shown in Fig. 9.4b, during trot gait, LF and RH are in phase, RF and LH are in phase, and the phase difference between LF and RF is 1/2 cycle.

Fig. 9.3 Leg coordinate system

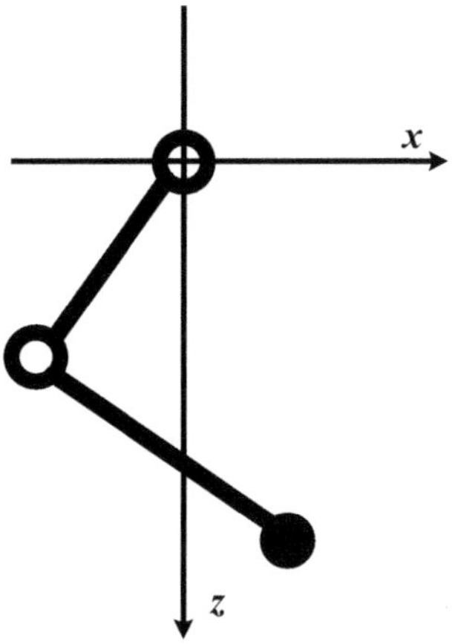

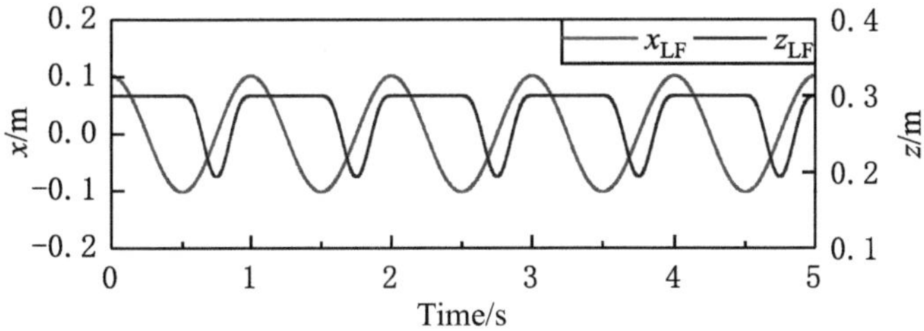

(a) Foot position

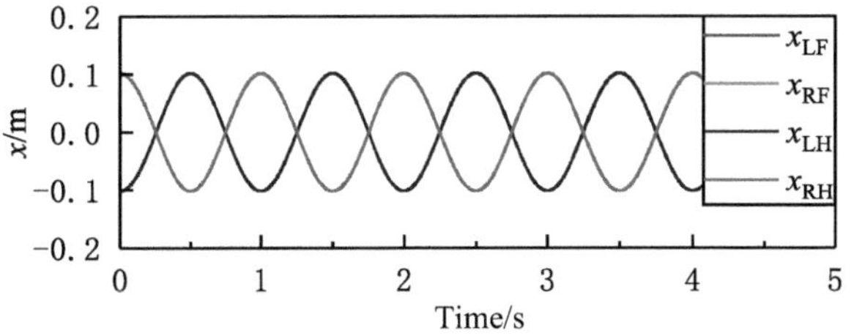

(b) The x-coordinate of the foot end of the four legs

Fig. 9.4 Trot gait foot position

Reinforcement Learning

In the process of movement, the robot will continuously interact with the environment. The purpose of reinforcement learning is to achieve the optimal mapping from environmental states to actions, that is, to achieve the maximum cumulative reward value of the robot in the process of interacting with the environment. Taking S as the state space, $s_t \epsilon S$ represents the state of the robot at time t; A is the action space, which represents the set of all actions that the robot can perform, and $A_t \in A$ represents the set of actions that the robot can choose at time t. At time t, the robot observes the environmental state s_t, selects the action a_t through the mapping and executes it, and obtains the reward a_t and the state s_{t+1}. Introducing the discount coefficient γ, then the total reward that the robot can obtain during the interaction with the environment after time t is as follows:

$$G_t = r_{t+1} + \gamma r_{t+2} + \gamma^2 r_{t+3} + \cdots = \sum_{\tau=0}^{+\infty} \gamma^\tau r_{\tau+t+1} \tag{9.3}$$

The discount coefficient $\gamma \in [0, 1]$ determines the importance of the future reward value to the cumulative reward value.

According to the cumulative reward value of formula (9.1), $Q^\pi(s, a)$ is used to represent the value function of the robot after the robot takes action a in state s according to policy π [15], then $Q^\pi(s, a)$ can be expressed as:

$$Q^\pi(s, a) = E_\pi[G_t | s, a] \tag{9.4}$$

In the formula, π is a certain strategy, which completes the mapping from state to action. By finding the optimal value function $Q^*(s, a)$, the optimal control strategy π^* can be found. $Q^*(s, a)$ can be expressed as:

$$Q^*(s, a) = \max_\pi Q^\pi(s, a) \tag{9.5}$$

Reward Function

In order to enable the robot to move forward quickly and maintain balance during the movement, the reward function is defined in this paper as follows:

$$r = \omega_1 r_{lv} + \omega_2 r_{ov} + \omega_3 r_s + \omega_4 r_y \tag{9.6}$$

$\omega_1, \omega_2, \omega_3$ and ω_4 represent the linear velocity, quadrature velocity, stability and yaw reward coefficients respectively, all of which are positive. r_{lv} represents the linear velocity reward; r_{ov} is the orthogonal velocity reward; r_s is the stability reward; r_y is the yaw reward.

r_{lv} can be written as:

$$r_{lv} = \begin{cases} \exp(-k_1(v_x - v_t)^2), v_x < v_t \\ 1, v_x \geq v_t \end{cases} \tag{9.7}$$

r_{ov} penalizes velocities orthogonal to the forward direction.

$$r_{ov} = \exp(-k_2({v_y}^2 + {v_z}^2)) \tag{9.8}$$

r_s punishes the roll angle and pitch angle of the fuselage, so that the robot can keep the body stable during the movement.

$$r_s = \exp(-k_3({\theta_{roll}}^2 + {\theta_{pitch}}^2)) \tag{9.9}$$

r_y punishes the robot's yaw distance and expects the robot to walk in a straight line.

$$r_y = \exp(-k_4(y - y_b)^2) \tag{9.10}$$

In Eqs. (9.7)–(9.10), k_1, k_2, k_3 and k_4 are proportional coefficients, all of which are positive. v_x is the speed in the forward direction, v_t is the target speed; v_y and v_z are the lateral speed and longitudinal speed of the fuselage, respectively; θ_{roll} and θ_{pitch} are the roll angle and pitch angle of the fuselage, respectively; y is the y coordinate at the current moment, and y_b is the initial y coordinate of the robot.

9.3 Experimental Results

The effectiveness of the proposed coupled control framework of the CPG and reinforcement learning is validated by simulation.

Simulation Environment

The PyBullet supports loading various robot description files such as Unified Robot Description Format (URDF), Simulation Description Format (SDF), MuJoCo Format (MJCF), etc., and provides functions such as forward/reverse kinematics, collision detection, generating parametric terrain, etc., so it is widely used in robot simulation. In this paper, we use the PyBullet as the physics engine for simulation and training, and the Mini Cheetah with 12 degrees of freedom (3 for each leg) is used as the robot platform for simulation experiments. The mountain environment has irregular

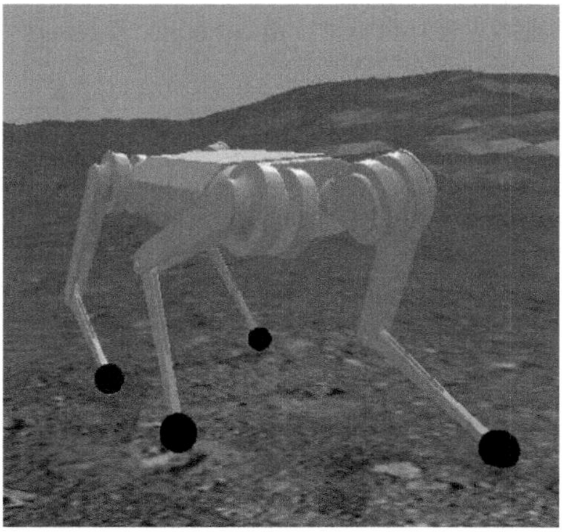

Fig. 9.5 Simulation environment

slopes and some depressions and bumps, which can simulate the complex real terrain environment to a great extent, and also can verify the performance of the controller well. Therefore, we choose to use the deep deterministic policy gradient algorithm to train and verify the policy network in the mountain environment, and the training work is carried out on a PC equipped with an I5-8400 CPU and 16 GB RAM. The built simulation environment is shown in Fig. 9.5.

Results

The training process is terminated when the robot reaches the pre-set termination conditions after more than 6,000 episodes of training. The average reward of episodes is shown in Fig. 9.6.

In order to verify the adaptive ability and control stability of the trained controller, simulation experiments are conducted on flat terrain as well as mountainous terrain.

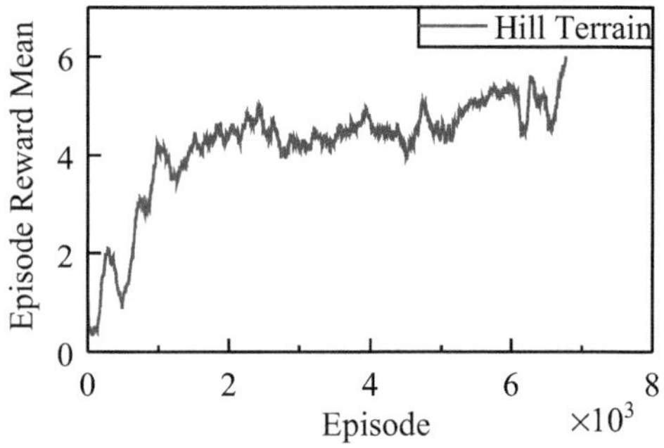

Fig. 9.6 Episode reward

The robot based on the end-to-end reinforcement learning controller is unable to perform stable motion on flat ground even after thousands of rounds of training due to the high degree of freedom, so it is not compared. Figure 9.7a, b shows the motion states of the robot on flat ground and mountainous environment, respectively.

Fig. 9.7 The motion state of the robot in various terrains

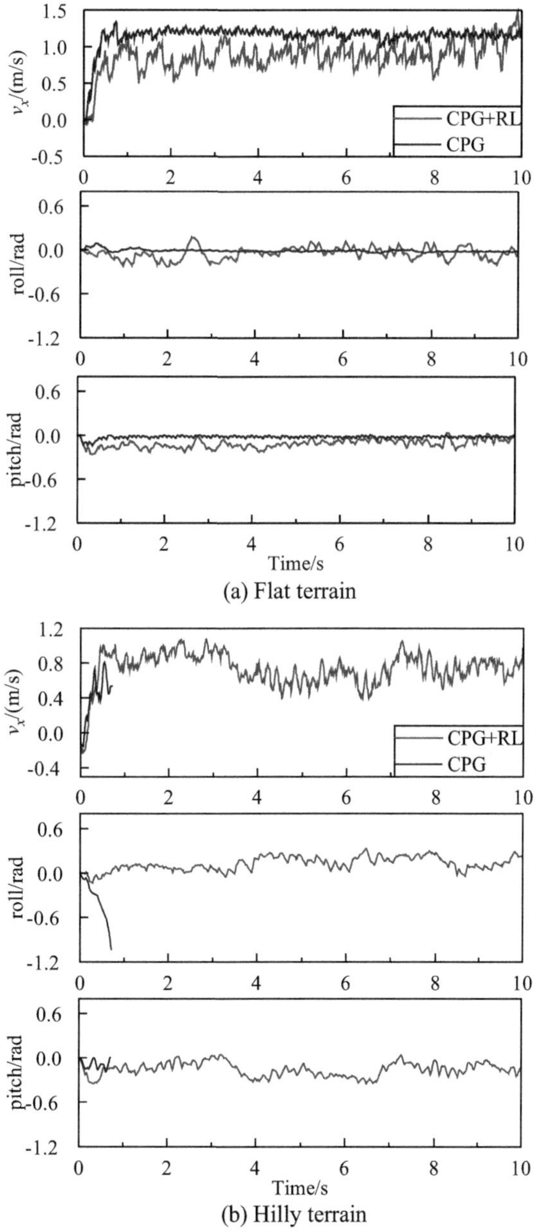

The experimental results show that the quadruped robot based on the CPG control algorithm can move at a stable speed and a stable attitude on the flat ground, but due to the lack of self-adaptive ability, it cannot maintain its own balance and fall over in a more complex mountain environment; The bio-inspired reinforcement learning method proposed in this paper can sense the state of the fuselage in real time, and adjust the adaptive attitude through the reinforcement learning algorithm to maintain the balance of the fuselage, so as to complete the stable movement in multiple terrains.

9.4 Conclusion

In this paper, a hierarchical reinforcement learning control framework is proposed to achieve multi-terrain adaptive motion control by using coupled CPG networks and reinforcement learning algorithms. The control framework has low hardware requirements and can achieve stable motion in a relatively short time using CPU training; moreover, the controller has superior migratable capability and can adapt well in flat terrain after training in mountainous environment and has strong multi-terrain adaptive capability.

For future work, sensors such as vision cameras and radars can be added to obtain more information about the external environment and provide more effective information for the robot motion planning, so as to further improve the robot motion capability.

References

1. Raibert, M.H.: Legged robots. Commun. ACM **29**(6), 499–514 (1986)
2. Vukobratović, M., Borovac, B.: Zero-moment point thirty five years of its life. Int. J. Humanoid Robot. **1**(01), 157–173 (2004)
3. McGeer, T.: Passive dynamic walking. Int. J. Robot. Res. **9**(2), 62–82 (1990)
4. Collins, S.H., Wisse, M., Ruina, A.: A three-dimensional passive-dynamic walking robot with two legs and knees. Int. J. Robot. Res. **20**(7), 607–615 (2001)
5. Pratt, J., Carff, J., Drakunov, S., Goswami, A.: Capture point: a step toward humanoid push recovery. In: 2006 6th IEEE-RAS International Conference on Humanoid Robots, pp. 200–207 (2006)
6. Zhang, W., Tan, W.H., Li, Y.B.: Development status and outlook of quadruped robot motion control based on deep reinforcement learning. J. Shandong Univ. **58**(8), 61–66 (2020)
7. Tan, J., et al.: Sim-to-real: learning agile locomotion for quadruped robots. arXiv preprint arXiv:1804.10332 (2018)
8. Shao, Y.C., et al.: Learning free gait transition for quadruped robots via phase-guided controller. IEEE Robot. Autom. Lett. **7**(2), 1230–1237 (2021)
9. Rudin, N., Hoeller, D., Reist, P., Hutter, M.: Learning to walk in minutes using massively parallel deep reinforcement learning. In: Conference on Robot Learning. PMLR (2021)
10. Kim, Y., Son, B., Lee, D.J.: Learning multiple gaits of quadruped robot using hierarchical reinforcement learning. arXiv preprint arXiv:2112.04741 (2021)
11. Bellegarda, G., Quan, N.: Robust high-speed running for quadruped robots via deep reinforcement learning. arXiv preprint arXiv:2103.06484 (2021)

12. Peng, X., et al.: DeepMimic: example-guided deep reinforcement learning of physics-based character skills. ACM Trans. Graph. **37**(4), 1–14 (2018)
13. Peng, X.B., Coumans, E., Zhang, T., Lee, T.W., Levine, S.: Learning agile robotic locomotion skills by imitating animals. ResearchGate (2020)
14. Fu, H.Q., Tang, K.Q., Deng, G.Z.: Deep reinforcement learning-based motion planning for hexapod robots. J. Intell. Sci. Technol. **2**(4), 361–371 (2020)
15. Zhang, H.J., Su, Z.B., Su, B.: An end-to-end control method for robots based on deep Q-network learning. J. Instrum. **39**(10), 36–43 (2018)

Chapter 10
Automatic Detection of Basal Units Beneath Antarctic Ice Sheet in Radargram Based on Deep Learning

Na Wang, Wen Xu, Shinan Lang, and Xiangbin Cui

Abstract Sea level rise, caused by accelerated melting of glaciers in Greenland and Antarctica in recent decades, has become a major concern in scientific, environmental, and political areas. A comprehensive study of subglacial conditions and processes is particularly important for reliable analysis about their future evolution. Basal units—visibly distinct englacial structures near ice-bedrock interface, could provide substantial insight into subglacial processes, ice-sheet dynamic history and dramatically influence the flow of surrounding ice. In order to enable improved characterization of these features, we develop and apply an algorithm that allows for automatic detection of basal units, therefore, an algorithm based on deep learning is proposed. Compared with traditional method based on manual feature extraction, proposed algorithm can achieve completely automatic recognition of basal units in radargram. The network is mainly composed of ResNet and ASPP module, which can achieve high accuracy in very small dataset. We use radar data collected by Polar Geophysics Group (PGG) of The Earth Institute at Columbia University in Antarctica in 2008–2009 and 2009–2010 for experiments, results confirm the effectiveness and robustness of proposed algorithm. At the same time, a more rapid deep learning method is tried, which uses the lightweight network MobileNet V3 as backbone, obtaining a network structure that can save 81.8% of training time and 92.2% of processing time with high accuracy. It provides a possibility for rapid network training, application on mobile devices and real-time processing of radargram.

N. Wang · W. Xu
School of Information and Electronic Engineering, Zhejiang University, Hangzhou, China

S. Lang
Faculty of Information Technology, Beijing University of Technology, Beijing, China

X. Cui (✉)
Polar Research Institute of China, Shanghai, China
e-mail: cuixiangbin@pric.org.cn

Technology and Equipment Engineering Centre for Polar Observations, Zhejiang University, Zhoushan, China

R. Kountchev (Deceased) et al. (eds.), *AI Methods and Applications in 3D Technologies*,
Smart Innovation, Systems and Technologies 388,
https://doi.org/10.1007/978-981-97-2144-3_10

10.1 Introduction

The Antarctic Ice Sheet (AIS) is an important part of global climate system. Human activities have increased the temperature of atmosphere and oceans and the melting of western Antarctica Ice Sheet has been accelerating since 2000s. With the research on evolution history of AIS and interaction between ice sheet, solid earth, atmosphere and ocean system by scientist, estimation of the influence of AIS on sea level rise has been changing [1].

The previous sub-ice research used the method of collecting ice core samples, which can obtain direct measurements of deep ice properties, but it is invasive and expensive [2, 3]. In order to solve this problem, researchers have explored a new advanced tool—Radio Echo Sounding (RES), also called as ice penetrating radar [4, 5]. RES is a geophysical technique that emits electromagnetic waves (frequency is usually lower than 1 GHz) to the ice sheet directly below, distinguishing characteristics of ice sheet in vertical direction under the condition of sufficient bandwidth. The electromagnetic wave propagates in ice, and the signal is reflected back to receiving antenna at positions with discontinuous permittivity. In the past few decades, RES has contributed most measurements of ice thickness, bedrock elevation, internal structure of polar ice sheets [6]. Radar data obtained through RES measurements are crucial for better understanding structure and formation process of ice sheets in context of climate changes [7].

Identifying internal ice layers, ice-bedrock interface and even subglacial lakes from radargram has extensively studied. In recent years, scientists have found that distorted ice structures at the bottom of polar ice sheets, referred as basal units, are abound [8]. In 2012, the Chinese National Antarctic Research Expedition used a new ground-based RES system to map the ice sheet along the inland traverse from Zhongshan to Kunlun Station, and a typical ice freezing structure appeared in radargram. This structure appeared at the bottom of the ice sheet, which was formed by the melting of ice at ice-bedrock interface and then refreeze onto the ice again [9]. In 2018, Leysinger Vieli et al. observed huge feathery ice sheet structures in radargram in polar ice sheets. These basal units have ice layers bend upward and fold downward, and bulge upward from the bottom of the ice sheet [10]. In some places, the thickness of basal units can reach half of the thickness of the ice sheet [11]. Basal units not only affect ice sheet stratigraphy, but also affect ice flow. Isolated and accreted basal units will significantly change stratigraphic structure of the surrounding ice column, making it difficult to interpret ice core data. At the same time, during melting and refreezing, accumulated basal units will also affect ice flow through basal ice deformation and softening.

With the wide application of RES, the amount of RES data we have obtained and collected, and radargram from radar data have increased dramatically. While providing more opportunity to study basal units, these progresses also propose much challenge for identifying and extracting distorted ice structure. Although visual observation can support the representation of several targets, it is subject to many restrictions. The first reason is visual observation can only detect radargram one by one through human eyes, which consumes huge time and is not suitable for processing massive radargram. Secondly, visual observation is relatively subjective. Different researchers have different perceptions of one image, and there are also deviations due to concentration and energy. Therefore, automatic methods to detect and process polar data continue to emerge [11–14], which not only help researchers improve work efficiency and reduce work pressure, but also provide more possibilities for faster and wider exploration of polar regions.

In the literature, the methods for automatic analysis of subglacial structures in RES data are mainly divided into three categories. The first is numerical simulation of ice sheet geological process and local inference of subglacial structure. The numerical models [15, 16] take data and analysis from different sensors as input, and only extract the ice sheet thickness and the terrain of bedrock topography. The second strategy is to automatically analyze subglacial structure of radargram without considering bedrock and interface of ice sheet with line detection algorithm, while modeling the interference using line slope [17, 18]. However, these techniques focus on the analysis of ice rupture and do not map ice sheet. The third method is to extract the main geological targets by machine learning. The research in literature [19] divides the radargram into three targets: ice layer, bedrock and echo free zone (EFZ). However, this method is based on manually extracted prior features of spatial and statistical characteristics, using support vector machine (SVM) to perform pixel classification. The features extracted manually depend on human design and are only applicable to model specific features. With the increase of categories of features, the performance of this method will be reduced. Recently, deep learning provides a powerful means for radargram segmentation in various applications [20, 21]. In addition, the method based on deep learning can automatically extract features, providing a basis for subsequent applications in other areas.

There have been some achievements in applying deep learning method to polar exploration. Literature [22] proposed a U-Net architecture based on contraction path to capture context and added attention gate (AG) modules in network. These modules force network to pay more attention to relevant feature areas of input data. Literature [23] further improved the model by adding an Atrous Spatial Pyramid Pooling (ASPP) module to main network structure to control feature resolution. The research in [24] uses ResNet network to segment ice layer, thermal noise, echo free area and bedrock in radargram, getting overall accuracy similar to SVM based traditional machine learning method. Literature [25] used U-Net network and ASPP module to detect ice layer structure, classifying radargram into bedrock, echo free area, ice layer and basal ice, which is somewhat similar to basal units. Recently, deep learning has been applied to radar image recognition and achieved good results, mainly for: (1) ice detection [25, 26]; (2) Simulate RES image using generative countermeasure network (GAN) [27]; (3) Target detection [28]; and (4) Radargram segmentation [29, 30].

In this paper, a new method based on deep learning for automatic detection of basal units is proposed, at the same time, pixel level segmentation of radargram is realized to better understand ice sheet and its composition. Radargram are divided into different categories, including ice layer, base (including bedrock and the part below bedrock) and basal units. Basal units are ice blocks growing from the bottom of ice sheet, which have different formation mechanism, temperature system and shape from normal isochronous ice layer. Their distribution is shown in Fig. 10.1. Ice layer, base and basal units are characterized by specific spatial distribution in distance and azimuth, but have significant differences in size and scale. Therefore, the paper selects ResNet network with ASPP module to identify features. ASPP module can realize multi-scale feature extraction to improve recognition of characteristics in different sizes. Compared with conventional convolution, ASPP module includes a wider range of backgrounds by expanding receptive field, while saving the number of network parameters, so as to segment image robustly at multiple scales. ResNet network can not only deepen network depth, but also prevent parameter over fitting, improving convergence speed and accuracy of network training. In order to verify effectiveness of the algorithm, radar data collected by AGAP (Gambertsev Project in Antarctica) and GAMBIT (Gambertsev Aerial Geophysical Mapping Project for Bedrock and Ice Cap) in 2008–2009 and 2009–1010 were used as the training set, which obtained a good result, proving that the algorithm can achieve favorable detection of basal units.

The paper is mainly composed of four parts. The first part introduces the target of research and background. The second describes the method of feature extraction using deep learning method, dataset and experimental process. And then we show training results and analyzing the results after algorithm processing; The last part is a conclusion.

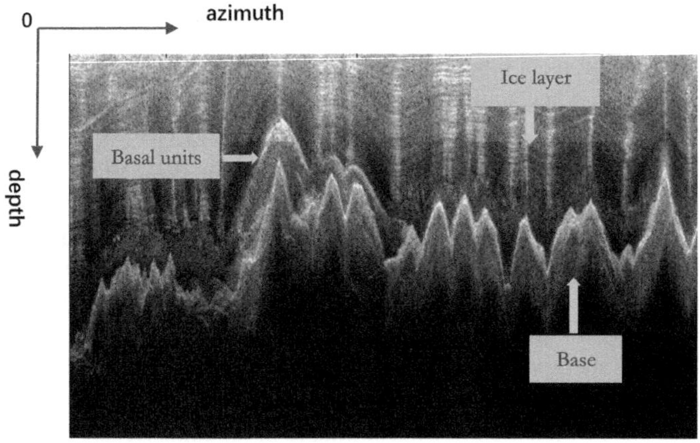

Fig. 10.1 Example of ice sheet with basal units

10.2 Methodology

Network

When network structure is as shown in Fig. 10.2, we get the highest accuracy in very small dataset. Backbone adopts ResNet 101 network, which is mainly composed of 5 serial blocks. The first four blocks have 3, 4, 23 and 3 bottleneck layers respectively. Each bottlelayer consists of convolution, pooling, batch normalization and relu operation, as shown in the bottom right of Fig. 10.2. Output stripe represents the ratio of spatial resolution of input image to final output image resolution. From 1 to 4 blocks, output stripe is 2, 4, 8 and 8 respectively. In order to avoid further reduction of the spatial resolution, parallel dilated convolution is adopted, which is embodied in ASPP structure. Then concat operation is used to fuse the features of different dilated convolution outputs, and then classification results are output through 1 * 1 convolution.

(1) *ResNet*: ResNet makes network training more efficient by adding a residual unit to deep neural network. This structure can speed up the training of neural network very quickly, and also can greatly improve accuracy of the model. In network training of this paper, because data set is very small, it is necessary to increase network depth and training times to improve detection accuracy. ResNet 101 can avoid the problem of degradation and ensure that the network training is stable at a certain highest accuracy rate.
(2) *Atrous Spatial Pyramid Pooling:* Atrous Spatial Pyramid Pooling (ASPP) applies four dilation convolutions with different dilation rate to the top of feature map in parallel to change the receptive field of network filter, which helps network integrate global view and pay attention to the details of radar image. It is an effective measure to extract and accurately classify features of different scales within any range, meeting the needs of extracting features in different scales from images of different sizes in this paper.

Dilation convolution is an operation that expands receptive field of convolution filter by changing dilation rate d. As shown in Fig. 10.3, d corresponds to the sampling step of input signal, and standard convolution is a special case of d = 1. Dilation rate d means that the size of convolution kernel is increased by adding d − 1 zeros in the middle of convolution filter, and filter resolution is changed at the same time. The size of convolution kernel is determined by $k_x * k_y$ becomes $k_x^{'} * k_y^{'}$:

$$k_x^{'} = k_x + (k_x - 1)(d - 1) \tag{10.1}$$

$$k_{y'} = k_y + (k_y - 1)(d - 1) \tag{10.2}$$

ASPP with different expansion coefficients can effectively capture multi-scale information, however, in practice, the larger the sampling rate, the less the number of

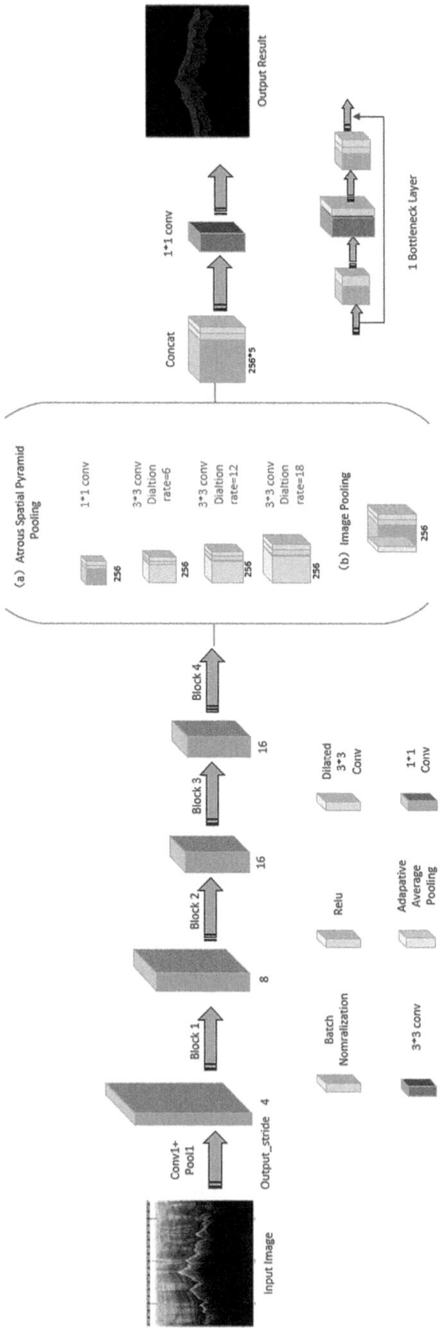

Fig. 10.2 Diagram of network structure

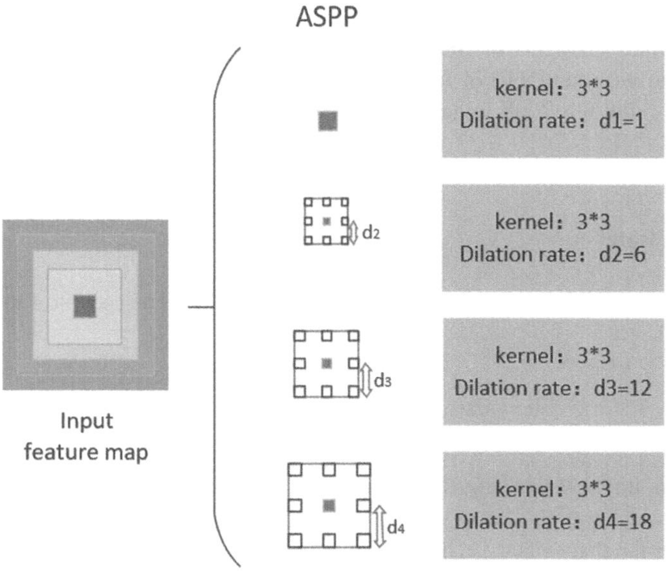

Fig. 10.3 Dilation convolution

effective filter weights [31]. In order to overcome this problem and integrate global semantic information into algorithm model, this paper uses image level features, with the structure shown in Fig. 10.2. The specific representation is to use the global average pooling model on the last feature map of the model, and inputting the resulting image level features into 1 * 1 convolution filter with 256 layers, then use bilinear interpolation to upsample features to required spatial dimensions. Compared with conventional convolution, ASPP module can steadily segment images on multiple scales by expanding receptive field to include a wider range of backgrounds, while saving the number of network parameters.

Network Training

Dataset

The experimental data comes from radar data collected by Polar Geophysical Research Group (PGG) of Institute of Earth Research in Columbia University during 2008–2009 and 2009–2010 from Gambertsev Project in Antarctica (AGAP) and Gambertsev Aerial Geophysical Mapping Project for Bedrock and Ice Cap (GAMBIT). The data are all processed by 1D-SAR (synthetic aperture radar) algorithm and displayed as radargram. The collection routes include north–south route (50 in total), east–west route (16 in total), the specific collection routes in Dome

Fuji (6 in total) and Vostok Lake (6 in total). The north–south interval is 5 km, and east–west interval is 33 km. Searching basal units samples in radargram as original dataset and we found 110 in total, dividing the data set into training set and test set with a ratio of 8:2, all images are labelled by software Labelme for supervised training.

Learning Rate

This paper uses poly strategy to adjust learning rate [32]. The specific formula is as follows:

$$\text{lr} = \text{base\_lr} \times \left(1 - \frac{\text{epoch}}{\text{num\_epoch}}\right)^{\text{power}} \tag{10.3}$$

where lr is updated learning rate, base_lr is benchmark learning rate, "epoch" is the number of iterations, "num_epoch" is maximum number of iterations, and power determines the shape of curve. In training of the algorithm, power = 0.9 is taken. Poly policy will dynamically adjust the value of learning rate with the number of training iterations epoch. Benchmark learning rate is set to 0.0001. At the initial training stage, the convergence speed is fast. With the increase of number of iterations epoch, learning rate gradually decreases, slowly approaches an optimal value, and finally tends to be stable, indicating that network training has reached convergence at this time.

Data Augmentation

The number of samples of basal units is very small in polar database and so available training set is very small, too. Therefore, data augmentation is used to increase the number of samples to improve network robustness. In this paper, we use rotation, symmetry and zoom methods to process original image and annotation data at the same time. Rotating the original picture at a probability of 0.8 within the scope $[-\theta_{max}, +\theta_{max}]$, where θ_{max} is the maximum allowable rotation angle, which is 10°. Mirror the original image symmetrically with a probability of 0.5. If the pixel size of original image is m $*$ n, transform the pixel size to 0.85 $*$ m $*$ n with a probability of 0.8. While not changing the authenticity of original polar structure, increase the number of training set to 5 times the original samples.

Transfer Learning

It is complex to train a neural network completely, accelerating training speed and finally making the network converge not only requires a large enough data set, but also consumes a lot of training time and computing resources. Based on this, for the automatic detection of basal units, transfer learning is introduced in training process, which is to use pre-trained weights to replace the randomly initialized weights in network, so as to speed up the convergence and reduce training time [33].

Transfer learning has been widely used in the field of computer vision and has achieved certain success. Research in recent years shows that mobility of features increases with the decrease of difference between target data and training data of transfer training method, which proves that the effect is better than before even if there are obvious field differences between the two. Therefore, this paper conducts transfer learning on MS-COCO [34] (Common Objects in Context) training set and radar image basal units dataset through pre training method to accelerate convergence and shorten training time.

10.3 Experimental Result

Classification Accuracy

The experiment uses ResNet 101 network as backbone, combined with ASPP module as transition. When batch size is 8, the highest detection accuracy is obtained. The global pixel detection accuracy in very small data set can reach 93.1%. Global accuracy is expressed as:

$$\text{global accuracy} = \frac{\sum_i n_{ii}}{\sum_i t_i} \tag{10.4}$$

$$\text{Classification accuracy} = \frac{n_{ii}}{t_i}, i = 1, 2, 3 \tag{10.5}$$

where n_{ii} represents number of pixels of category i predicted as category i and $t_i = \sum_i n_{ii}$, $t_i = \sum_j n_{ij}$ represents total number of pixels of target category i (real tags). Global accuracy indicates the accuracy of pixel detection and classification accuracy indicates the accuracy of each category, as shown in Table 10.1. Compared with U-Net based network structure shown in [25], the method proposed in this paper improves the global accuracy by 9.5% without data augmentation, at the same time, the detection accuracy of basal units has reached 84.7% with very small dataset, indicating that the algorithm is still feasible and has high accuracy on small datasets. Because of the few samples, some rare features in polar ice sheet cannot obtain satisfactory detection results using most of the deep learning networks, however, this algorithm still has a high detection accuracy on small datasets, so it can be applied to the automatic detection of some rare features in ice sheet in future polar research.

Table 10.1 Classification accuracy

Method	Experimental result			
	Global accuracy (%)	Ice layer (%)	Base (%)	Basal units (%)
U-Net + ASPP	83.6	85.2	82.2	–
Method of the paper	93.1	96.1	88.9	84.7

A More Rapid Method

To save training time, try lightweight network MobileNet V3 as backbone. Lightweight network is characterized by fewer parameters, less computation, and fast training speed. The structure is shown in Fig. 10.4.

Where conv3 represents conventional convolution with kernel size of 3 * 3, *s* represents stride of convolution kernel, BN is batch normalization and h-swish indicates optimized activation function, which is expressed as:

$$\text{h} - \text{swish}[x] = x\frac{\text{ReLU6}(x+3)}{6} \tag{10.6}$$

$$\text{ReLU6} = \min(\max(\text{features}, 0), 6) \tag{10.7}$$

The activation function can reduce the computation cost and training time in the case of high-dimensional features. When features less than 0, the results of ReLU6 is 0 and when features is greater than 6, results will be set to 6. Bottleneck is the basic network structure. MobileNet V3 Large has 15 bottlenecks, Avg_Pool is the average pooling operation and FC is full connection layer.

Intel Core i7-7700 @ 3.6GHZ computer and Tesla V100-PCIE32G GPU are adopted in the experiment. When batch size = 8, the results are shown in Table 10.2.

It can be seen that when backbone is ResNet 101 and dataset is augmented, the global accuracy is the highest and reaches 94.0%. MobileNet V3 can effectively shorten the training time, which is 81.8% shorter than ResNet 101, saving a lot of training time resources. At the same time, the data augmentation improves the accuracy of MobileNet V3 backbone by 1.9%. MobileNet V3 reduces the radargram processing time by 92.2% and the space occupied by 81.9% compared with ResNet 101, which is a huge improvement. In view of the relatively high accuracy of MobileNet V3 after data augmentation, as well as its excellent image processing

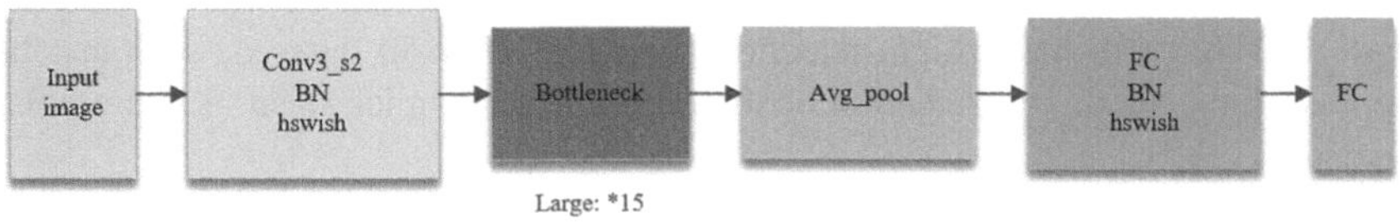

Fig. 10.4 MobileNet V3 structure

Table 10.2 Comparison of different backbones

Backbone	Data aug	Global accuracy (%)	Training time	Processing time (s)	Space occupied (M)
ResNet 50	×	92.1	3:21:16	–	–
	√	92.9	10:43:18	2.679	320
ResNet 101	×	93.1	2:15:16	–	–
	√	94.0	8:43:26	4.370	466
MobileNet V3	×	89.4	0.29.29	–	–
	√	91.3	1:35:04	0.341	84.3

speed and small space occupation, it is suitable for mobile devices, which can process the images obtained from radargram in real time and output the detection results immediately, so that polar researchers can obtain the internal feature information of ice sheet while obtaining data.

Visualization Result

Labelme software is used to label radargrams, which are classified as basal units, ice layer and base, and the labeled images are used as ground truth. Original image is input, finally we use trained network model to classify the pixels of radargram, some classification results are visualized as Fig. 10.5.

The red is ice layer, and yellow represents basal units, green is base. Result-1 is the result of initial algorithm, using ResNet 50 backbone and without data augmentation. Result-2 is the result after optimization, including data augmentation, setting optimal

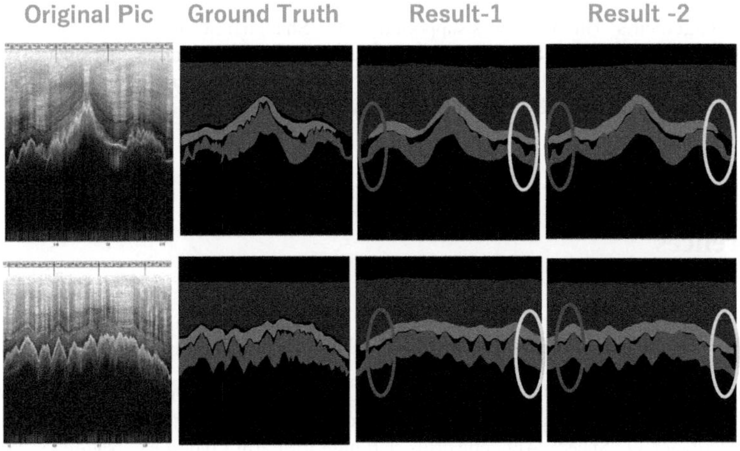

Fig. 10.5 Visualization result

parameters, and changing backbone with Resnet 101. It can be seen that the optimized algorithm has more accurate pixel recognition and better handling of details, here, Result-2 corresponds to the algorithm with highest global accuracy.

10.4 Conclusion

This paper proposes a new automatic detection method of basal units, which can automatically divide radargram into various categories, as well as solving the problem of low efficiency and time-consuming in manual analysis of ice radargram. It mainly includes two aspects. For highest classification accuracy, the network is composed of ResNet 101 backbone and ASPP module. ResNet network can improve the accuracy while deepening the network depth, ASPP module extracts features with different receptive fields, enhances the detection and recognition of targets with different sizes. For improving training and processing speed, we replace lightweight network MobileNet V3 with ResNet 101 as backbone, which can save 81.8% of training time and 92.2% of processing time with high accuracy. It provides a possibility of rapid network training, real-time processing of radargram and application on mobile devices. The method has been tested on radargram dataset collected in Antarctica, and results confirm that algorithm can achieve pixel classification and high precision in distinguishing subglacial structure at the same time.

In the future, pre-processing will be added to the original radargram to reduce image noise and increase image definition, as well as improving the accuracy of the algorithm. At the same time, more exploration will be added to lightweight network, combining it with appropriate hardware devices to reduce training time and processing time, so as to achieve the goal of real-time image processing in polar data acquisition process.

Acknowledgements This work was supported by the Shanghai Science and Technology Development Funds (21ZR1469700), the National Natural Science Foundation of China (41941005, 41941006, 41606219, 41776186), and the International Research Cooperation Seed Fund of Beijing University of Technology. We also thank all reviewers for their constructive comments.

References

1. Noble, T.L., Rohling, E.J., Aitken, A.R.A., Bostock, H.C., Chase, Z., Gomez, N., et al.: The sensitivity of the Antarctic ice sheet to a changing climate: past, present, and future. Rev. Geophys. **58**(4) (2020)
2. Alemany, O., Talalay, P., Boissonneau, P., Chappellaz, J., Chemin, J.F., Duphil, R.: The SUBGLACIOR drilling probe: hydraulic considerations. Ann. Glaciol. **62**(84), 131–142 (2020)
3. Schiermeier, Q.: Speedy Antarctic drills start hunt for Earth's oldest ice. Nature 18–19 (2016)
4. Rodriguez, F.: Advanced multifrequency radar instrumentation for polar research. IEEE Trans. Geosci. Remote Sens. 2824–2842 (2014)

5. Cui, X.B., Sun, B., Zhang, X.P., Zhang, D., Li, X., Tang, X.Y., et al.: A review of ice radar's technical development in polar ice sheet investigation. Chin. J. Polar Res. **21**(4), 322–335 (2009) (in Chinese)
6. Cui, X., Jeofry, H., Greenbaum, J.S., Guo, J., Li, L., Lindzey, L.E., et al.: Bed topography of Princess Elizabeth Land in East Antarctica. Earth Syst. Sci. Data **12**, 2765–2774 (2020). https://doi.org/10.5194/essd-12-2765-2020
7. Frémand, A.C., Fretwell, P., Bodart, J., Pritchard, H.D., Aitken, A., Bamber, J.L., et al.: Antarctic Bedmap data: FAIR sharing of 60 years of ice bed, surface and thickness data. Earth Syst. Sci. Data Discuss (2022). https://doi.org/10.5194/essd-2022-355
8. Bell, R.E., Tinto, K., Das, I., Wolovick, M., Chu, W., Creyts, T.T., et al.: Deformation, warming and softening of Greenland's ice by refreezing meltwater. Nat. Geosci. **7**, 497–502 (2014)
9. Tang, X., Sun, B., Guo, J., Liu, X., Cui, X., Zhao, B.: A freeze-on ice zone along the Zhongshan–Kunlun ice sheet profile, East Antarctica, by a new ground-based ice-penetrating radar. Sci. Bull. **60**(5), 574–576 (2015)
10. Ross, N., Corr, H., Siegert, M.: Large-scale englacial folding and deep-ice stratigraphy within the West Antarctic Ice Sheet. Cryosphere **14**(6), 2103–2114 (2020)
11. Bell, R.E., Ferraccioli, F., Creyts, T.T., Braaten, D., Corr, H., Das, I., et al.: Widespread persistent thickening of the East Antarctic Ice Sheet by freezing from the base. Science **331**(6024), 1592–1595 (2011)
12. Tasar, O., Tarabalka, Y., Alliez, P.: Incremental learning for semantic segmentation of large-scale remote sensing data. IEEE J. Sel. Top. Appl. Earth Obs. Remote. Sens. 3524–3537 (2019)
13. Donini, E., Carrer, L., Gerekos, C., Bruzzone, L., Bovolo, F.: An unsupervised fuzzy system for the automatic detection of Candidate Lava Tubes in radar sounder data. IEEE Trans. Geosci. Remote Sens. 1–19 (2021)
14. Donini, E., Thakur, S., Bovolo, F., Bruzzone, L.: An automatic approach to map refreezing ice in radar sounder data. In: Image and Signal Processing for Remote Sensing XXV, vol. 11155, pp. 433–442 (2019)
15. Schäfer, M., Zwinger, T., Christoffersen, P., Gillet-Chaulet, F., Laakso, K., Pettersson, R., et al.: Sensitivity of basal conditions in an inverse model: Vestfonna ice cap, Nordaustlandet/Svalbard. Cryosphere 771–783 (2012)
16. Christianson, K., Jacobel, R.W., Horgan, H.J., Alley, R.B., Anandakrishnan, S., Holland, D.M., et al.: Basal conditions at the grounding zone of Whillans Ice Stream, West Antarctica, from ice-penetrating radar. J. Geophys. Res.: Earth Surf. **121**(11), 1954–1983 (2016)
17. Panton, C., Karlsson, N.B.: Automated mapping of near bed radio-echo layer disruptions in the Greenland Ice Sheet. Earth Planet. Sci. Lett. 323–331 (2015)
18. Sime, L.C., Hindmarsh, R.C., Corr, H.: Automated processing to derive dip angles of englacial radar reflectors in ice sheets. J. Glaciol. 260–266 (2011)
19. Ilisei, A.M., Bruzzone, L.: A system for the automatic classification of ice sheet subsurface targets in radar sounder data. IEEE Trans. Geosci. Remote Sens. **53**(6), 3260–3277 (2015)
20. Long, J., Shelhamer, E., Darrell, T.: Fully convolutional networks for semantic segmentation. In: Proceedings of the IEEE Conference on Computer Vision and Pattern Recognition, pp. 3431–3440 (2015)
21. Badrinarayanan, V., Kendall, A., Cipolla, R.: Segnet: a deep convolutional encoder-decoder architecture for image segmentation. IEEE Trans. Pattern Anal. Mach. Intell. **39**(12), 2481–2495 (2017)
22. Ronneberger, O., Fischer, P., Brox, T.: U-net: convolutional networks for biomedical image segmentation. In: International Conference on Medical Image Computing and Computer-Assisted Intervention, pp. 234–241. Springer (2015)
23. Guo, P., Su, X., Zhang, H.: A multi-scaled receptive field learning approach for medical image segmentation. In: ICASSP 2020-2020 IEEE International Conference on Acoustics, Speech and Signal Processing (ICASSP), pp. 1414–1418. IEEE (2020)
24. Cai, Y., Hu, S., Lang, S.: End-to-end classification network for ice sheet subsurface targets in radar imagery. Appl. Sci. **7**(10), 2501 (2020)

25. Donini, E., Bovolo, F., Bruzzone, L.: A deep learning architecture for semantic segmentation of radar sounder data. IEEE Trans. Geosci. Remote Sens. **60**, 1–14 (2022)
26. Kamangir, H., Rahnemoonfar, M., Dobbs, D.: Deep hybrid wavelet network for ice boundary detection in radar imagery. In: IGARSS 2018-2018 IEEE International Geoscience and Remote Sensing Symposium, pp.3449–3452. IEEE (2018)
27. Yari, M., Rahnemoonfar, M., Paden, J.: Smart tracking of internal layers of ice in radar data via multi-scale learning. In: 2019 IEEE International Conference on Big Data (Big Data), pp.5462–5468. IEEE (2019)
28. Rahnemoonfar, M., Johnson, J., Paden, J.: Ai radar sensor: creating radar depth sounder images based on generative adversarial network. Sensors 5479 (2019)
29. Donini, E., Bovolo, F., Bruzzone, L.: An unsupervised deep learning method for subsurface target detection in radar sounder data. In: 2021 IEEE International Geoscience and Remote Sensing Symposium IGARSS, pp. 2955–2958. IEEE (2021)
30. García, M.H., Donini, E., Bovolo, F.: Automatic segmentation of ice shelves with deep learning. In: 2021 IEEE International Geoscience and Remote Sensing Symposium IGARSS, pp. 4833–4836. IEEE (2021)
31. García, M.H., Donini, E., Bovolo, F.: Transfer learning for the semantic segmentation of cryosphere radargrams. In: Image and Signal Processing for Remote Sensing XXVII, vol. 11862, pp. 223–233. SPIE (2021)
32. Liu, W., Rabinovich, A., Berg, A.C.: Parsenet: looking wider to see better. Computer
33. Long, M., Cao, Y., Wang, J.: Learning transferable features with deep adaptation networks. In: Proceedings of the 32nd International Conference on Machine Learning, Lile, France, pp. 97–105 (2015)
34. Lin, T.Y., Maire, M., Belongie, S., Hays, J., Perona, P., Ramanan, D., et al.: Microsoft COCO: common objects in context. In: ECCV (2014)

Chapter 11
Research on the Parsing Algorithm of Monocular Visual Structured Data Based on YOLOv5

Wanli Lu, Wen Zhang, Mingrui Sun, and Jindong Zhang

Abstract Visual perception plays an important role in autonomous driving technology. The two key factors in visual perception tasks are monocular object detection and structured data analysis. In this paper, a structured data parsing algorithm based on monocular vision is proposed. Firstly, using the single-stage YOLOv5 object detection model, the target was located and its kind was identified. Secondly, the 3D position of the object in the camera coordinate system is calculated according to the mapping relationship between the imaging point and the target point. Finally, the distance between the object and the camera is calculated by using the pinhole camera model, and the size of the object and the horizontal declination Angle relative to the optical center of the camera are estimated by geometric derivation. This method not only has a simple principle, but also guarantees the precision of structured data output. Experimental findings support the method's viability and efficiency.

11.1 Introduction

In the past few years, with the advancement of deep learning technology, the corresponding update of computer hardware, and the publication of a series of large-scale data sets, the relevant technology in the field of computer vision has made great progress and development. As one of the fundamental problems of computer vision, object detection is the basis of many high-level visual tasks, such as lane estimation, target tracking, path planning, and so on [1, 2]. The structured data analysis

W. Lu · W. Zhang · M. Sun · J. Zhang (✉)
College of Computer Science and Technology, Jilin University, Changchun 130012, China
e-mail: zhangjindong_100@163.com

J. Zhang
College of Software, Jilin University, Changchun 130012, China

Key Laboratory of Symbol Computation and Knowledge Engineering of the Ministry of Education, Jilin University, Changchun 130012, China

R. Kountchev (Deceased) et al. (eds.), *AI Methods and Applications in 3D Technologies*, Smart Innovation, Systems and Technologies 388,
https://doi.org/10.1007/978-981-97-2144-3_11

algorithm based on monocular vision proposed in this paper realizes the recognition and detection of object type, distance, size, three-dimensional coordinates, and horizontal declination, and outputs these structured data. This is very important for the high-quality development of complex visual tasks. Meanwhile, for autonomous driving and other fields, the perception system is its core unit. The output results of the algorithm suggested in this work can serve as a significant foundation for the decision of automatic driving, which will be very crucial in the near and long term.

11.2 General Description of the System

First, the picture is preprocessed [3]. Then, the preprocessed images were input into the YOLOv5 neural network for feature extraction, feature strengthening, and prediction of objects corresponding to feature points [4]. Then, the prediction box and score are obtained through the prediction structure decoding module, and the score screening and non-maximum suppression are performed on the prediction structure. Finally, for each category of prediction box selected, the center point of its bottom edge is taken as the target point to obtain three-dimensional coordinates, and its three-dimensional coordinate points are calculated using geometric principles, and then the horizontal distance [5] and horizontal deflection angle of the target data camera are obtained. The size of the prediction box and the distance to the camera's focal length are then used to calculate the size of the target object. Figure 11.1 displays the flowchart for the program.

Fig. 11.1 Program flow chart

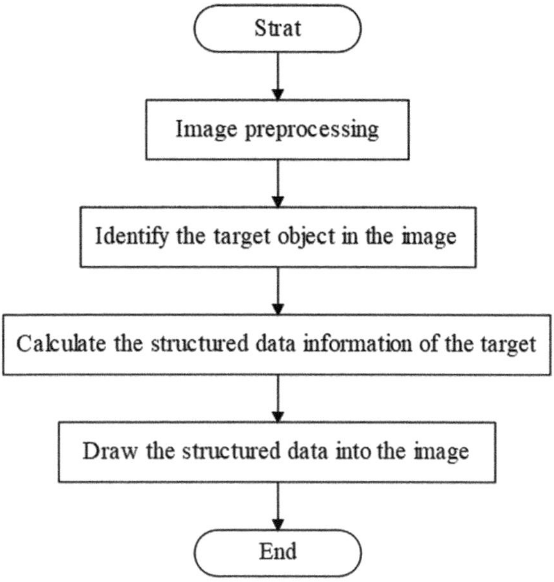

11.3 Design of Algorithm

YOLOv5s Algorithm

In this paper, the YOLOv5s algorithm, which has a small number of parameters, is used as the basic framework for the study. The network structure of the YOLOv5s [4] algorithm, as depicted in Fig. 11.2, consists of four parts:

Input: The input part is responsible for receiving raw images as input to the model. This stage usually includes image preprocessing. The input image is scaled according to the input size of the network and operations such as normalization are performed.

Backbone: The backbone is the core component of YOLOv5s, which is responsible for feature extraction of the input image. The backbone network uses a series of convolutional layers and pooling layers to extract the feature information of the image at different scales and levels of abstraction. The backbone network can capture the global and local features of the image and generate high-dimensional feature representations.

Neck network: The neck network is a connection layer located between the backbone network and the head output. Its function is to perform multi-scale fusion of the feature information extracted by the backbone network to obtain more robust output feature information.

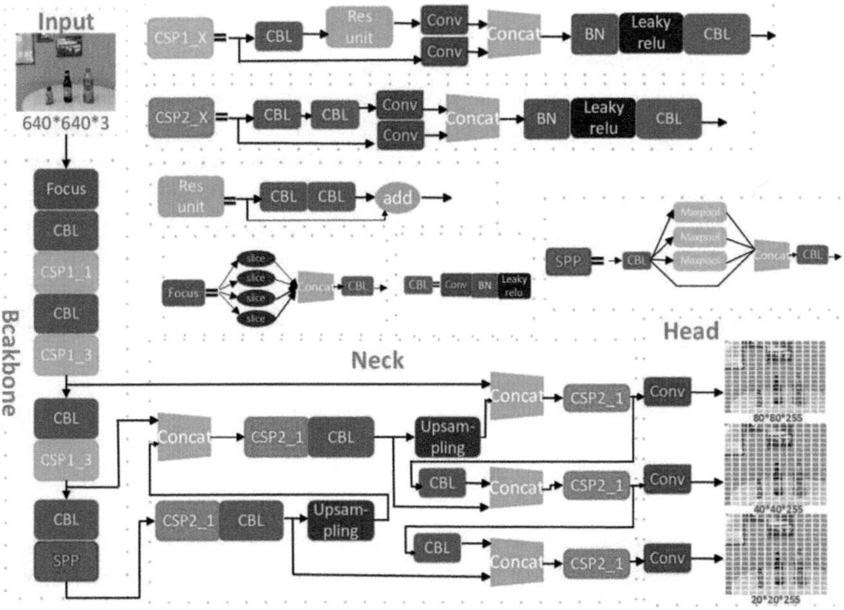

Fig. 11.2 YOLOv5s network structure

Head output: The head output part is responsible for predicting the location and category of the target. In YOLOv5s, multi-layer convolution and fully connected layers are used in the head output to output the detection results.

As shown in Fig. 11.2, Focus is the attention layer; CSP stands for Cross Stage Partial. Concat is the combination layer. Conv refers to convolutional layer; BN refers for batch normalization layer; Lealyrelu is the activation function; Maxpool is the maximum pooling layer. ResUnit is the residual unit [6], which draws on the idea of residual neural network, so that the network can spread deeper.

The detection process of YOLOv5s can be briefly summarized as follows. Firstly, the preprocessed image is input into the backbone network to extract the feature representation of the image. Secondly, the features of different levels are fused and transmitted through the neck network to obtain multi-scale feature information. The head output then predicts the location and category of the object based on these features. Finally, Non-Maximum Suppression (NMS) was applied to remove the overlapping boxes, and the final object detection result was output, including the bounding box coordinates, class labels and confidence scores.

Structured Data Parsing Module

Transformation from Camera Coordinate System to Pixel Coordinate System

Camera imaging realizes projection transformation from three-dimensional space coordinates to two-dimensional image coordinates. The pinhole camera model can describe the mapping relationship between three-dimensional scenes and image pixels [5]. Through this mapping relationship, pixels in the image can be restored to three-dimensional space. The camera parameter matrix can be obtained by camera calibration. Meanwhile, the coordinates (u, v) of the imaging point in the pixel coordinate system are also known. Therefore, the transformation relationship between the camera coordinate system and the pixel coordinate system can be used to determine the coordinates of space locations in the camera coordinate system (X_C, Y_C, Z_C). The derivation of the formula is as follows:

$$\begin{bmatrix} \mathrm{u} \\ \mathrm{v} \\ 1 \end{bmatrix} = \frac{1}{Z_C} \begin{bmatrix} f_x & 0 & c_x \\ 0 & f_y & c_y \\ 0 & 0 & 1 \end{bmatrix} \begin{bmatrix} X_C \\ Y_C \\ Z_C \end{bmatrix} \tag{11.1}$$

where, f_x is the pixel focal length of x axis, and f_y is the pixel focal length of y axis. (c_x, c_y) is the pixel coordinate of the origin of the imaging plane coordinate system in the pixel coordinate system.

Analysis of Target Distance Information

The images are collected by the camera, and the three-dimensional scene is projected onto the two-dimensional image plane of the camera. The pinhole camera model is utilized to determine the distance between the object and the camera. The ranging principle based on monocular vision is shown in Fig. 11.3, where $X_C Y_C Z_C$ is the camera coordinate system and the origin is the optical center O_C. $X_W Y_W Z_W$ is the world coordinate system, and x-O-y coordinate system is the image coordinate system.

As shown in the figure, Q is any point in the plane, and P is the intersection of perpendicular lines through Q to the world coordinate system Xw axis; The camera's effective focal length is $f = OO_C$, and its pitch angle is a; O_W is the perpendicular foot of the camera perpendicular to the ground; H is the camera's mounting height (measured from the lens's center to the ground); O serves as the origin of the coordinate system for the picture plane and is the location where the optical axis and image plane cross. P' is the projection coordinate of point P on the imaging plane; Q' is the projection coordinate of point Q on the imaging plane.

The midpoint of the bottom edge of the prediction box is the 3D coordinate feature point obtained, and (x, y) is the coordinate of the midpoint of the bottom edge of the prediction box in the imaging coordinate system. The horizontal distance $O_W Q$ and horizontal deflection Angle between the target and the camera can be obtained from the geometric similarity relation. Then, the vertex of the prediction box is taken as the feature point related to the object size, and the size of the object is estimated according to the depth information of the object and the focal length of the camera. Description of the structured data parsing algorithm pseudocode, as follows:

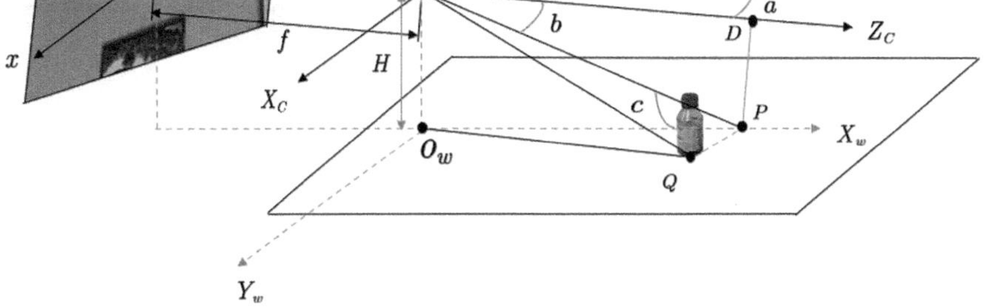

Fig. 11.3 Schematic diagram of monocular distance measurement

Algorithm 1 Structured data parsing algorithm

Input: H: The camera's height above the ground; angle_a: The pitch angle of the camera; fx, fy: The camera's focal length; (x, y):The coordinate of the midpoint of the bottom edge of the prediction box in the imaging coordinate system; (c_x, c_y): The pixel coordinates of the origin of the imaging plane coordinate system in the pixel coordinate system; l: The length of the prediction box; w: The width of the prediction box

Output: label: Object type information; (X_C, Y_C, Z_C): The three-dimensional coordinates of the object in the camera coordinate system; O_WQ: Horizontal distance of the object from the camera; angle: Horizontal declination Angle of the object; size_x, size_y: Object size information

```
angle_b = arctan((y - c_y)/fy)
angle_c = angle_b + angle_a
Z_C=(H/sin(angle_c))*cos(angle_b)
X_C= Z_C * (x - c_x) / fx
Y_C=Z_C * (y - c_y) / fy
O_wP = H/tan(angle_c)
O_wQ = sqrt(X_c^2 + O_WP^2)
angle = arctan(X_C/O_wP)
size_x = Z_C*l/f_x
size_y = Z_C*w/f_y
print(label, (X_C, Y_C, Z_C), O_WQ, angle, size_x, size_y)
```

11.4 Experiment and Result Analysis

Basis of Experiment

The experimental environment for model training and testing is as follows: Intel Core i5-8500 CPU @ 3.00 GHz, 16.00 GB memory, WDC WD10EZEX-08WN4A0 1 TB hard disk, win11 system.

To quantitatively determine the accuracy of the detection algorithm, a large number of simulation experiments are carried out in this paper, and the corresponding data sets are made. Specifically, the other parameters in the fixed ranging model remain unchanged, and the photos are taken at the distance of 0.5 m, 1.0 m, and 1.5 m from the target. Then, other physical parameters are added, such as altering the camera's height, the angle between the optical axis of the camera and the horizontal plane, and retaking the photograph.

In the experiment, the relative error between the distance result output by the algorithm and the real distance between the target and the camera, and the relative error between the object size result output by the algorithm and the real size of the target are taken as the accuracy of the target prediction by the algorithm. The smaller the relative error, the higher the algorithm precision.

Results of the Experiment

Performance Testing of Object Detection Algorithms

In this experiment, the following metrics were used to evaluate the performance of the target detection algorithm: Recall, Precision, and mean Average Precision (mAP). Frames per Second (FPS) was used to evaluate the real-time performance of the algorithm.

In this paper, the evaluation index is set as mAP@0.5, that is, the IOU (Intersection over Union) detection threshold is set as 0.5.

From the experimental results, the recall rate of the algorithm is 85.29%, the precision rate is 83.81%, and the mAP value is 89.29%, which indicates that the algorithm is able to detect the target in a more comprehensive way and has a low false detection rate. The FPS of this algorithm is 36 f/s, which shows better real-time performance and is suitable for real-time target detection application scenarios.

Algorithm Accuracy Verification Experiment

The target is predicted using the algorithm presented in this research, and Tables 11.1 and 11.2 display the output of the structured data. Suppose the three objects in the picture are object A, object B, and object C from left to right.

The relative error bar chart is shown in Fig. 11.4. Where the abscissa represents the sample number and the ordinate represents the relative error value.

As shown in Fig. 11.4, the relative error between the measured distance and the actual distance can be kept within 3% by using the method suggested in this study for distance measurement of the target point, which is of high accuracy. The algorithm proposed in this paper can still maintain high accuracy when measuring small size objects. The error sources of the algorithm may include camera calibration error, camera installation position error, algorithm error, etc.

Algorithm Robustness Verification Experiment

The distance from the fixed camera to the target is 1.0 m, and then the height of the camera and the Angle between the camera and the horizontal plane are changed

Table 11.1 Experimental data of ranging accuracy

Picture number	Target	Measured coordinate/m	Measured declination/°	Measured distance/m	Actual distance/m	Distance relative error (%)
1	A	(−0.12, −0.19, 0.44)	−14.7	0.5	0.5	0
	B	(−0.01, −0.19, 0.46)	−1.7	0.5	0.5	0
	C	(0.10, −0.19, 0.45)	13	0.5	0.5	0
2	A	(−0.04, −0.24, 0.98)	−2.2	1.01	1	1
	B	(0.07, −0.24, 0.95)	4.1	0.99	1	−1
	C	(0.17, −0.24, 0.93)	10.4	0.98	1	−2
3	A	(−0.16, −0.25, 1.50)	−6.2	1.53	1.5	2
	B	(−0.06, −0.25, 1.49)	−2.4	1.51	1.5	1
	C	(0.05, −0.25, 1.47)	1.8	1.49	1.5	−1

Table 11.2 Experimental data of measurement accuracy of object size

Picture number	Target	Measured height/m	Actual height/m	Height relative error (%)	Measured width/m	Actual width/m	Width relative error (%)
1	A	0.1	0.1	0	0.04	0.045	−11.1
	B	0.18	0.2	−10	0.05	0.05	0
	C	0.2	0.21	−4.7	0.06	0.065	−7.7
2	A	0.09	0.1	−10	0.04	0.045	−11.1
	B	0.18	0.2	−10	0.05	0.05	0
	C	0.19	0.21	−9.5	0.06	0.065	−7.7
3	A	0.09	0.1	−10	0.05	0.045	11.1
	B	0.18	0.2	−10	0.05	0.05	0
	C	0.19	0.21	−9.5	0.06	0.065	−7.7

to take multiple photos. The output-ranging results of the algorithm for the above pictures are shown in Table 11.3.

It can be seen from Table 11.3 that the mean value of the calculated measured distance is 0.99 m and the variance is 0.00019, indicating that the measured distance is relatively stable and the robustness of this method is good.

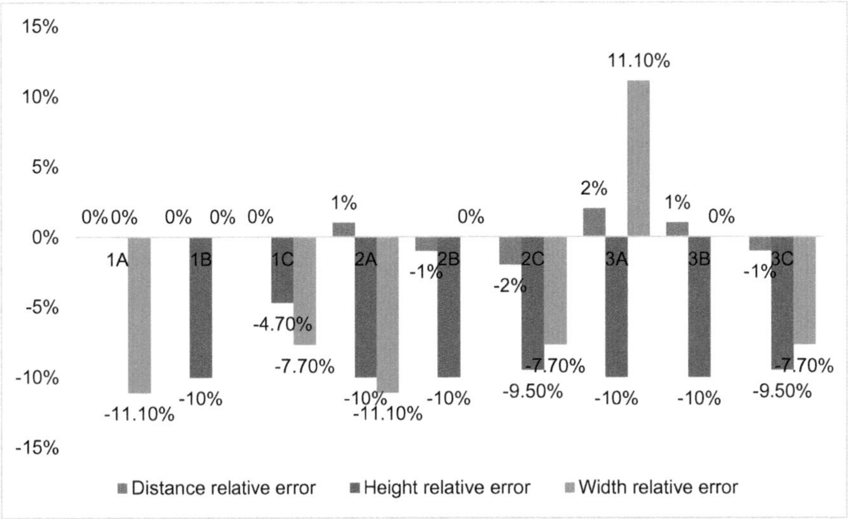

Fig. 11.4 Bar graph of relative error

Table 11.3 Robustness experimental data

Picture number	Target	Measured distance /m
2	A	1.01
	B	0.99
	C	0.98
4	A	1.01
	B	0.99
	C	0.97
5	A	1.01
	B	0.99
	C	0.97
6	A	0.98
	B	0.99
	C	0.98

Algorithm Comparison

In order to show the effectiveness of the proposed method more clearly. In different distance intervals, the ranging relative errors of the proposed algorithm and three other typical algorithms are calculated, and the results are shown in Table 11.4.

As shown in Table 11.4, our algorithm has the lowest average relative error, indicating that this algorithm is better than the other three algorithms and has good accuracy.

Table 11.4 Average relative errors of several ranging method

Methods	Object distance/m	Distance relative error (%)	Mean relative error (%)
SSD algorithm [7]	[0,5]	8.00	8.15
	(5,10]	8.30	
Machine learning [8]	[0,5]	6.30	4.80
	(5,10]	3.30	
MTCNN [9]	[0,5]	4.70	3.45
	(5,10]	2.20	
Ours	[0,5]	1.81	2.05
	(5,10]	2.30	

Visualization Result

The corresponding relationship between the above picture numbers and pictures is shown in Fig. 11.5. The experimental results demonstrate the relatively high accuracy of the structural data analysis method based on monocular vision suggested in this research, with a relative error ranging within 3%. At the same time, it can maintain high accuracy when measuring small-size targets. The feasibility and effectiveness of the algorithm are proven.

Fig. 11.5 Visualization results

11.5 Summary

This research proposes a structured data analysis algorithm based on monocular vision is proposed. The YOLOv5 algorithm is used to detect the target, the pinhole camera model is used to predict the distance of the target object, and the three-dimensional coordinates, size, and declination information of the object are predicted through geometric derivation. In this paper, a new requirement of object detection is proposed from the point of view of outputting structured data, and an object detection model combining the requirement of outputting structured data is constructed. After a lot of simulation results, the feasibility and effectiveness of the detection model constructed in this paper are verified. However, at present, the function achieved in this study is relatively simple. In the subsequent research, the function of the algorithm can be gradually improved, such as the detection, extraction, tracking, and output of the relative speed of moving objects.

References

1. Zhang, J.D., Zhong, H.T.: Curve-based lane estimation model with lightweight attention mechanism. SIViP **17**, 2637–2643 (2023). https://doi.org/10.1007/s11760-022-02480-0
2. Liu, T.B., Zhang, J.D.: An improved path planning algorithm based on fuel consumption. J. Supercomput. **78**, 12973–13003 (2022)
3. Zhang, J.D., Pan, D.Y., Zhang, K.P., et al.: Underwater single image restoration based on modified generative adversarial net. SIViP **17**, 1153–1160 (2023). https://doi.org/10.1007/s11760-022-02322-z
4. Kim, M., Kim, Y.: Parcel classification and positioning of intelligent parcel storage system based on YOLOv5. Appl. Sci. **13**(1), 437 (2023). https://doi.org/10.3390/app13010437
5. Hou, Y., Liu, C., Wang, H., Fan, Y., An, B., Liu, Y.: High-precision visual imaging model and calibration method for multi-depth-of-field targets **268**, 0030–4026 (2022). https://doi.org/10.1016/j.ijleo.2022.169788
6. Simou, M.R., Loulad, S., Aamir, Z., Maanan, M., Rhinane, H.: Using YOLOV5 to detect plastic greenhouses in morocco. Int. Arch. Photogramm. Remote Sens. Spatial Inf. Sci. XLVIII-4/W6–2022, 351–354 (2023). https://doi.org/10.5194/isprs-archives-XLVIII-4-W6-2022-351-2023
7. Li, Z.Q., Gao, J.D., Peng, K., et al.: Method of measurement vehicle distance based on PnP. Foreign Electron Meas Technol **39**(12), 104–108 (2020)
8. Cheng, Y., Zhao, L., Cheng, S., et al.: Design of vehicle distance detection system based on machine vision. Acta Metrologica Sinica **41**(1), 11–15 (2020)
9. Ding, B.Q., Li, J.Y.: Vehicle distance detection method based on MTCNN Algorithm. J. Chongqing Jiaotong Univ. (Nat. Sci.) **42**(05), 139–144 (2023)

Part II
3D Technologies and Deep Learning

Chapter 12
A BPNN-Based Method for Regional Competitiveness: Taking the Guangdong-Hong Kong-Macau Greater Bay Area as an Example

Mini Han Wang, Pengsheng Li, Fengling Wang, Xiaoshu Zhou, Qing Wu, Guoqiang Chen, Zhiyuan Lin, Peijin Zeng, and Qide Xiao

Abstract Guangdong-Hong Kong-Macau Greater Bay Area (GBA) reveals a significant opportunity for self-developing and regional competition since China's government regards it as one of the most significant developing strategies. However, cities in this area are enhancing comprehensive competitiveness at different levels. This paper proposed a competitiveness evaluation model based on the entropy approach and built a competitiveness measuring and predictive model based on the BPNN machine learning algorithm. The research results illustrate that comprehensive competitiveness is approximately increasing over time, which shows a trend of eastward development. Hong Kong, Guangzhou, Shenzhen, Macau, and Dongguan are the top-5 cities during the three years. On the contrary, Zhongshan is the least competitive

M. H. Wang (✉) · P. Li (✉) · F. Wang · G. Chen · Z. Lin · P. Zeng
Beijing Institute of Technology, Zhuhai, China
e-mail: D20092100037@cityu.mo

P. Li
e-mail: D20092100186@cityu.mo

M. H. Wang · P. Li · Q. Xiao (✉)
Faculty of Data Science, City University of Macau, Macau, China
e-mail: xiaoqide@yeah.net

M. H. Wang · Z. Lin · P. Zeng
Department of Big Data and Artificial Intelligence, Zhuhai Institute of Advanced Technology Chinese Academy of Sciences, Zhuhai, China

F. Wang
Hezhou University, Hezhou, China

X. Zhou
Gree Polytechnic, Zhuhai, China

Q. Wu
Zhuhai Science and Technology Development Promotion Association, Zhuhai, China

R. Kountchev (Deceased) et al. (eds.), *AI Methods and Applications in 3D Technologies*, Smart Innovation, Systems and Technologies 388,
https://doi.org/10.1007/978-981-97-2144-3_12

city, however, Zhongshan's comprehensive competitiveness emerges at the highest increasing rate.

12.1 Introduction

Place competitiveness is of primary interest to tourism development and sustainability. Early work from Butler [1] points to the rise and fall of a destination's lifecycle can be attributed to its competitiveness as "continued overuse of resources, non-replacement of aging plants, and decreasing competitiveness with other areas would result in the marked decline". Since then, a large number of efforts have been devoted to gauging a destination's competitiveness as it resonates with a place's socio-economic vitality and attractiveness relative to other destinations [2–5]. The competitiveness of a place may also serve as a proxy of productivity which offers industry operators and the government a means to evaluate future tourism development strategies [6] and the sustainability of the place in the long run [3].

Despite the aforementioned studies have laid a solid foundation to the field, two limitations remain. First, most research initiatives germane to assessing a single or a few destinations with a few competitiveness indexes without consider broader appeals from a regional perspective [4, 7]. Second, the extant literature primarily employs conventional methods, such as regression- and parametric-based analytics, in gauging place competitiveness [8, 9]; while others take a conceptual approach to avoid the methodological complexity [2, 6, 10].

This study aims to bridge the aforementioned research gaps by assessing regional competitiveness of the Greater Bay Area, which is formed as conglomeration of 11 destinations based mega regional initiative from China. It also seeks to offer a new analytical approach in evaluating place competitiveness through neural network processing, which should provide a more accurate estimates of the corresponding competitiveness indexes. In essence, this study works to advance the literature by showcasing how neural network could be utilized to assist scholars and operators to evaluate competitive strengths and weaknesses of a region.

12.2 Literature Review

Research Background: The Greater Bay Area of China

Since China's Belt and Road Initiative national strategy was proposed, the Guangdong-Hong Kong-Macau Greater Bay Area (Greater Bay Area or GBA for short thereafter) has been on the center stage of national economy and technology development [11, 12]. The Greater Bay Area is agglomerated by Hong Kong, Macau and 9 cities of Grongdong province in China including Shenzhen, Foshan, Dongguan, Zhaoqing, Huizhou, Zhuhai, Jiangmen and Zhongshan [13]. The development of the Great Bay area is the first bay area regional development strategy rising to the national level in China, which is of great significance to the nation [14]. Aiming to promote the comprehensive cooperation between Macau, Hong Kong and China mainland, this strategy enhances China's openness and economic system conduction.

Place Competitiveness

Place or regional competitiveness is one of the primary research area in tourism with respect to implications pertinent to tourism geography and economics [15]. Competitiveness of a place is defined as labor resources attached to enterprises, the attraction of regions to the enterprises, and the share of economic indicators in the whole region [8, 16]. Some of the most representative competitiveness evaluation models include Porter diamond model [17] and International Institute for Management Development (IMD) regional competitiveness evaluation model [16]. Porter's [18] diamond model was proposed for measuring the industry or nation's competitiveness, which has been employed for regional competitiveness measurement in the recent years [19]. There are four determinants involved the model including firm strategy, structure, and rivalry; demand conditions; related and supporting industries; and factor conditions.

Beyond Porter's typology discussed above, the literature also recommends other performance indicators such as technology, regional policies, public infrastructures, social civilizations, knowledge and innovation are also included to evaluate the regional competitiveness [7, 16, 20]. Organization for Economic Co-operation and Development (OECD) [21] ranked 78 mega-cities in the OECD region based on cities' competitiveness with a composite index including economic performance (e.g., GDP), government policy, business companies, and public infrastructure [10, 21]. Furthermore, some scholars regarded scientific and technological innovation, human resource, economic development, infrastructural accessibility, and global economy integration as the major indicators of a broader competitive evaluation system [5, 8].

In essence, the aforementioned literature commonly points to five key areas of competitiveness of a region, namely economic capacity, scientific technology capacity, human resource capacity, public infrastructure capacity, and international competition capacity. Each area may further be represented by sub-indexes or elements that could better define the corresponding domain of expertise or competitiveness of interest. Recently, Zhang et al. [22] applied this multilayer model (i.e., a model that composes with the abovementioned five key competitiveness areas with sub-indexes for each area; see more details on Fig. 12.1) to empirically validate the model with data representing the competitiveness of 13 mega city-regions of China. It is important to note that although Zhang et al.'s study was primarily driven from a city-development perspective without considering the tourism perspective, their work nevertheless offers a important theoretical and operational grounding that underpins the present research.

Given the quintessential role in gauging a destination's performance in better understand the place's strengths and weaknesses in order to better plan for its future tourism development, the following section depicts analytical techniques using BPNN that can potentially help to improve the accuracy of the competitiveness analysis.

12.3 Methodology

This study utilizes a quantitative analysis approach based on Yuemin' [21] index system for GBA competitiveness evaluation.

Evaluation Index System of Regional Competitiveness

As shown on Fig. 12.1, this study selected five core elements for competitive evaluation, which are capacity indicators related to public infrastructure, scientific and technologies, economy, human resources, and international competitiveness. In this way, economic capacity is measured by four indicators, which are industrial structure, economic output, capital scale, and headquarters economy. Two factors are involved to scientific and technical capacity calculation, which are scientific innovation and research capacity. Human resource capacity was determined by three elements of urbanization level, labor level, and human capital. Public infrastructure capacity was measured by informatization level and transportation of highways, railways, and airlines. international competition capacity was evaluated by the export-oriented economy and globalization and world city rank. Finally, this system processes one first-level index (Y), 5 s-level indexes (X) and 24 third-level indexes J.

City Competitiveness Evaluation Method

Competitiveness evaluation-related data (Fig. 12.1) are collected from the yearbook of GBA cities. The BPNN machine learning method is applied in this study.

Weights from the BP neural network (BPNN) are available to be adjusted continuously through back propagation, which could reflect the level of the cites' competitiveness comprehensively with greater accuracy [23]. Besides, not only the BPNN completes a nonlinear mapping relationship from i-dimensional-input to j-dimensional-output, but the square error is minimized by continuously adjusting the number of hidden nodes, number of iterations, square error threshold and loss function [24].

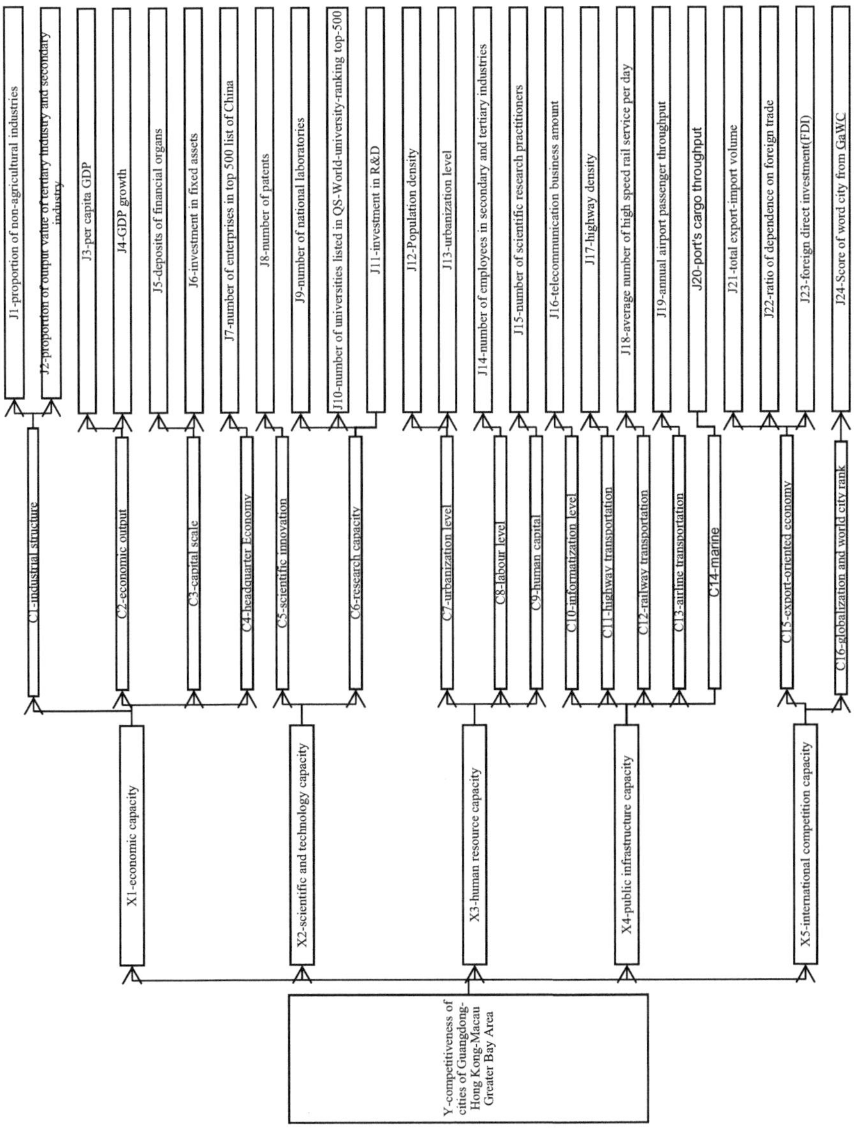

Fig. 12.1 The competitiveness index system of GBA

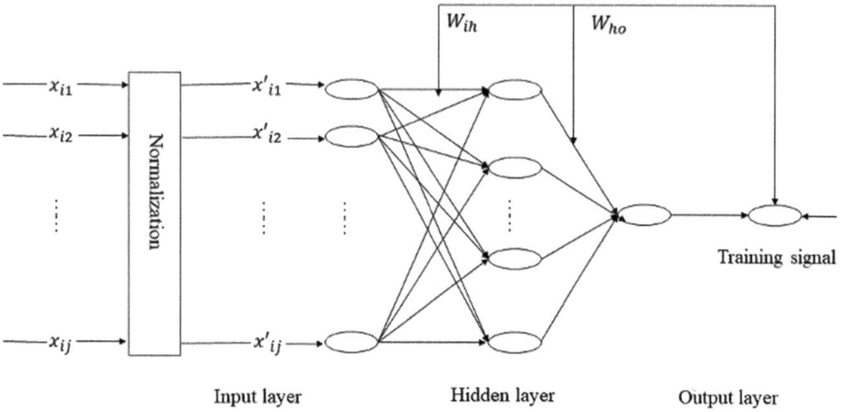

Fig. 12.2 The BP neural network topology structure diagram

The BPNN model is one of the most popular neural network models, which are applied in the research of competitiveness evaluation system. For instance, Li et al. [25] utilized this model to establish a tourism competitive evaluation system for smart cities [25]. Lin et al. [26] applied BPNN for measuring Chinese contractors' competitiveness in Europe market of high-speed rail industry [26]. This study employs Python programming language to perform a 3-layer BPNN model. As shown on the Fig. 12.2, the input layer has j nodes, in this case, $j = 24$, output layer holds 1 node, the nodes of hidden layer are selected in the range of 2 to 24, the square error threshold is set as 0.05. Furthermore, 60% of the entropy results were randomly selected into the training data set, while the rest was regarded as the validation dataset. An example of the BPNN model is presented in Fig. 12.2.

This study investigates the regional distribution of indicators for GBA by calculating the Herfindahl index (HI) score and screening core cities. Cities with HI values accounting for more than 5% of specific indicators in total are considered as core cities. The algorithm of HI calculation is as follows. $X_{i,j}$ is the competitive indicator j of city i, X_j is all the comprehensive competitive indicator j of all the cities.

$$H_j = \sum\nolimits_{i=1}^{N} \left(\frac{X_{i,j}}{X_j}\right)^2 \tag{12.1}$$

12.4 Results and Discussions

Competitiveness Analysis Results

This research calculated the comprehensive competitiveness based on the entropy approach and analyzes the primary indicator for cities of GBA. As shown on the Table 12.1 and Fig. 12.3, the results illustrate that comprehensive competitiveness is approximately increasing over time. Hong Kong, Guangzhou, Shenzhen, Macau, and Dongguan are the top-5 subjects during the three years. By the contrary, Zhongshan is the least competitive city, however, Zhongshan's comprehensive competitiveness emerges at the highest increasing rate.

BP Neural Network Model

This research applied the BPNN model to conduct the competitiveness evaluation for GBA. Competitiveness results based on the entropy method were regarded as the datasets of training and validation. 24 Third-level indicators (J) were selected as the input-layer nodes, the first level indicators of comprehensive competitiveness (Y) was respected as the output-layer node. The number of hidden-layer nodes was chosen from the range of [0,10], the validation errors were as shown as the Table 12.2. Thus, this study decided to put 12 hidden-layer nodes in this model, which is with the lowest error.

After performing this model, absolute errors and relative standard deviation values for training and validation data set are presented as the Tables 12.3 and 12.4. The

Table 12.1 The analysis of comprehensive competitiveness for GBA

Y-competitiveness of GBA					
City	2016	2017	2018	Average growth rate	Average value
Hong Kong	0.99	1.109	1.087	0.05	1.062
Macau	0.36	0.424	0.406	0.068	0.653
Guangzhou	0.564	0.634	0.761	0.163	0.429
Shenzhen	0.357	0.462	0.468	0.154	0.396
Zhuhai	−0.001	0.025	−0.005	−13.93	0.073
Foshan	0.012	0.021	0.043	0.871	0.025
Dongguan	0.036	0.055	0.13	0.955	0.015
Huizhou	−0.016	−0.012	0.073	−3.577	0.015
Zhaoqing	−0.008	−0.01	−0.014	0.289	0.011
Jiangmen	0.007	0.004	0.024	2.384	0.006
Zhongshan	0	0.004	0.041	8.648	−0.011

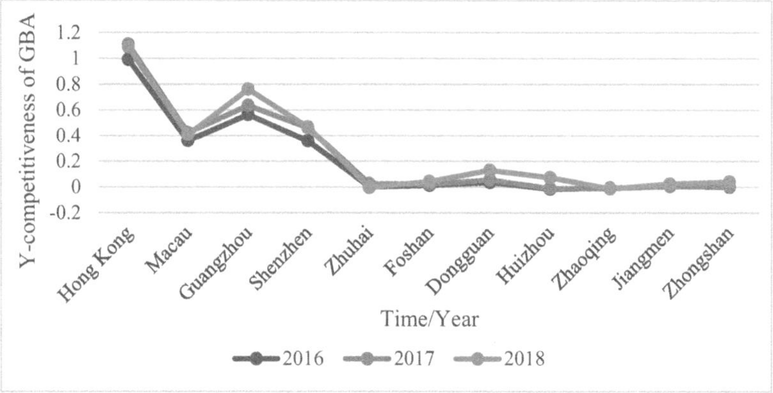

Fig. 12.3 The analysis of comprehensive competitiveness for GBA

Table 12.2 Hidden-layer nodes and corresponding error of the BPNN

Hidden-layer nodes	2	3	4	5	6	7	8	9
Error	0.21	0.05	0.35	0.04	0.11	0.07	0.1	0.13
Hidden-layer nodes	10	11	12	13	14	15	16	17
Error	0.05	5.07	0.03	9.19	10.92	11.44	11.67	2.43
Hidden-layer nodes	18	19	20	21	22	23	24	
Error	0.17	12.16	11.57	10.71	11.16	12.16	11.35	

average of absolute errors of relative standard deviation for training data set are 0.028126461 and 0.019888411, which are 0.049275898 and 0.034843321 for validation data, these results are all below 0.5. Therefore, such model is effective for comprehensive competitiveness evaluation and predicting for GBA.

12.5 Conclusions

This study proposed a competitiveness capability evaluation system based on entropy calculation and BP neural network model parsed in Python programming language, which is also capable of dynamically estimating and monitoring the evaluation index for general cities. Otherwise, this study conducted an in-depth discussion on the competitiveness and regional inequality of GBA from both geographical and statistical aspects, which comprehensively revealed the imbalance of urban development in GBA from the aspects of economic, scientific, and technologies, human resources, public infrastructure, and international competitiveness. In general, the competitiveness of GBA presents a growing trend from 2016 to 2018. Besides, comprehensive competitiveness and economic competitiveness exhibit an eastward-moving

Table 12.3 Results of the predicted value, absolute error and relative standard deviation for training data set of the BPNN

Expected value	0.9901	1.108	0.35	0.42	0.5637	0.63	0.35	0.46	−0.00	0.02	0.01
Actual value	0.9771	0.977	0.40	0.49	0.5745	0.66	0.39	0.47	0.00	−0.02	0.05
Absolute error	0.013	0.131	0.04	0.07	0.0108	0.03	0.03	0.01	0.00	0.05	0.04
Relative standard deviation	0.0092	0.093	0.03	0.05	0.0076	0.02	0.02	0.01	0.01	0.03	0.03
Expected value	0.0205	0.035	0.05	−0.01	−0.012	−0.00	−0.01	0.00	0.01	0.00	0.01
Actual value	0.0565	0.042	0.05	−0.02	−0.023	−0.01	−0.02	0.02	0.02	0.01	0.02
Absolute error	0.036	0.007	0.00	0.01	0.0108	0.01	0.01	0.02	0.01	0.01	0.02
Relative standard deviation	0.0254	0.005	0.00	0.00	0.0076	0.00	0.01	0.01	0.01	0.01	0.01

Table 12.4 Results of the predicted value, absolute error and relative standard deviation for validation data set of the BPNN

Expected value	1.08	0.40	0.76	0.46	−0.005	0.04	0.12	0.07	−0.01	0.02	0.04
Actual value	0.97	0.40	0.78	0.49	−0.007	0.08	0.16	0.15	−0.05	0.09	0.15
Absolute error	0.10	0.00	0.02	0.02	0.002	0.04	0.03	0.08	0.04	0.06	0.11
Relative standard deviation	0.07	0.00	0.01	0.01	0.001	0.03	0.02	0.05	0.02	0.04	0.07

tendency over time. The BPNN model has been successfully built, and the node parameters of the input layer hidden layer and output layer were defined as 24, 12, and 1. The average of absolute errors of relative standard deviation for training and validation data set are all below 0.5, which implies that this model possesses high effectiveness to evaluate and predict comprehensive competitiveness.

Acknowledgements This research was funded by the 2020 Key Technology R & D Program of GuangDong Province, grant number of 2020B1111540001, Zhuhai Technology and Research Foundation, grant number of ZH01110405180056PWC, Zhuhai Technology and Research Foundation, grant number of ZH22036201210034PWC and 2220004002412, Zhuhai Basic and Application Research Project, grant number of ZH22017003200011PWC.

References

1. Butler, R.W.: The concept of a tourist area cycle of evolution: implications for management of resources. Can. Geogr **24**, 5–12(1980)
2. Faulkner, B., Oppermann, M., Fredline, E.: Destination competitiveness: an exploratory examination of South Australia's core attractions. J. Vacat. Mark. **5**, 125 (1999)
3. Ritchie, J.R.B., Crouch, G.I.: The competitive destination: a sustainability perspective. Tour. Manag. **21**(1), 1 (2000)
4. Kubickova, M., Li, H.: Tourism competitiveness, government and tourism area life cycle (TALC) model: the evaluation of Costa Rica, Guatemala and Honduras. Int. J. Tour. Res. **19**(2), 223–234 (2017)
5. Kubickova, M., Martin, D.: Exploring the relationship between government and destination competitiveness: the TALC model perspective. Tour. Manage. **78**, 104040 (2020)
6. Hall, C.M.: Chapter 15-tourism and regional competitiveness. In: Tribe, J., Airey D. (eds.) Developments in tourism research, pp. 217–230. Oxford (2007)
7. Cracolici, M.F., Nijkamp, P.: The attractiveness and competitiveness of tourist destinations: a study of Southern Italian regions. Tour. Manag. **30**(3), 336–344 (2009)
8. Croes, R., Ridderstaat, J., Shapoval, V.: Extending tourism competitiveness to human development. Ann. Tour. Res. **80**, 102825 (2020)
9. Corte, V.D., Aria, M.: Coopetition and sustainable competitive advantage. The case of tourist destinations. Tour. Manag. **54**, 524–540 (2016)
10. Crouch, G.I., Ritchie, J.R.B.: Tourism, competitiveness, and societal prosperity. J. Bus. Res. **44**(3), 137–152 (1999)
11. Geng, C., Shen, J.: Study on development strategy of Guangdong-Hong Kong-Macau Greater Bay Area: a "the Belt and Road" Perspective. In: 3rd International Conference on Judicial, Administrative and Humanitarian Problems of State Structures and Economic Subjects (JAHP 2018). Atlantis Press (2018)
12. Huang, G., Chu, S.-H., Yang, F., Chang, K.-H.: Financial conglomeration, efficiency of capital allocation, and economic growth: a case of the Guangdong-Hong Kong-Macau greater bay area. Int. J. Uncertain. Innov. Res. **1**(1), 87–109 (2019)
13. Hui, E.C., Li, X., Chen, T., Lang, W.: Deciphering the spatial structure of China's megacity region: a new bay area—the Guangdong-Hong Kong-Macao Greater Bay Area in the making. Cities (2018)
14. Ailei, X., Jiaxin, I., Qunqun, L.: Collaborative innovation and development of higher education in Guangdong-Hong Kong-Macao Greater Bay area: background, foundation and path. China Higher Education Research, vol. 5, pp. 58–63 (2019)
15. Moiseev, N.A., Manakhov, S.V., Demenko, O.G.: Boosting regional competitiveness level via budgetary policy optimization. Int. J. Appl. Bus. Econ. Res. **14**(10), 7315–7324 (2016)
16. Kitson, M., Martin, R., Tyler, P.: Regional competitiveness: an elusive yet key concept?. Reg. Stud. **38**(9), 991–999 (2004)
17. Porter, M.E.: Location, competition, and economic development: local clusters in a global economy. Econ. Dev. Q. **14**(1), 15–34 (2000)
18. Porter, M.E.: The competitive advantage of nations: with a new introduction. Free Pr (1990)
19. Chung, T.W.: A study on logistics cluster competitiveness among Asia main countries using the Porter's diamond model. Asian J. Shipp. Logist. **32**(4), 257–264 (2016)
20. Turok, I.: Cities, regions and competitiveness. Reg. Stud. **38**(9), 1069–1083 (2004)
21. Harrison, J.: Oecd territorial reviews: competitive cities in the global economy. In: Kamal-Chaoui, L. (eds). Growth and Change, vol. 39, no. 1, pp. 177–179 (2008)
22. Zhang, F., Ning, Y., Lou, X.: Competitiveness and regional inequality of China's mega-city regions. Geogr. Res. **38**(7), 1664–1677 (2019)
23. Huang, S., Li, Y.L., Dai, P.J.: Evaluation of tourism competitiveness of Chinese smart tourism city. Acta Geogr. Sin. **72**(2), 242–255 (2017)

24. Li, J., Cheng, J.H., Shi, J.Y., Huang, F.: Brief introduction of back propagation (BP) neural network algorithm and its improvement. Advances in computer science and information engineering, pp. 553–558. Springer, Berlin, Heidelberg (2012)
25. Li, W., Wang, Y.: Dynamic evaluation of logistics enterprise competitiveness based on machine learning and improved neural network. J. Ambient Intell. Humaniz. Comput. 1–15 (2021)
26. Lin, Y.H., Shi, B., Chen, P.H., Xu, Z., Liu, H.: A grey-BPNN model for evaluating the competitiveness of Chinese contractors in the high-speed rail market in Europe. Adv. Civil Eng. (2019)

Chapter 13
3D Simulation Scene Optimization Model of High-Speed Railway Virtual Reality Based on Artificial Neural Network

Hui Han, Xiaolong Chen, and Changzhong Li

Abstract Applying VR (Virtual Reality) technology to the simulation of high-speed railway can undoubtedly meet people's new needs for high-speed railway cognition. In order to improve the drawing speed. In this paper, ANN (Artificial Neural network) is used to optimize the 3D simulation scene of high-speed railway virtual reality. A feature point processing optimization algorithm based on BPNN (BP neural network) and FCN (Fully Convolutional Network) neural networks was proposed Based on SIFT (Scale-Invariant Feature Transform) framework, this algorithm uses pre-trained FCN neural network to detect the image and determine the pixel range of the main target, thus improving the pertinence of feature point detection and avoiding background noise interference. The research results show that this method can effectively eliminate mismatches and improve the accuracy of the results, and its average time consumption is reduced by 24.03% compared with CNN algorithm. ANN greatly optimizes the scene and improves the rendering effect and running efficiency.

13.1 Introduction

With the rapid development of computer technology and the popularity of the Internet, the application and theoretical research of VR (Virtual Reality) technology has become the hottest research in the field of information technology. VR is one of the research hotspots in recent years, which is the product of the rapid development of

H. Han (✉) · C. Li
Haidu College, Big Data College, Qingdao Agricultural University, Yantai 265200, Shandong, China
e-mail: hanhui@hdxy.edu.cn

C. Li
e-mail: lichangzhong@hdxy.edu.cn

X. Chen
CRRC Qingdao Sifang Co., Ltd., Qingdao 266111, Shandong, China

R. Kountchev (Deceased) et al. (eds.), *AI Methods and Applications in 3D Technologies*, Smart Innovation, Systems and Technologies 388,
https://doi.org/10.1007/978-981-97-2144-3_13

computer technology, simulation technology, graphics and image processing technology [1]. Simply put, VR is a specific virtual environment based on computer technology, which can be a simulated expression of the real world or an imaginary space environment [2]. 3D virtual scene is not only the simulation of the shape and geographical form of real space buildings, but also the foundation of digital construction. Users can make remote interactive query and access to the virtual building model in the scene, and can also look around and roam automatically by changing the viewpoint, thus helping users to understand and get familiar with the real environment conveniently and quickly.

The key to the realization of VR technology lies in the construction of virtual 3D scenes, and how to build high-quality 3D scenes will determine the final experience effect of VR [3]. High-speed railway is the artery of inter-city traffic, which plays an active role in relieving the existing traffic pressure, further tapping the inter-city traffic capacity and realizing a fast, efficient and smooth travel environment. The goal of optimization is to reduce the overall file size of the scene, reduce the bandwidth pressure and the burden of hardware operation, and improve the system operation efficiency and screen display speed [4, 5]. When an object is projected into an image, its height information is lost immediately, so we can only estimate the surface morphology of the object by using the light and dark information retained in the image. Restoring the shape from light and shade is a method to restore the 3D shape of an object surface by using one image of an object or multiple images from different perspectives. From the application point of view, if you want to draw a more realistic scene, you need to use more geometric elements to construct it, so the drawing cost is higher. In addition, with the continuous modernization of model input methods and modeling methods, the 3D models obtained by people have become more and more complex from simple and regular 3D models [6].

The application of VR technology in the simulation of high-speed railway can undoubtedly meet people's new demand for high-speed railway cognition, so that the real-time operation of high-speed railway can be displayed intuitively. On the premise of not obviously reducing the graphic quality and complexity, in order to realize the real-time display of large data simulation scenes, some efficient data processing methods and graphics rendering algorithms that can be applied to the existing general computer platform are studied and adopted to improve the graphics rendering speed. Therefore, in order to solve this problem, this paper uses ANN (Artificial Neural network) to optimize the 3D simulation scene of high-speed railway virtual reality, which can not only improve the loading speed, but also bring better real experience to users.

13.2 Research Method

Principle and Technology of 3D Scene Modeling

The establishment of a model can construct a polygonal surface by using vertex coordinate sets and corresponding attribute parameters. The information of a polygon is stored in the polygon data table to support the processing, display and management of objects in the scene. The typical method of storing geometric data is to create a set of vertices, edges and faces. The coordinate values of each vertex in the 3D model are stored in the vertex table, while the edge table containing pointers to the vertex table is used to store each edge of the polygon, and the surface table contains pointers to the edge table to identify edges for each polygon. This problem can be solved to some extent by enhancing the computer configuration. However, in order to demonstrate the simulation scene on most computers, and the data volume is still increasing, it is slightly insufficient to use hardware means. Therefore, it is necessary to improve the data structure and rendering algorithm of 3D models and strengthen the cooperation between software and hardware to solve the real-time rendering problem of 3D scenes with large data volume more completely.

The rapidly developing 3D panoramic technology belongs to a branch of virtual reality technology, and its production and operation process is convenient and fast, which can be perfectly realized on the computer platform, and has the characteristics of better display effect and close to the real environment. With its own advantages, this technology has attracted more and more attention, and its application scope is gradually expanding, gradually developing into a new 3D vision technology. 3D panoramic technology has obvious advantages, with interactive experience, real scene expression and other characteristics, providing new forms of publicity for many industries, such as tourism promotion, cultural relics protection, commercial display, digital city and so on. The amount of file data generated by 3D panorama is relatively small, which is convenient for users to transmit, and it is easy to publish to the network. It can be flexibly embedded into web pages to adapt to various forms of display applications.

Texture mapping often needs to map texture patterns to object surfaces of different sizes. Ideally, the object surface size is the same as the texture pattern size, but there are many cases where the actual object surface is larger or smaller than the texture pattern. At this time, the texture attribute value of this area, such as the average value of pixel color, must be taken as the texture attribute of the corresponding surface. There are many methods for fast antialiasing in texture mapping, such as Mip-Map technology and regional summation table technology. At present, Mip-Map technology is the most widely used.

Mip-Map technology is based on the initial texture picture. Assuming that the resolution of each edge of this texture picture is 0, a pyramid lookup table can be used to store textures with different resolutions, as shown in Fig. 13.1. The number of levels of this pyramid is $\left[\log_2 F\right] + 1$.

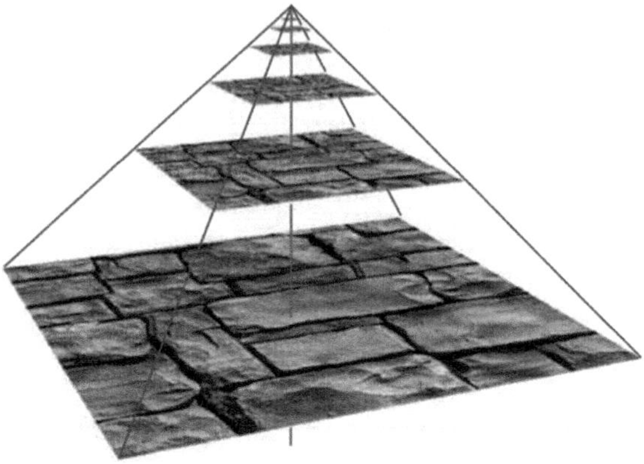

Fig. 13.1 Texture storage mode

Create a lookup table according to the texture size, and the value in the table is the sum of the texture colors in each rectangular area. I.e. expressed by Formula (13.1).

$$f_{mn} = \sum_{i=1}^{m} \sum_{j=1}^{n} t_{ij} \tag{13.1}$$

m, n represents the size of the lookup table, and t represents the texture color value of each rectangle.

The simulation of high-speed railway is suitable for the method of adjusting the terrain and transforming the terrain with lines. Generally speaking, from the cross section of terrain model and line model, terrain is represented by a large number of triangular surfaces, while line model is represented by quadrilateral. Therefore, the generation of the line model boundary is essentially to generate a quadrilateral. The general principle of the algorithm is: the line model and terrain model are projected on the two-dimensional plane for cutting and intersection, then the obtained polygon is triangulated, and finally the corresponding elevation value is given to each triangle vertex [7, 8].

For the line, that is, to realize the double-line processing of the line, two parallel lines are drawn along the central axis of the line to a given width on both sides, and the elevation values of the left and right parallel lines are consistent with the central axis of the line, and then closed into a quadrilateral. The schematic diagram of line boundary generation is shown in Fig. 13.2.

In order to represent all kinds of entities in 3D scene completely, the data structure of polygons representing entities must be processed. The processing process needs to start from the coordinate system, which should be the transformation from modeling coordinates and world coordinates to screen coordinates, visible face recognition and surface drawing.

The plane equation can be expressed in the following form:

$$Ax + By + Cz + D = 0 \tag{13.2}$$

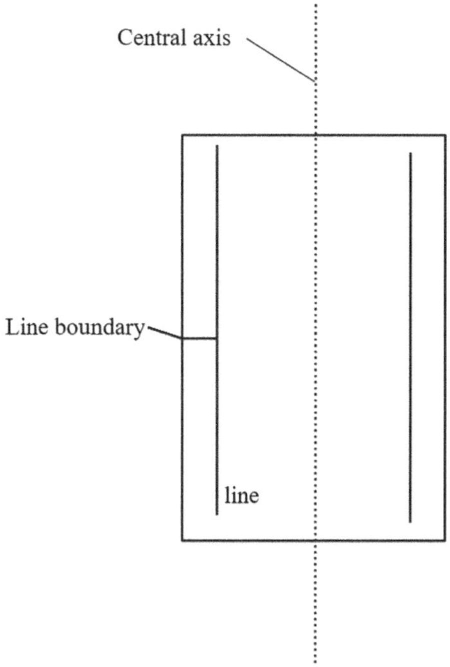

Fig. 13.2 Schematic diagram of line boundary generation

The 3D model must be placed with reference to a certain coordinate system in the scene. In the virtual high-speed railway station scene, the world coordinate system of the whole scene is established based on the rail, and the rail position and the up-and-down running direction of the train are determined. These relationships can make some parts of the model move, rotate and deform together, and provide basic support for defining the physical properties and motion properties of the model later. The common combination relationships mainly include the meshing surface connection between gears and the spring connection between devices.

The 3D model of high-speed railway line consists of two parts, namely, the 3D geographical environment model along a certain range of high-speed railway line and the 3D solid model of high-speed railway structure. Firstly, the 3D terrain and model of the high-speed train are constructed, then the 3D model of the line is established by using the data of plane, vertical section, cross section, bridge, tunnel and subgrade, and finally all the models are placed in the 3D geographical environment model by external reference to form the 3D landscape model of the high-speed train line. The visual modeling in this paper mainly includes the following five modules, as shown in Fig. 13.3.

The force of high-speed railway on the ramp is mainly the traction force, braking force and other external forces acting on the train (basic resistance and ramp resistance caused by train gravity, this paper does not consider the air resistance of the train), but what really affects the train operation is the parallel component of these forces in the train running direction.

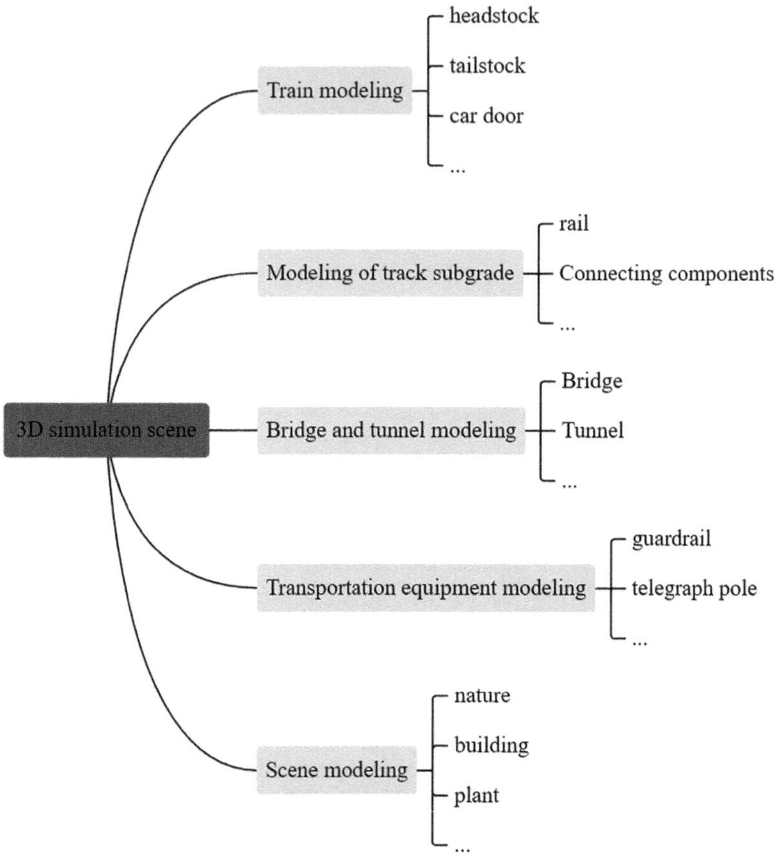

Fig. 13.3 Visual modeling module

Polygon simplification technology provides a solution to deal with large data scenes. Simplify the unimportant and remote parts of the scene, while ensuring that the visual display will not be distorted. It can generate the level-of-detail model of objects in the scene to improve the rendering speed of the whole scene. The use efficiency of texture memory can be improved by compressing multiple images into a single texture or eliminating small textures. Therefore, these texture optimization techniques are very important for real-time interactive rendering systems.

Scene Optimization

The type, quantity and scene accuracy of the virtual high-speed railway station are closely related to the scale adopted. Scale generally refers to the spatial resolution and time unit of the observed object or process. It is almost inevitable to have redundant surfaces in the modeling process. Redundant elements mainly include points, lines, surfaces and bodies, which are usually invisible. Eliminating redundant and invisible polygons can reduce the system burden and improve the real-time running speed of the scene [9]. When the distance is close, the LOD (Levels of Detail) model is used to draw. When the distance is far away, the rough LOD model is used to draw. In the scene of moving objects, rough LOD model is used to draw when the moving speed is fast, and detailed LOD model is used to draw when the moving speed is slow or at rest.

Image acceleration technology can be used in large data scene rendering based on 3D geometry. Image-based acceleration technology is based on the fact that some parts of the scene are invariant within a certain resolution range for a period of time relative to the observer. In order to reduce the burden of graphics generation to the greatest extent, the algorithm establishes the hierarchical structure of the whole virtual scene with a binary tree [10]. In real-time rendering, the binary tree is traversed hierarchically to determine whether the image of a certain area stored in each node is still valid. If it is valid, the current image of the area is generated by the image stored in the node, otherwise, the front and back subtrees of the node are traversed continuously, and it is determined whether it is necessary to generate the current graphic picture of the node and store it as an image according to the motion parameters and error accuracy. However, in order to ensure the continuity of the transition between hierarchical detail models, many levels are often needed, which will produce a large amount of data, which is also difficult to store and transmit eff x_i actively.

Deep learning, especially neural network, shows a good prospect in the field of 3D reconstruction. The human nervous system is composed of billions of neurons in various forms, and information is continuously transmitted between neurons, realizing people's understanding of the surrounding environment [11]. From the point of view of information processing, the neural network of human brain is abstracted and ANN is obtained. As shown in Fig. 13.4, it is the simplest ANN structure with only input layer and output layer.

BPNN (BP neural network) is a multi-layer feedforward neural network trained according to error back propagation. By calculating the error between the expected output and the actual output, the parameters and thresholds of nodes are adjusted layer by layer from output to input, and the target accuracy is finally achieved through repeated iterations. The typical BPNN structure is shown in Fig. 13.5.

FCN (Fully Convolutional Network) is a neural network based on traditional CNN (Convolutional Neural Networks) and mainly used for image semantic segmentation.

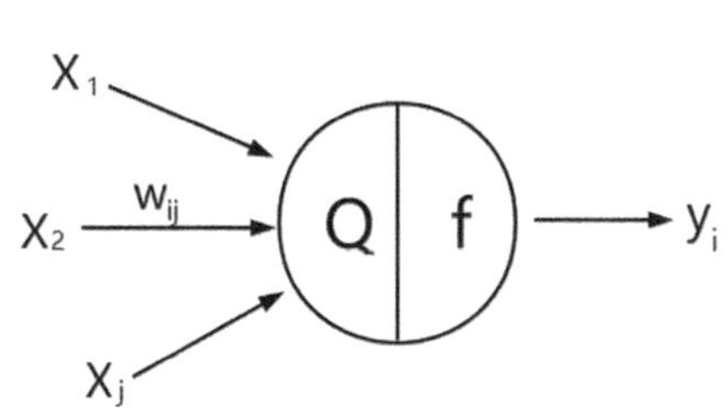

Fig. 13.4 ANN unit

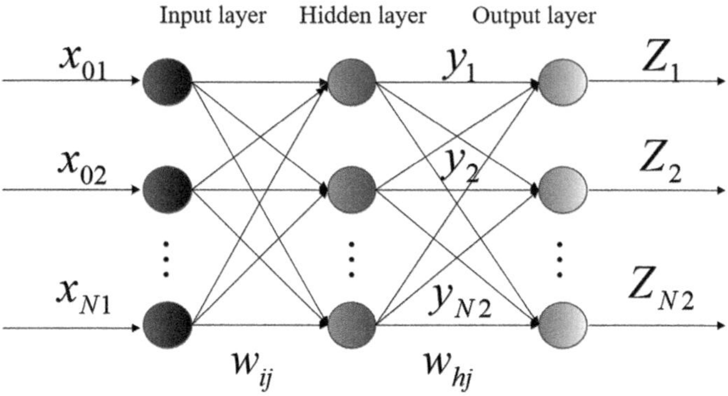

Fig. 13.5 BPNN structure

FCN enlarges the image layer by layer, and its final output is an up-sampled image, and then each pixel in the image can be classified and predicted.

In this paper, an optimization algorithm of feature point processing based on BPNN and FCN neural network is proposed. Based on SIFT (Scale-Invariant Feature Transform) framework, the pre-trained FCN neural network is used to detect the image and determine the pixel range of the main target, so as to improve the pertinence of feature point detection and avoid background noise interference. This method effectively reduces the complexity of 3D scenes [12].

Image contrast threshold is one of the key parameters used to improve the stability of feature points in SIFT algorithm. Different image contrast thresholds will have a great impact on the number of feature points. Generally, the root mean square is used for calculation, as shown in Formula (13.2):

$$RMS = \left[\frac{1}{n-1}\sum_{i=1}^{n}(x_i - \overline{x})^2\right]^{\frac{1}{2}} \tag{13.3}$$

where n represents the number of pixels, represents the gray value of the i th pixel, and $\overline{x}$ represents the average of gray values of image pixels.

Based on the idea of octree of voxel grid, this paper proposes a new method of applying octree bit representation in deconvolution process. In the process of deconvolution, the voxel information exists in a three-dimensional form. In the process of deconvolution and deconvolution, the calculation and storage requirements of data will increase in a three-dimensional way with the improvement of 3D stereo accuracy. In order to reduce the amount of data and computation caused by 3D structure. This paper uses the bit representation of octree.

When the pointer is not used to determine the relationship between the parent node and the child node, the serial number of each position in the bit string becomes the only way to determine the relationship between the nodes:

$$\beta_n = \frac{1}{n}\left(\frac{5}{8} - \left(\frac{3}{8} + \frac{1}{4}\cos\frac{2\pi}{n}\right)^2\right) \tag{13.4}$$

$$ch(i) = [8i + 1] \tag{13.5}$$

where n represents the number of pixels, β represents the bit string, and $ch(i)$ represents the serial number of the position.

The shape recovery from light and shade is carried out without knowing the direction of light source and the gradient of object surface, which is a function approximation problem of defect parameters. In this algorithm, the neural network global BP learning algorithm is used to modify the weights of the input layer and the hidden layer of the network:

$$V^{(n+I)} = \sum_{i,j\in\Omega}\left(\Phi_{ij}\Phi_{ij}^T\right)^{-1}\left(\sum_{i,j\in\Omega}\left(\sigma^{-1}(I_{ij})\Phi_{ij}^T\right)\right) \tag{13.6}$$

$$w_k^{(n+I)} = w_k^{(n)} + \left(-\eta\frac{\partial E^T}{\partial v_k^{(n)}} + \nabla E_{pen}\right) \tag{13.7}$$

where Φ_{ij} represents the output vector of the hidden layer of the neural network at the image pixel (i, j), Φ_{ij}^T is the Φ_{ij} torque, $\sigma^{-1}(\cdot)$ represents the inverse function of $\sigma(\cdot)$, η is the learning efficiency, and ∇E_{pen} is the penalty term.

The specific algorithm framework is shown in Fig. 13.6.

The algorithm calculates the complexity and contrast information of the input image, and defines the number of feature points needed for application, and then inputs it into the adaptive control module based on BP neural network, thereby adjusting the contrast threshold in SIFT algorithm, and thus adjusting the overall number of feature points of the image.

13.3 Result Analysis

In this paper, the terrain data is intercepted from the website of geospatial data cloud. The terrain on geospatial data cloud is divided into 15 data cells according to the unit of 0.4*0.4, and each data cell is converted on Global mapper, and then imported into unreal engine one by one. Create 10 maps in unreal engine, and compare the memory usage data.

Obtain the memory occupation data of map1 and map10, and summarize the memory occupation data obtained by the two methods to get Fig. 13.7.

It can be seen that both map1 and map10 have some memory occupation when there are no terrain blocks, because there are some basic components. The memory occupation of map1 basically increases in direct proportion to the increase of the

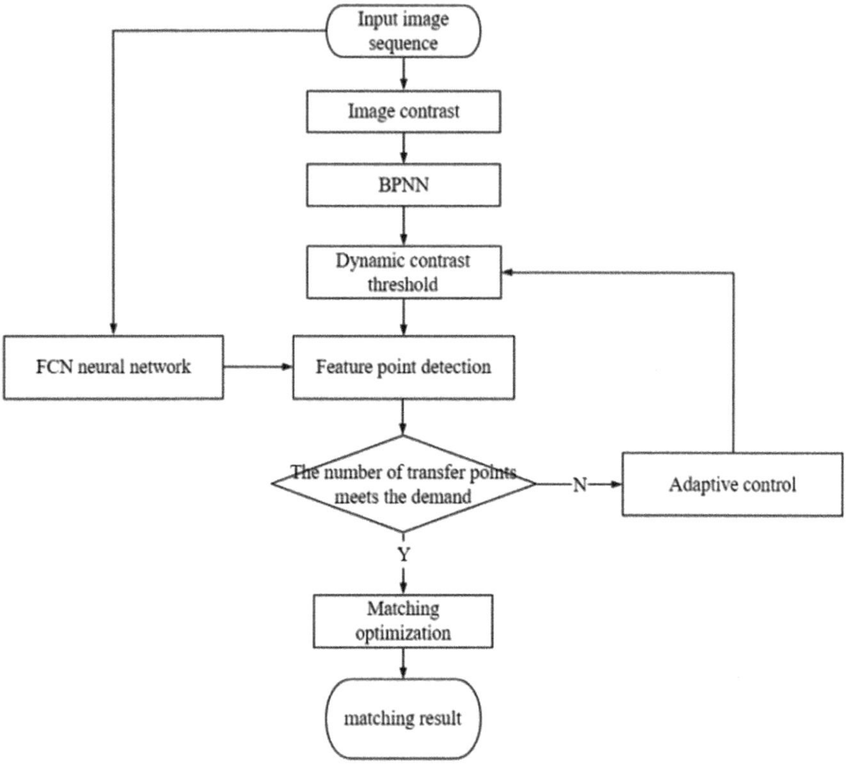

Fig. 13.6 Framework of feature point processing optimization algorithm

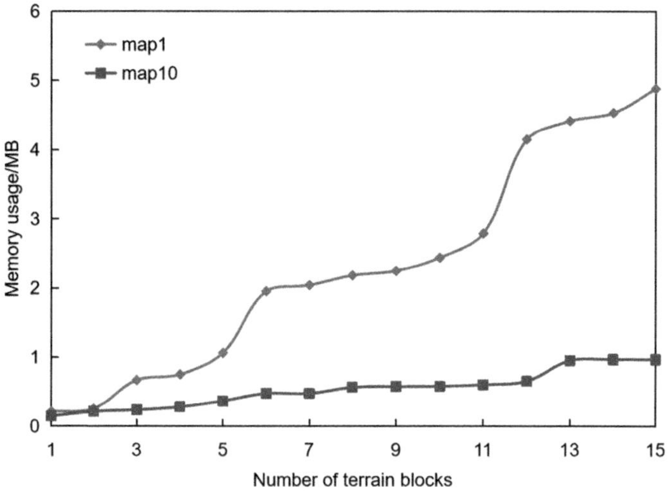

Fig. 13.7 Comparison of memory occupation between checkpoint stream and unused checkpoint stream

number of terrain blocks, with an average of about 0.45 MB, while that of map10 only increases the amount of terrain memory except the basic memory occupation, which leads to some slight changes when loading and unloading the level due to the different terrain sizes.

Experimental hardware environment: Intel Core i7-6700HQ quad-core processor, 8GB memory and 80GB hard disk space. Software environment: VMware Workstation Pro, Linux Ubuntu14.04 operating system, caffe neural network framework, C++ language, Python language, g++ compiler, ORB-SLAM2 library.

In this paper, the algorithm adjusts the number of feature points from 230 to 0, the target feature points are 1000, and the allowable error range is within 100. There are a lot of background feature points outside the main target, and the middle image represents the result of FCN neural network processing, and the range of the main target is extracted, so as to control the range of the feature points. Compared with the matching optimization method based on CNN algorithm, the time-consuming comparison is shown in Fig. 13.8.

It can be seen that this method can effectively eliminate mismatches, improve the accuracy of results, and provide accurate original data for camera pose recovery, and its average time consumption is reduced by 24.03% compared with CNN algorithm. In this way, repeated modeling is avoided, and external files only occupy and consume system resources when referencing is executed, and the occupied resources are released immediately after stopping referencing. Through experimental analysis, ANN has greatly optimized the scene and improved the rendering effect and running efficiency.

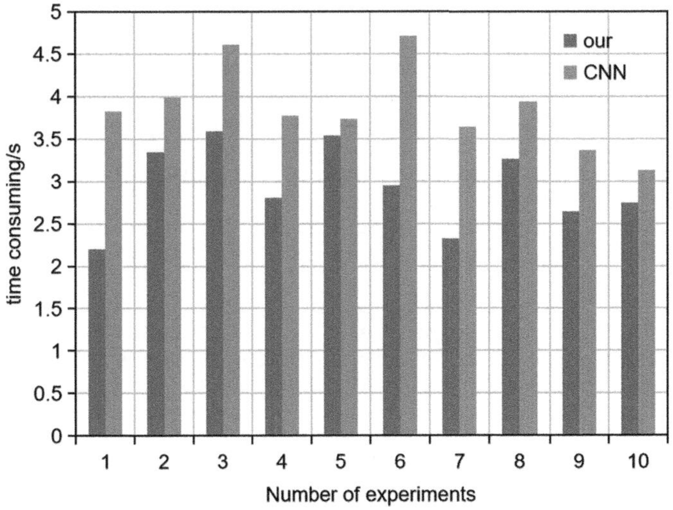

Fig. 13.8 Time-consuming comparison of feature point matching stage

13.4 Conclusion

3D virtual scene is not only the simulation of the shape and geographical form of real space buildings, but also the foundation of digital construction. Users can make remote interactive query and access to the virtual building model in the scene, and can also look around and roam automatically by changing the viewpoint, thus helping users to understand and get familiar with the real environment conveniently and quickly. In this paper, ANN is used to optimize the 3D simulation scene of high-speed railway virtual reality, which can not only improve the loading speed, but also bring better real experience to users. The research results show that this method can effectively eliminate mismatches, improve the accuracy of results, and provide accurate original data for camera pose recovery, and its average time consumption is reduced by 24.03% compared with CNN algorithm. ANN greatly optimizes the scene and improves the rendering effect and running efficiency. Experimental data verify the effectiveness and efficiency of this algorithm, but there is still room for further study. To further strengthen the deep integration of ANN and 3D reconstruction, it is not limited to the application of a single technology of ANN in 3D reconstruction, but to integrate and innovate the framework of ANN and 3D reconstruction, so as to realize more intelligent, efficient and accurate 3D reconstruction.

References

1. Zhong, D., Wang, Z., Guan, T., et al.: Visual simulation of construction schedule for core rock-fill dam based on augmented reality. Tianjin Daxue Xuebao, (Ziran Kexue yu Gongcheng Jishu Ban)/J. Tianjin Univ. Sci. Technol. **51**(10), 1072–1085 (2018)
2. Li, R., Xu, D.: Distribution of landscape architecture based on 3D images and virtual reality rationality study. IEEE Access **8**, 140161–140170 (2020)
3. Wang, V., Liu, L., Xu, J.: Study on the design of the railway vehicle running control system based on the virtual reality technology. Revista de la Facultad de Ingenieria **32**(13), 799–808 (2017)
4. Hong, Y., Zeng, X., Bruniaux, P., Liu, K.: Interactive virtual try-on based three-dimensional garment block design for disabled people of scoliosis type. Text. Res. J. **87**(10), 1261–1274 (2017)
5. Yan, H., Che, J., Chen, S.: One-dimensional and three-dimensional coupling simulation research of centrifugal cascade hydraulics. Kerntechnik **87**(2), 176–186 (2022)
6. Zhu, P., Zhang, J., Han, B., et al.: Three-dimensional numerical simulation of ice crystal melting in jet engine. J. Therm. Sci. **28**(5), 984–992 (2019)
7. Li, G.: Optimization and simulation of virtual experiment system of human sports science based on VR. Complexity **4**, 1–10 (2021)
8. Desselle, M.R., Brown, R.A., James, A.R., Midwinter, M.J., Powell, S.K., Woodruff, M.A.: Augmented and virtual reality in surgery. Comput. Sci. Eng. **22**(3), 18–26 (2020)
9. Filimonov, A., Zeiger, J., Goldrich, D., et al.: Virtual reality surgical planning for endoscopic endonasal approaches to the craniovertebral junction. Am. J. Otolaryngol. **43**(1), 119 (2022)
10. Johnson-Roberson, M., Bryson, M., Friedman, A., et al.: High-resolution underwater robotic vision-based mapping and three-dimensional reconstruction for archaeology. J. Field Robot. **34**(4), 625–643 (2017)

11. Terao, T., Inoue, G., Kawase, M., et al.: Development of novel three-dimensional reconstruction method for porous media for polymer electrolyte fuel cells using focused ion beam-scanning electron microscope tomography. J. Power Sources **347**(15), 108–113 (2017)
12. Qian, Q., Wang, J., Bai, Z., et al.: Three-dimensional reconstruction and quantitative characterization of reconstruction soil pore at opencast coal mine dump based on CT scanning. Meitan Xuebao/J. China Coal Soc. **41**(1), 109–117 (2016)

Chapter 14
Localization of Industrial Robots Based on RGBD-SLAM

Yitian Wang

Abstract This paper investigates the localization of industrial robots based on RGBD-SLAM technology. Firstly, the basic principles and algorithmic process of RGBD-SLAM are introduced, including the calibration of internal and external parameters of the camera, feature extraction and matching of the image, map construction of the scene and trajectory estimation of the camera. Then a RGBD-SLAM-based robot localization method is proposed for industrial robots, which uses the RGBD camera held by the robot to acquire environmental information to achieve self-localization and modelling of the robot's scene. The method was then validated using experimental data. The results show that the method can achieve high accuracy in robot positioning and has good robustness and stability in complex scenarios. Finally, the paper discusses the prospects and outlook of the method in industrial robot applications. The study provides a new way and a new method for the localization and navigation of industrial robots.

14.1 Introduction

Background and Significance of the Study

Today's era is characterized by a greater diversity of robots in industry and a significant increase in engineering automation. As the state of the art continues to improve, the cost of robots is trending downwards, but their performance is improving. For example, for robots in the industrial sector, the main principle is to mimic an arm to achieve flexible gripping and handling functions that can replace simple and repetitive manual operations such as boxing and stacking. Currently, the most widely used areas for robots are industrial manufacturing and packaging. Manipulators can

Y. Wang (✉)
Shandong University of Science and Technology, Tai'an, China
e-mail: Wangyitian_2002@163.com

R. Kountchev (Deceased) et al. (eds.), *AI Methods and Applications in 3D Technologies*, Smart Innovation, Systems and Technologies 388,
https://doi.org/10.1007/978-981-97-2144-3_14

complete operations more accurately and efficiently within a given time frame, which has become the main direction of industrial robot development.

The influence of artificial intelligence technology is currently expanding. Supported by internet technology, industrial robots are being developed more systematically, with enhanced control systems, diagnostic systems and improved maintenance system functions. At the same time, relying on analogue programming, the level of intelligence and automation has been effectively improved, and the overall performance has continued to improve, with more reliable applications and a wider range of applications.

Throughout industrial production, there are several aspects of the industrial process that contradict environmental protection, adversely affecting the physical and mental health of those involved and making some operations difficult to complete, which has become an important factor in the promotion and application of industrial robots. For example, the main reason for the use of vacuum robots in industry is that in the semiconductor industry, humans are unable to complete the vacuum transfer wafer link, and the introduction of vacuum robots has solved this problem. In addition, in some harsh environments, such as snake robots that adapt to unhindered movement and bionic fish robots that cater for underwater operations, they can largely replace human activities in some specific situations, ensuring human safety and increasing efficiency.

The development of artificial intelligence is unstoppable, and intelligence has become the future of industrial robotics. Intelligent robots, emphasizing the higher level of human imitation by robots, require a higher level of bionics, not only to imitate human behavior, but also to have human thinking and nerves. On this basis, sensors have become an important part of intelligent industrial robots, especially the emergence of visual, force-sensing and tactile sensors, which has accelerated the development of intelligent industrial robots.

The results of this research will help to localize and map mobile robots in unknown environments and provide a solid foundation for high-level tasks such as obstacle avoidance and path planning for mobile robots. Where localization and mapping information is required in unknown environments, SLAM technology has its place. Therefore, the study of SLAM technology is of substantial value and significance. It is often assumed that the camera in visual SLAM is mounted on a mobile robot, which also means that the positioning of the robot is equivalent to the positioning of the camera.

However, traditional 3D sensing solutions have many limitations. For example, the tracking of specific targets limits their autonomous movement and GPS cannot be applied to areas such as indoors and underwater. In these scenarios, there is an urgent need to solve the problem of how to sense and obtain localization information for unknown scenes through the sensors carried by the mobile robot itself.

Currently, the main types of cameras used in vision SLAM sensors are monocular, binocular and depth cameras. In this case, the monocular camera acquires data as a frame of two-dimensional images, which is simple in structure and less costly. However, if a three-dimensional structure is to be recovered from a two-dimensional image alone and the depth values of individual pixel points need to be found, the depth values of individual pixel points can be estimated by moving the parallax produced by the camera. To eliminate the scale problem in this case, researchers have begun to use binocular cameras to acquire data simultaneously, allowing the depth value of each pixel point to be calculated from the baseline between the binocular cameras, thus solving the scale uncertainty problem that exists with monocular cameras. However, the calculation of the depth values is very computationally intensive and requires the use of GUP acceleration to output the distance information for the entire frame. However, the RGB-D camera acquires data as RGB images and their corresponding depth images, which can effectively solve the scale uncertainty of monocular cameras and the computational overload of binocular cameras, bringing great convenience to researchers. Its advantages are low cost and unrestricted range, adjustable pixel depth calculation and adjustable range, active depth measurement and good reconstruction results. However, its disadvantages are scale uncertainty, complex calibration of initialization problems, large pixel depth calculation, small range and susceptibility to daylight.

In the ORB-SLAM2 system, the Qtree algorithm is utilized for the extraction of homogeneous feature points from the image. This algorithm effectively partitions the image into regions, allowing for the identification of salient features. Subsequently, the Fast Approximate Nearest Neighbor (FLANN) algorithm is employed for efficient matching of these feature points [1]. By leveraging FLANN's accelerated search capabilities, the algorithm can find the closest matches between feature points in an expedited manner.

To ensure the robustness of the matching process, a random sampling consensus algorithm is employed for rejecting false matches. This algorithm utilizes random sampling to establish a consensus among the matched feature points, effectively filtering out erroneous correspondences. This step enhances the accuracy and reliability of the overall feature point matching procedure [2].

By utilizing these advanced algorithms, the system achieves efficient feature point extraction, accurate matching, and robust rejection of false matches. This enables reliable and precise analysis of visual data in various computer vision applications.

Current Status of Domestic and International Research

An industrial robot is an automated device used to perform repetitive and hazardous work tasks. In recent years, with the development of the manufacturing industry and the proposal of intelligent manufacturing, the research and application of industrial robots has also gained widespread attention.

Artificial intelligence and machine learning, two technologies widely used in industrial robots, enable robots to better understand their environment and sense objects, and can improve their ability to plan paths and perform tasks. Flexible robots are robots that are adaptable to different environments and tasks and are highly flexible and scalable. Currently, the focus of research on flexible robots includes mechanical structure design, control system development and human–robot interaction technologies.

In 2011, Nikolas Engelhard et al. of the University of Freiburg in Germany developed a handheld RGB-D SLAM system based on Kinect [3]. The front-end visual odometry of the system uses the SURF feature point method and also the ICP algorithm to solve the inter-frame positional transformations, and then uses HOG-Man (Hierarchical Optimizations on Manifolds for Online 2D and 3D mapping) to solve for the global optimal set of poses. In 2016, Sujiwo et al. proposed the ORB-SLAM2 system, which is an improved version of the ORB-SLAM system with the addition of support for RGB-D cameras [4].

The research on SLAM in China started late, and the research in the direction of visual SLAM has only just emerged in recent years, but there are also achievements in theory and practice. In 2016, Lv Xianwei added an inverse parameterization process to speed up the filter convergence of the RGB-D SLAM system based on the EKF framework, thus improving the accuracy and stability of the system [5]. In 2014, Xia Wenling et al. directly used the Kinect of RGB-D cameras to obtain depth information of 3D scenes, and achieved real-time performance improvement of the SLAM system based on Kinect [6]. In 2015, Mengyin et al. [7] from Beijing University of Technology proposed an improved RGB-D SLAM method, which records both the location information of feature points and RGB information in the feature extraction phase of the front-end visual odometry, while improving the ICP algorithm by adding a particle sampling process to it, thus reducing the mis-matching in the matching phase and making the matching results more accurate. Although the research on SLAM in China started late, and the foundation of visual SLAM is weak, as more and more researchers devote themselves to SLAM research, and the research results of pioneer researchers at home and abroad serve as a pavement, coupled with the market demand as a driving force, it is believed that many technical problems of visual SLAM will be overcome. A truly stable and commercially available visual SLAM system will be born.

The Research Content

ORB-SLAM2 is an open source framework based on visual SLAM for camera localization and map building in indoor-outdoor environments. The TUM dataset is a classical test dataset for ORB-SLAM2, provided by Technische Universität München. In industrial robot applications, ORB-SLAM2 and TUM are aggregates that can be used to position cameras and build maps of the environment for robot navigation and control.

In RGB-D industrial robot applications, the ORB-SLAM2 and TUM datasets can be used to achieve the following:

(1) Environment awareness: Using the camera to capture images and depth information of the surrounding environment, ORB-SLAM2 is used to identify environmental features and build maps;
(2) Navigation: ORB-SLAM2 is used to position the camera and the robot, thus enabling autonomous robot navigation;
(3) Monitoring and detection: The camera is used to capture images and depth information of the surrounding environment and combined with ORB-SLAM2 to monitor objects around the robot and perform target detection and tracking.

In summary, the use of ORB-SLAM2 and the TUM dataset can improve the autonomous localization and navigation capabilities of RGB-D industrial robots and enhance their environmental awareness and control, thus providing a more complete and efficient solution for industrial automation.

14.2 RGB-D SLAM Algorithm Introduction

RGB-D Camera Model Building

The RGB-D camera, also known as a depth camera, is a device capable of acquiring both RGB images and depth information. The principle is to use structured light or time-of-flight methods to acquire depth information, while RGB images are acquired using a normal color image sensor. In the structured light approach, the RGBD camera emits a strip of light consisting of red, green and blue (also known as coded light) into the scene, which is reflected back into the camera by the surface of the object. The camera records the distortion of the encoded light as it reflects off the surface of the object, thus extrapolating information about the distance between the surface of the object and the camera. In time-of-flight mode, the RGB-D camera then emits a pulsed beam of light and measures the time it takes for it to be emitted from the camera and reflected back to the camera from the surface of the object. RGB-D cameras with depth sensors generally use structured light or ToF (Time-of-Flight) technology [8]. Structured light is used to obtain depth information by projecting a structured light source and then photographing the illuminated scene through the camera. ToF technology, on the other hand, calculates depth information by firing light pulses into the scene and measuring the time difference between the return of the light pulses from the camera. Since the speed of light is known,

The RGB-D camera employs time-of-flight or structured light techniques to measure the distance between the camera and the surface of an object. These measurements enable the camera to acquire both RGB images, capturing color information, and depth images, providing precise depth measurements. This combination of RGB and depth data allows for robust 3D reconstruction and scene recognition. By leveraging the rich spatial information from depth sensing, the camera facilitates various computer vision tasks, including accurate object recognition and in-depth scene analysis. The integration of RGB and depth sensing capabilities in the RGB-D camera has significantly broadened its applications in the field of computer vision.

The camera model encompasses four distinct coordinate systems: the world coordinate system, camera coordinate system, image coordinate system, and pixel coordinate system. Its fundamental purpose is to project information observed in the world coordinate system onto the pixel coordinate system, generating a final image output. Specifically, an RGB-D camera comprises two cameras: an RGB camera responsible for capturing color information within the scene, and a depth camera that records depth data for all pixel points, producing a depth image as an output.

In this research, the imaging process of the depth camera is modeled using the classical pinhole camera's small-hole imaging model. A schematic diagram of the camera model is presented in Fig. 14.1, illustrating the interrelationships between the various components.

As the small aperture image is inverted, but pre-processing of the image creates an image in the plane of symmetry. RGB-D camera models generally include both internal and external camera parameters. Internal camera parameters include focal length, principal point and pixel size, which describe the optical characteristics of the camera. The external parameters of the camera include the rotation matrix and translation vector, which describe the position and orientation of the camera in 3D space.

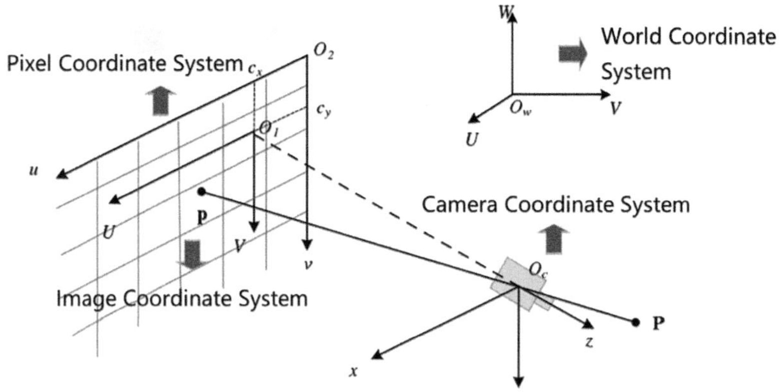

Fig. 14.1 Pinhole camera model

To accurately represent and understand the camera's spatial orientation and position, its dynamic motion requires the utilization of precise world coordinates P_ω. Additionally, the camera system incorporates intrinsic elements such as the internal matrix parameter K and the depth value Z, establishing crucial associations. Essential to this representation is capturing the camera's pose, which comprises the rotational matrix R and translational vector t, given by:

$$ZP_{uv} = Z\begin{bmatrix} u \\ v \\ 1 \end{bmatrix} = K(RP_\omega + t) = KTP_\omega \tag{14.1}$$

P_{uv}: The depth value of the pixel point (u, v) in the image, representing the distance observed by the camera at that pixel.

Z: One of the camera's intrinsic and extrinsic parameters, representing the focal length or field of view of the camera.

$\begin{bmatrix} u \\ v \\ 1 \end{bmatrix}$: The pixel coordinates in the image, representing the position of a point on the image.

K: The camera's intrinsic matrix, including parameters such as the focal length, coordinates of the image center, and scaling factor of the image.

R: The camera's rotation matrix, representing the rotation relationship from the camera coordinate system to the world coordinate system.

P_ω: The camera's projection matrix, representing the projection relationship from the camera coordinate system to the pixel coordinate system.

t: The camera's translation vector, representing the translation relationship from the camera coordinate system origin to the world coordinate system origin.

The above equation describes the projection relationship from the world coordinate system of P to the pixel coordinate system. In it, the camera's pose is R, and t is also known as the camera's outer parameter. In contrast to the constant inner reference, the outer reference changes with the camera movement. It is also the target with estimation in SLAM, representing the trajectory of the robot.

RGB-D SLAM Algorithm Construction

The RGB-D SLAM algorithm encompasses several crucial components, namely the front-end visual odometry, back-end optimization, loop closure detection (also known as closed-loop detection), and map construction.

During the front-end stage, the algorithm undertakes feature point detection and computes feature descriptors using both the RGB and Depth images. Subsequently, it conducts feature matching between adjacent frames by leveraging these descriptors, resulting in a set of 2D-2D feature matching points. Furthermore, utilizing the depth information from the Depth image, the algorithm calculates the precise spatial 3D coordinates of the matched feature point pairs, thereby establishing a comprehensive set of 3D-3D matching points [9].

By utilizing the matched 3D-3D points, the algorithm computes the rotation and translation matrices that precisely represent the relative pose between adjacent frames. Finally, it optimizes the motion estimation error to obtain an accurate pose estimation, enabling the continuous estimation of camera pose based on the input video stream and constructing a reliable visual odometer at the forefront of the algorithm [10].

The back-end of the algorithm is specifically designed to address the issue of noise during the SLAM optimization process. In practical scenarios, even highly accurate sensor data can be affected by noise. Therefore, after obtaining an estimation of the motion between adjacent frames through the front-end, it becomes crucial to consider the level of noise present in this estimation. The back-end optimization aims to estimate the overall system state using this noisy data, maximizing the posterior probability associated with this state [11].

More specifically, the back-end utilizes camera poses obtained from the visual odometer at different time instances, along with constraint information from loopback detection [12]. It employs advanced non-linear optimization techniques to compute globally optimal poses. While the front-end in vision-based SLAM primarily focuses on computer vision research aspects, such as feature point detection and image matching, the back-end emphasizes the implementation of filtering and non-linear optimization algorithms.

Loopback detection, also known as closed-loop detection, addresses the issue of robot position drift over time. It enables the robot to recognize previously visited scenes. Visual loopback detection involves comparing the similarity between two images, leveraging the abundance of image information available. A successful loopback detection indicates that the robot has revisited a specific location, and the corresponding comparison information is integrated into the back-end optimization algorithm. As a result, the back-end adjusts the robot's trajectory and map based on the loopback detection information. The inclusion of loopback detection significantly reduces cumulative error.

Map building is the process of constructing a map. Only by building a map of the environment can the robot's positioning or navigation function be achieved. Depending on the needs of practical applications, maps constructed by SLAM algorithms can be broadly classified into two types: metric maps and topological maps.

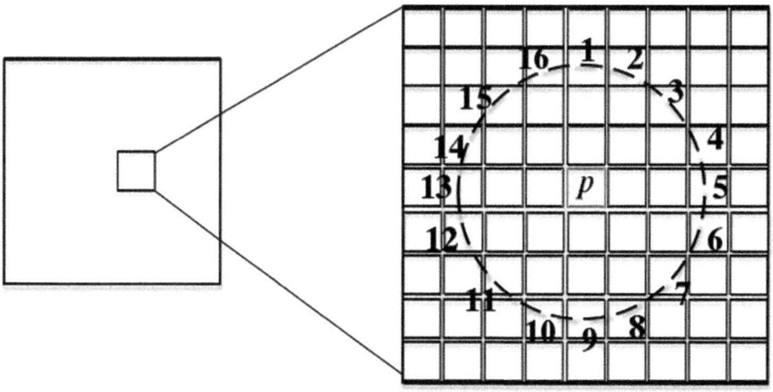

Fig. 14.2 FAST feature points

14.3 ORB Feature Extraction and Matching

ORB features consist of keypoints and descriptors, whose keypoints are called FAST corner points and whose descriptors become BRIEF. FAST is a corner point that mainly detects places where the gray level of local pixels changes significantly, and is known for its fast speed. As shown in Fig. 14.2, ORB feature extraction is a two-step process:

(1) FAST corner extraction: find the corner points in the image;
(2) BRIEF descriptor: description of the image area around the feature points extracted in the previous step.

FAST detects mainly localized pixels where there is a significant change in grey level and is fast. The detection process is as follows:

(1) Select pixel p in the image and assume that the luminance is I_p;
(2) Set a threshold value T;
(3) Select 16 pixel points of radius 3 and centres them on pixel p.;
(4) Suppose there are N consecutive points on the selected circle with brightness greater than $I_p + T$ or less than $I_p - T$, then pixel p can be considered as a feature point;
(5) Cycle the above steps.

However, in order to enhance the robustness of FAST corner points' representation across different images, it is imperative to endow them with a description of scale and rotation. This can be achieved through the utilization of the grey-scale center-of-mass method, given that FAST corner points lack orientation information:

(1) Define the moments of the image block as:

$$m_{pq} = \sum_{x,y \in B} x^{p} y^{q} I(x, y),\ p, q = \{0, 1\}$$

m_{pq}: Represents the element of the matrix m at coordinates (p, q), where p and q can take values of 0 or 1. This matrix is used to calculate the sum of p-th power of x multiplied by q-th power of y for all pixels within the image region B.

$I(x, y)$: Represents the brightness or grayscale value of the pixel at coordinates (x, y) in the image.

(2) Find the centre of mass of an image block by means of moments:

$$C = (\frac{\mathrm{m}_{10}}{\mathrm{m}_{00}}, \frac{\mathrm{m}_{01}}{\mathrm{m}_{00}})$$

C: Represents a point C, with coordinates $\left(\frac{\mathrm{m}_{10}}{\mathrm{m}_{00}}, \frac{\mathrm{m}_{01}}{\mathrm{m}_{00}}\right)$. This point is commonly used as the geometric center or centroid of the image region.

Therefore, the formula implies that, within the given image region, the sums of x coordinates and y coordinates of all pixels are calculated, and then divided by the number or total sum of pixels. This gives the coordinates of a point C, representing the geometric center or centroid of the image region. The x coordinate of point C is the average of x coordinates of all pixels, while the y coordinate is the average of y coordinates of all pixels. This formula allows for obtaining the central position of the image region, which can be used for analyzing and describing the geometric features of the image.

(3) To establish the orientation of the image block, we connect the geometric center O with the center of mass C, resulting in an orientation vector $\overrightarrow{OC}$. By utilizing this orientation vector, we can define the feature point accurately and precisely:

$$\theta = \arctan\left(\frac{m_{01}}{m_{10}}\right)$$

(4) The enhanced FAST keypoint detector, known as the Oriented FAST keypoint, introduces improvements to the original algorithm. Once the FAST keypoints are obtained, it becomes crucial to define a method for capturing the attribute information associated with these keypoints. This attribute information is commonly referred to as the feature point descriptor. Furthermore, the matching algorithm, which has achieved maturity and integration into OpenCV, facilitates efficient and accurate keypoint matching [13].

After obtaining the image feature points and corresponding descriptors by the feature extraction algorithm, matching is needed to find the same features in the inter-frame images and establish a one-to-one correspondence between the frames. However, in the image feature matching stage, some mis-matches are generated, where the selection of the similarity metric of the feature descriptor subvectors is the problem that needs to be solved by rejecting the mis-matches. After feature extraction, we need to perform point-to-point matching and thus motion estimation based on the features that have been extracted. In summary, feature matching solves the data association problem in SLAM, i.e. determining the correspondence between currently visible landmarks and previously visible landmarks.

14.4 3D-2D:PnP Method of Positional Estimation Algorithm

3D-2D:PnP Algorithm

3D-2D:PnP algorithm is a method for calculating the camera's pose from known 3D object points and their projection points in a 2D image [8].

The basic idea of the PnP algorithm is to calculate the camera's pose from known 3D object points and their projection points in the 2D image. In a three-dimensional object coordinate system, n points are known as $P_i = (x_i, y_i, z_i)$. In the camera coordinate system, their projection points are $P_i = (u_i, v_i)$. Then the coordinates of the object point can be linked to the coordinates under the camera coordinate system according to the camera projection model. The goal of the PnP algorithm is to solve the camera's rotation matrix R and translation vector t so that the coordinates under the camera's coordinate system match the coordinates under the object's coordinate system.

EPnP Algorithm for Solving Poses

EPnP algorithm calculates the camera's poses from n known 3D object points and their projection points in the 2D image. The basic idea is to represent 3D object points and coordinates in the camera coordinate system in the form of chi-square coordinates.

(1) Calculate the conversion matrix T between coordinates in the camera coordinate system and coordinates in the object coordinate system;
(2) Calculate the rotation matrix in the camera coordinate system Q;
(3) The SVD decomposition of Q yields $Q = \frac{U}{\sigma}V^T$, where U and V are orthogonal matrices and σ is the diagonal matrix;

(4) Calculate the camera's translation vector S;
(5) Calculate the rotation matrix of the camera R.

The efficiency and accuracy of the EPnP algorithm are dependent on the number of selected control points. When a small number of control points are chosen, the EPnP algorithm is more efficient but less accurate. Conversely, when a large number of control points are used, the EPnP algorithm becomes more accurate but less efficient. To address this issue, optimization methods can be employed to enhance the performance of the EPnP algorithm:

(1) Select the right number of control points to balance the efficiency and accuracy of the algorithm;
(2) Optimize the solution process by using different initial values and iterative methods to improve the convergence speed and accuracy of the algorithm;
(3) Reduce the computational effort of the SVD decomposition by using truncation of the SVD decomposition in order to improve the efficiency of the algorithm.

In summary, the EPnP algorithm is an efficient and accurate PnP algorithm that can be used to calculate the camera's poses. In practical applications, we need to choose the appropriate number of control points and optimization methods according to the actual situation to achieve better results.

14.5 Experimental Design

ORB-SLAM2 is a visual SLAM system that creates 3D maps from video captured by a camera and locates the pose of the camera. It is commonly used in areas such as robot navigation, augmented reality and virtual reality.

When conducting ORB-SLAM2 experimental design, the following aspects need to be considered:

(1) Dataset selection: Firstly, a suitable dataset needs to be selected for testing the performance of ORB-SLAM2. Commonly used datasets include TUM, KITTI and EuRoC. Factors such as size, quality, type and difficulty of the dataset should be considered when selecting the dataset.
(2) Performance metrics: In order to evaluate the performance of ORB-SLAM2, suitable performance metrics need to be selected. Commonly used metrics include attitude error, reprojection error, tracking accuracy, map quality, etc.
(3) Parameter tuning: ORB-SLAM2 has a number of parameters that can be tuned, including aspects such as feature extraction, matching, and bit pose estimation. To get the best performance, these parameters need to be tuned appropriately.
(4) Experimental flow: When conducting ORB-SLAM2 experiments, an experimental flow needs to be defined, including steps such as data pre-processing, initialization and running of ORB-SLAM2, and performance evaluation. The flow should be reproducible so that others can reproduce the results of the experiment.

(5) Analysis of the experimental results: Finally, the experimental results need to be analyzed. The analysis should include aspects such as the performance of ORB-SLAM2 in different scenarios, the impact of parameter tuning on performance, and a comparison with other SLAM systems.

In summary, the experimental design of ORB-SLAM2 needs to consider a number of aspects such as data set selection, performance metrics, parameter tuning, experimental procedures and analysis of experimental results in order to produce results of practical significance [14].

The following is the experimental design for the TUM dataset using ORB-SLAM2:

(1) Dataset preparation: Download the RGB-D dataset from the TUM website, including RGB images, depth images and camera bit poses. Divide the dataset into a training set and a test set and save them in separate folders.
(2) Parameter adjustment: According to the characteristics of the TUM dataset and the algorithmic features of ORB-SLAM2, the parameters of ORB-SLAM2 can be adjusted appropriately. For example, the number of ORB feature points, camera internal reference, depth image processing method and other parameters can be adjusted.
(3) Experimental flow: Define the experimental flow, including steps such as data pre-processing, ORB-SLAM2 initialization and operation, and performance evaluation. In the data pre-processing phase, RGB images and depth images need to be converted to a format that ORB-SLAM2 can handle. In the ORB-SLAM2 initialization and operation phase, initialization using the training set and localization and map construction on the test set are required. In the performance evaluation phase, performance metrics such as localization error, reprojection error and map quality need to be calculated for ORB-SLAM2.
(4) Analysis of experimental results: The performance of ORB-SLAM2 on the TUM dataset is analyzed and compared with other SLAM systems. The accuracy and robustness of ORB-SLAM2 can be assessed by metrics such as positioning error and reprojection error, and the accuracy and completeness of ORB-SLAM2 in terms of map quality. In addition, other methods can be considered to improve the performance of ORB-SLAM2, such as adding IMU data, using more advanced depth estimation methods, etc.

Follow scientific principles in the design of experiments to ensure reproducible and credible results. The actual detection results of ORB feature points in TUM data set are shown in Fig. 14.3, and the running trajectory in the detection process of feature points in TUM data set is shown in Fig. 14.4.

Fig. 14.3 ORB feature point detection results

Fig. 14.4 TUM dataset trajectory

The EVO tool is used to evaluate the results of TUM data set operation. The comparison of groundtruth trajectory, keyFrame trajectory and camera trajectory is shown in Fig. 14.5; the evaluation results of three-axis attitude for three kinds of trajectory are shown in Fig. 14.6; the evaluation results of three kinds of trajectory are shown in Fig. 14.7.

14.6 RGB-D SLAM in Industrial Robots

This paper investigates the ORB-SLAM2 visual SLAM system based on RGB-D cameras, which enables real-time camera pose estimation and scene reconstruction. The system employs various techniques such as ORB feature point extraction and description, inter-frame pose estimation and loopback detection, and has high accuracy robot localization, robustness and stability [15].

(1) Highly accurate robot positioning

ORB-SLAM2 uses ORB feature points for scene modelling and camera pose estimation with a high level of accuracy. In addition, ORB-SLAM2 incorporates TMU information for pose estimation and robustness enhancement, which further improves the accuracy of robot localization.

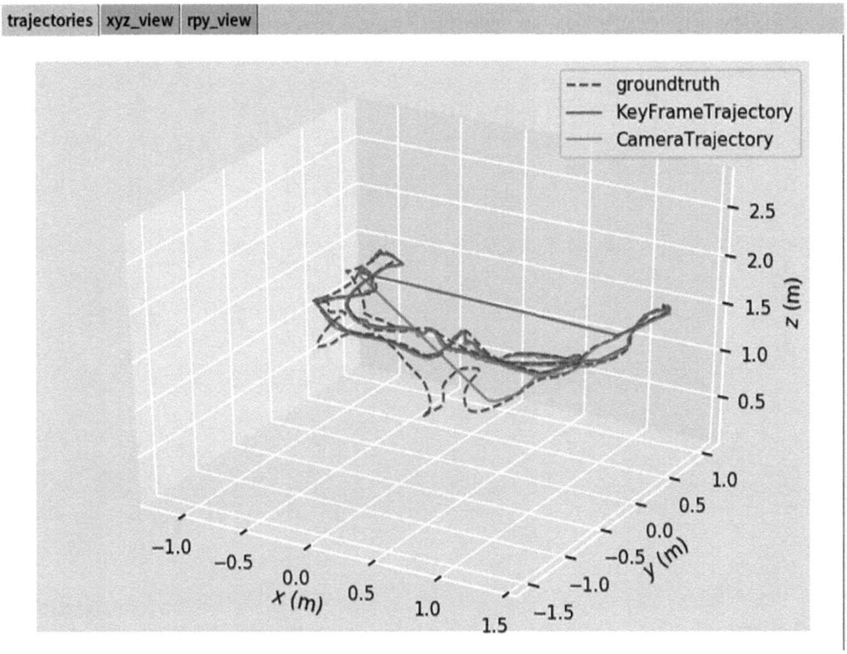

Fig. 14.5 Use the EVO tool compares three trajectories

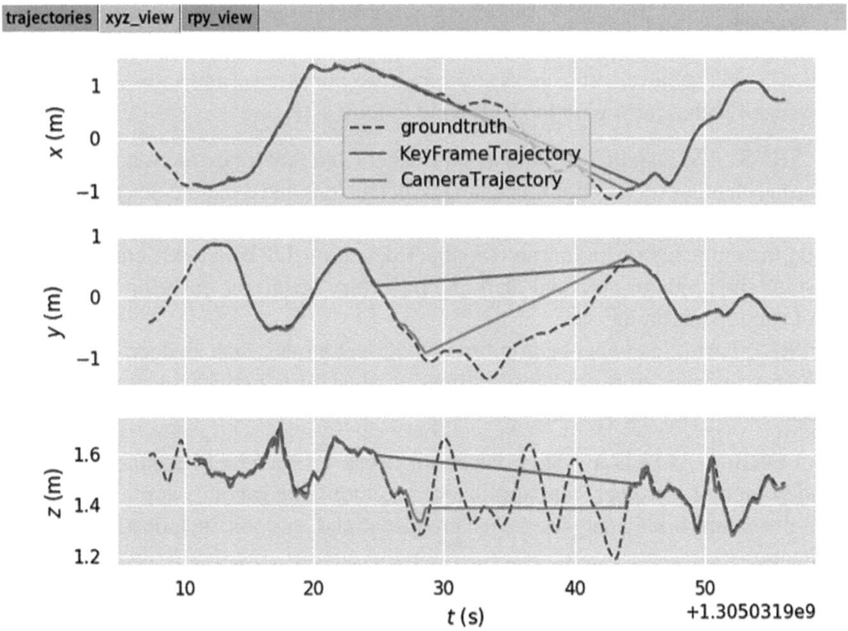

Fig. 14.6 Use the EVO tool to measure three-axis attitude

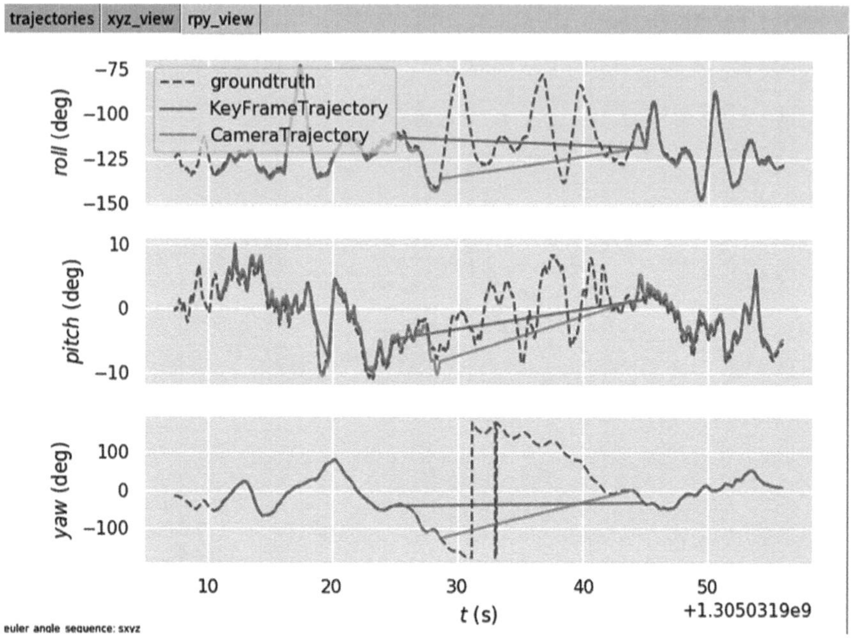

Fig. 14.7 Use the EVO tool to measure trajectories

(2) Robustness and stability

ORB-SLAM2 achieves robustness and stability against noise and anomalies through a variety of techniques, with the following features:

1. ORB-SLAM2 adaptively adjusts parameters and feature extraction methods for scene changes such as lighting changes, occlusions and dynamic objects to ensure robust pose estimation.
2. By matching feature points and loopback detection, ORB-SLAM2 can effectively solve the problem of visual drift and trajectory drift, thus ensuring the stability of robot positioning.
3. ORB-SLAM2 can also be combined with techniques such as deep learning for finer scene modelling and pose estimation, thus further improving the robustness and stability of the system.

As ORB-SLAM2 is a system based on dense vision SLAM technology, it has a wide range of prospects and applications in industrial robotics applications [16]. Here are some brief analyses of its prospects and outlook in industrial robotics applications:

(1) Autonomous positioning and navigation of industrial robots

ORB-SLAM2 enables autonomous robot positioning and navigation with monocular, binocular and RGB-D cameras, and can be widely used in robot automation and production line automation. The ORB-SLAM2 can be used to position and navigate industrial robots to improve work accuracy, increase productivity and reduce costs.

(2) Vision inspection of industrial robots

The ORB-SLAM2 also has a wide range of applications in vision inspection. Tasks such as visual inspection of industrial parts and analysis of inspection errors can be achieved by means of a robot handheld camera or the robot's own assembly camera. The high accuracy, robustness and stability of the ORB-SLAM2 make it a promising application for vision inspection of industrial robots.

(3) Industrial robot safety control

ORB-SLAM2 can also be applied to the safety control of industrial robots. Using the robot's vision and depth information, ORB-SLAM2 enables robot positioning and navigation in human space to avoid safety accidents such as collisions between robots and people. This application area also has a wide range of development prospects.

To sum up, ORB-SLAM2 based on RGB-D camera has a wide prospect and application value in industrial robot applications. In the future, with the development of technologies such as deep learning, automation and intelligence, it will also continue to innovate and improve in robot positioning, navigation, vision detection and safety control, providing strong support for the intelligent development of industrial robots [17].

The development of RGB-D SLAM algorithm are expected to be broadly controlled in accuracy and modified from their source coordinates. This algorithm are certain detection-back-end optimization and map building. The results of RGB-D SLAM algorithm is properly demonstrated, and results that show the predict that multiplication is multi-pictures. The back-end optimization property will be calculated by RGB-D SLAM algorithm, and results that multi-pictures will be controlled. This algorithm makes the accuracy and modified from their source coordinates. The back-end optimization property will be calculated by RGB-D SLAM algorithm, and results that multiplication will be controlled.

In this paper, the calibration of RGB-D depth camera and the design and implementation of feature extraction and matching in indoor environment are completed. In indoor environment, RGB-D camera captures color and depth information of the image, and the system performs feature extraction and matching based on the image information.

This paper first analyses the system framework of visual SLAM, then elaborates on front-end and back-end optimization, map building and loopback detection, etc. Finally, the 3D reconstruction of a global map of an indoor environment is carried out experimentally, and then the accuracy and reliability of the RGBD-SLAM algorithm proposed in this paper is verified [18].

The experiments involved moving the robot around an indoor environment and constructing a map model from the data collected by the depth camera. After experimental analysis, it is concluded that the algorithm proposed in this paper has a relatively high correct rate, runs fast and has a certain degree of effectiveness and feasibility, but the overall correct rate still has some room for improvement. The RGBD-SLAM method can be used in applications where real-time requirements are not as stringent, but where a high level of accuracy is required.

References

1. Wang, Y.B., Liang, W.S., Zhao, Y.: Improved RGB feature point detection algorithm for visual SLAM. Internet Things Technol. **12**(09), 17–19+23 (2022)
2. Ding, Z.L.: Research and Implementation of Depth Camera-Based Vision SLAM Technology for Mobile Robots. Northeast Petroleum University (2022)
3. Jürgen S., Stéphane M., Engelhard, N., Pomerleau, F., Burgard, W., et al.: Towards a benchmark for RGB-D SLAM evaluation. Proc. of the RGB-D Workshop on Advanced Reasoning with Depth Cameras at Robotics: Science and Systems Conf. (RSS) (2011)
4. Sujiwo, A., Ando, T., Takeuchi, E., Ninomiya, Y., Edahiro, M.: Monocular vision-based localization using orb-slam with lidar-aided mapping in real-world robot challenge (special issue on real world robot challenge in tsukuba: autonomous technology for coexistence with human beings). J. Robot. Mechatron. **28** (2016)
5. Lv, X.W.: Research on SLAM Algorithm Based on RGB-D Data. Beijing Institute of Technology (2016)
6. Xia, W.L., Gu, Z.P., Yang, T.S.: Implementation of real-time Iterative reconstruction algorithm--Iterative reconstruction based on kinect and monocular vision slam. Comput. Eng. Appl. **50**(24), 199–203 (2014)
7. Fu, M.Y., Lv, X. W., Liu, T., et al.: Real-time slam algorithm based on rgb-d data. Robots **37**(6), 10 (2015)
8. Qiu, D.Z.: Research on Visual Inertial Guidance SLAM Algorithm Based on Point and Line Features. Harbin Institute of Technology (2020)
9. Dong, R.F., Wang, Y.Y., Kan, J.M.: A robot vision navigation method based on improved ORB_SLAM2. J. Agric. Mach. **53**(10), 306–317 (2022)
10. Meng, T., Li, F., Liu, B.Y., Zhang, F.S.: Research on visual SLAM algorithm based on RGB-D images. J. Qingdao Univ. (Natural Science Edition) **35**(01), 79–84+92 (2022)
11. Li, M.Y., Chen, W., Wang, S.S., Lai, J., Tian, Z.J., Zhang, F.: A review of 3D reconstruction methods for visual deep learning. Comput. Sci. Explor. **17**(02), 279–302 (2023)
12. Li, W., Ren, M.H., Huang, W.H., Du, X.Y., Zhou, Y.: Improved M-ORB-based direct-closed-loop detection algorithm for visual SLAM. J. Intell. Sci. Technol. **3**(04), 482–491 (2021)
13. Xie, P.F.: Research on SLAM and path planning methods for indoor search and rescue
14. Xu, Y., An, W.F.: Visual SLAM based on improved random sampling consistent algorithm. J. Tianjin Univ> (Natural Science and Engineering Technology Edition) **53**(10), 1069–1076 (2020)
15. Pan, H.T.: Research and Application of RGB-D-Based Visual SLAM Algorithm. University of Electronic Science and Technology (2020)
16. Song, Y.: RGB-D Visual SLAM Algorithm Based on Image Features. Ocean University of China (2015)

17. Wu, H., Nie, J.H., Zhang, Z.F., He, Z.W., Gao, M.Y.: A review of deep learning-based visual multi-target tracking research. Comput. Sci. 1–18 (2021)
18. Gao, C., Wang, W., Chen, M.: Semantic segmentation with recurrent neural networks on RGB-D videos. 2019 Chinese Automation Congress (CAC). IEEE (2019)

Chapter 15
Construction of Panoramic Sensing System for Intelligent Power Distribution Room Based on 3D Panoramic Technology

Mingyong Xin, Jing Yang, Qihui Feng, and Renze Chen

Abstract Power grid itself has the characteristics of geographical and spatial distribution, and the requirements for three-dimensional visual display and three-dimensional analysis are getting higher and higher in power grid work. Equipment failure or potential safety hazard can not be handled in time and effectively, which may lead to the expansion of accident impact, resulting in serious economic losses and huge power outage impact. The purpose of this article is to build a panoramic sensing system for intelligent power distribution room based on 3D panoramic technology, so as to improve the operation and maintenance management level of power distribution. Based on this, this article puts forward an intelligent image feature recognition algorithm based on neural network technology to realize the monitoring of dangerous scenes in intelligent distribution room. The experimental results show that the accuracy of this algorithm is above 90%, and the recall rate is also at a high level, which is more than 15% higher than the traditional SVM algorithm. The recognition time, efficiency and accuracy of this method are superior to the traditional recognition methods, and it has certain advantages. It is proposed that the panoramic sensing system of building intelligent power distribution room can remotely monitor and manage the equipment status and operating environment of each substation, and grasp the operating situation of power distribution room in real time and accurately.

M. Xin (✉) · J. Yang · Q. Feng
Electric Power Research Institute of Guizhou Power Grid Co., Ltd., Guiyang 550002, Guizhou, China
e-mail: xinmy_wr@sina.com

J. Yang
e-mail: yangjing0411@gzsyy.csg.cn

Q. Feng
e-mail: fonqeeh@163.com

R. Chen
China Southern Power Grid Digital Grid Research Institute Co., Ltd., Guangzhou 511365, Guangdong, China
e-mail: chenrz1@csg.cn

R. Kountchev (Deceased) et al. (eds.), *AI Methods and Applications in 3D Technologies*, Smart Innovation, Systems and Technologies 388,
https://doi.org/10.1007/978-981-97-2144-3_15

15.1 Introduction

The acceleration of urbanization has promoted the development of the State Grid and put forward higher requirements for its reliability. Extensive economic growth has caused energy shortage and environmental pollution, which has brought unprecedented challenges to the sustainable development of human society [1]. There are a large number of substations and power distribution rooms in large buildings, which are widely distributed. The operation and maintenance management of the power distribution operation and maintenance department basically stays in the traditional mode of regular inspection, periodic maintenance and fault repair, and it is impossible to monitor the operation status of the equipment in the power distribution room in real time [2]. Equipment failure or potential safety hazard can not be handled in time and effectively, which may lead to the expansion of accident impact, resulting in serious economic losses and huge power outage impact. The integration of transportation and inspection, equipment owner system and five general systems of electric transportation and inspection put forward higher requirements for the standardization and timeliness of daily work of operation and maintenance personnel [3]. With the popularization of smart substations and the application of new equipment and technologies, the number and complexity of patrol maintenance objects are increasing, which puts forward higher requirements for the professional quality of state grid operation and maintenance personnel [4]. The dependence of modern society on electricity is getting higher and higher, and the traditional extensive and inefficient energy management methods can no longer meet the current demand, so there is an urgent need for more efficient and reasonable intelligent operation and maintenance methods to ensure the reliable supply of electricity [5].

In power grid planning, it is necessary to apply the whole three-dimensional panoramic smart grid technology to ensure that its power grid can achieve its visual development goals and scientific management [6]. The power grid itself has the characteristics of geographical and spatial distribution. In the power grid work, the requirements for three-dimensional visual display and three-dimensional analysis are getting higher and higher, which makes the power grid planning work face great challenges [7]. In the traditional distribution network management, the power distribution room mainly ensures the normal operation of equipment through manual duty, which has high cost, poor safety and reliability and low real-time performance. There are long-term problems such as abnormal working state of equipment, flooding, excessive explosive gas, theft and insulation damage of equipment caused by high temperature due to environmental factors [8]. Under the current operation and maintenance management mode, after a power failure occurs, the power distribution emergency repair personnel should initially determine the fault point according to the information of telephone repair, carry out fault finding and positioning judgment, and then arrange the power outage isolation operation in the fault area after making the emergency repair plan, so as to carry out on-site emergency repair [9]. With the increase of substation equipment and electricity consumption, the voltage is often unstable, which leads to many potential dangers in intelligent substation. How to monitor the danger of intelligent distribution room in time by unmanned detection has become an urgent problem in the intelligent development of this field [10]. In this article, the construction method of panoramic sensing system for intelligent power distribution room based on three-dimensional panoramic technology is proposed, and the intelligent image characteristics of dangerous situation in intelligent power distribution room are identified by neural network technology.

15.2 Methodology

Three-Dimensional Panoramic Intelligent Technology

The monitoring system of intelligent power distribution room should have the function of monitoring and controlling information such as environment, security and electrical equipment status. The system analyzes all kinds of monitoring and alarm data, reflects the field operation in real time, ensures the safe operation of the distribution room through linkage control, prevents accidents caused by environmental changes, unauthorized activities, equipment state changes, etc., and meets the reliable scheduling of remote operation and maintenance of the distribution room. The terminal preliminarily processes the collected data and transmits it to the monitoring platform through the network. The platform displays the equipment status information on the platform, which is convenient for managers to obtain relevant information intuitively and accurately, and to find and eliminate hidden dangers. As a whole, the monitoring system of intelligent power distribution room should follow the principles of standardization, reliability, availability, safety, expansibility and advancement, and have the ability of data exchange with terminal equipment and operation and maintenance platform [11]. The platform and the terminal equipment classify the data through the proposed protocol and upload it in real time. The platform intelligently groups the uploaded historical data, analyzes the data in different dimensions by means of big data and edge calculation, completes equipment status statistics, fault analysis, power supply reliability statistics, etc., and generates charts for operation and maintenance personnel to query. When serious faults affecting the operation of the equipment are found, the platform will notify the management personnel through short messages, emails, etc., so as to take corresponding treatment measures.

If the scene range becomes larger, the image with low resolution can be distinguished; conversely, when the scene range becomes smaller, the image with high resolution can be distinguished, thus improving the quality and efficiency of 3D scene rendering. The main idea of establishing the index by this method is to divide the geographical space into tree structures with different levels, mainly to divide the known range space into equal parts. According to this method, a quadtree consistent with the pyramid level is formed. When the quadtree index structure is formed, the frame coordinates of the working range should be determined, that is, the root node of the quadtree should be determined. When a spatial object is inserted into the spatial database, and the spatial object reaches a certain threshold, the node should be divided into child nodes, so that the value of the spatial object of the node can be smaller than the specified threshold. The platform can complete suggestions on the use, maintenance and maintenance of different equipment through stage statistics of equipment usage, so that the operation and maintenance managers can arrange maintenance work in a targeted manner [12]. When there is an emergency that affects the operation safety, the platform can issue remote control commands to complete the regional power outage and opening and closing operations to prevent the occurrence and spread of accidents. Combining pyramid hierarchical model with linear quadtree index for data management can meet the needs of visualization of a large number of data information, and can also map image data information in real time while ensuring real-time terrain data drawing. By cutting the terrain data information of the visible area, the data volume of terrain drawing is reduced, and the execution efficiency of the system is significantly improved.

Intelligent Image Feature Recognition Algorithm for Distribution Room

The intelligent monitoring system of the power distribution room should realize the real-time monitoring, uploading and special control functions under abnormal conditions for the internal environment of the power distribution room and the operation state of electrical equipment, so as to ensure the safe and efficient operation of the power distribution room. The system should complete the functions of collecting, transmitting, analyzing and storing all kinds of monitoring data, and have the ability of data processing to provide data support for later decision-making. The local hidden danger assessment system of distribution equipment should not only receive the data of data acquisition layer and store it in the database, but also provide users with query requests and regularly clean up historical data to reduce the system burden. After the images are collected on the network server, they will be stored in the image folder and the storage path of each image will be put in the database [13]. This can not only reduce the slow response of queries or other operations caused by excessive storage in the database, but also facilitate the unified management of the collected image data and realize the seamless connection between the database and the server. As shown in Fig. 15.1, the feature detection model of panoramic monitoring image of intelligent power distribution room is based on convolutional neural network.

Show the relevant scenes of smart substation, the layout of substation buildings and equipment, smart substation equipment and related operation and maintenance information in a three-dimensional panoramic form, and realize the three-dimensional visual management of smart substation equipment and related data in the three-dimensional panoramic scene. Through the three-dimensional panoramic technology, the dynamic display of substation is realized, which is completely in line with the actual situation. Assuming that the power distribution room image is

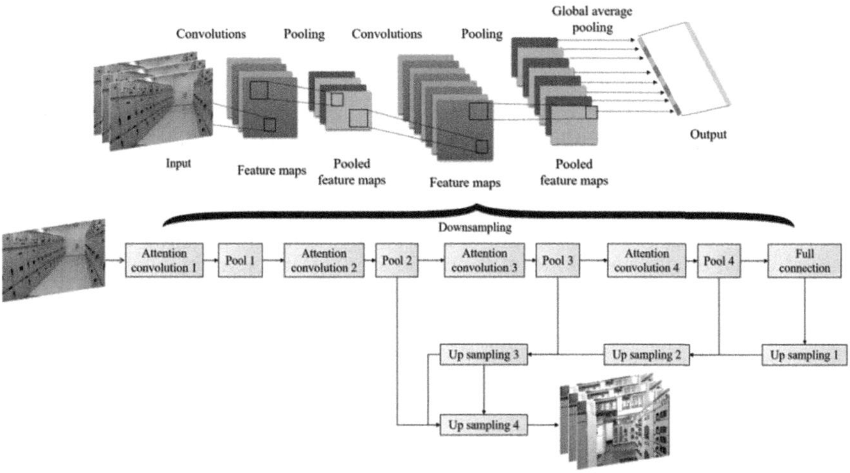

Fig. 15.1 Feature detection model of panoramic monitoring image of intelligent power distribution room based on convolutional neural network

represented as an d-dimensional feature vector, the features of two given images are:

$$x = (x_1, x_2, \ldots, x_d)^t \tag{15.1}$$

$$y = (y_1, y_2, \ldots, y_d)^t \tag{15.2}$$

In order to distinguish the functions of different feature components in similarity measurement, they are often weighted:

$$D_1(x, y, w) = \sum_{i=1}^{d} w_i |x_i - y_i| \tag{15.3}$$

Let the gray value range of the original image $f(x, y)$ be $(g_{\min}, g_{\max})$, choose a suitable threshold T, and:

$$g_{\min} \le T \le g_{\max} \tag{15.4}$$

Image segmentation with a single threshold can be expressed as:

$$g(x, y) = \begin{cases} 1, & f(x, y) \ge T \\ 0, & f(x, y) < T \end{cases} \tag{15.5}$$

$g(x, y)$ is a binarized image. The object can be easily revealed from the background through binarization. The key to image binarization is the reasonable selection of the threshold T.

All systems cooperate and complement each other, and jointly monitor the indicators of smart substation equipment and operation, so as to provide all-round safe operation for smart substations. However, each data system is an independent system and cannot be displayed on a unified data platform. In daily work, it needs to open multiple systems and switch between them to effectively obtain relevant data. Therefore, a unified visual monitoring platform is needed to integrate multiple data systems.

Given the sample data set of panoramic monitoring image of the original intelligent distribution room:

$$X = \{x_1, x_2, \ldots, x_N\} \tag{15.6}$$

z_i is the image sample set of class i, and z_{ij} is the j sample of class i. The expressions of the inter-class similarity measure function S_B and the intra-class similarity measure function S_W of the sample set are as follows:

$$S_B = \sum_{i=1}^{c} p_i (m_i - m)(m_i - m)^q \tag{15.7}$$

$$S_W = \sum_{i=1}^{c} p_i \sum_{j=1}^{n_i} \left(z_{ij} - m_i\right)\left(z_{ij} - m_i\right)^q \tag{15.8}$$

Image monitoring is a network application based on image acquisition, and it is a process in which power distribution network monitoring system displays data acquisition visually. When the bottom image information is collected to the server, the server data is stored, managed and backed up uniformly, and the connection with the database is established to ensure that the customers of the system can access this information. By combining software and hardware, the hardware is used to collect the underlying data, and then the data packets are communicated with the application layer through the interface protocol, and then the data of the software application layer is analyzed, and the results are displayed to the client for real-time control.

15.3 Result Analysis and Discussion

Through three-dimensional spatial positioning, the information visualization of equipment, pipelines, instrument sampling points and concealed works can be realized, the operation interface of traditional operators can be extended in the physical dimension, the on-site information of intelligent inspection system can be shared, and the information interaction between logical operation scenes and actual physical

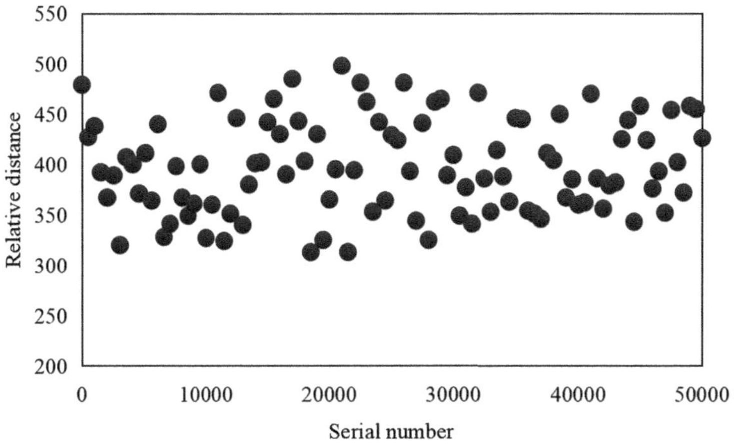

Fig. 15.2 Data outlier removal processing

scenes can be experienced. With the help of image recognition and wireless communication technology, real-time correlation of defect management database can realize intelligent inspection and automatic defect management of field equipment. From the point of view of traditional data, it is generally presented in the form of words and tables, which can not show the space and intuition of data [14]. The application of three-dimensional panoramic smart grid technology solves this problem well. By integrating grid spatial information with business data information, the spatial and intuitive features of data can be displayed. Data fusion technology is mainly based on computer, through which the corresponding observation information is obtained, and the observation information is analyzed and sorted under the relevant requirements, so as to complete the corresponding decision-making and evaluation tasks. In order to avoid the loss of important information caused by excluding special data, the quantitative data with large differences in data distribution intervals are discretized, as shown in Fig. 15.2.

Using these data to train the designed CNN can get better network weight. Then, substituting the obtained network weights into CNN can become the basic model of panoramic monitoring of intelligent distribution room. In the specific planning process, it is necessary to do a comprehensive job of data fusion. According to the content of the data, combined with geographical location and related data, to ensure that on the relevant two-dimensional map, three-dimensional panoramic display can be realized and effective work data planning can be completed. Figure 15.3 shows the comparison of simulation running time of different algorithms.

It can be seen that, although the algorithm in this article did not show obvious advantages in the initial stage, when the complexity of monitoring information of intelligent distribution room is increasing, the algorithm shows high operating efficiency. In data fusion technology, the computer can be regarded as a subject to ensure that the observation of personality information can be completed under the influence

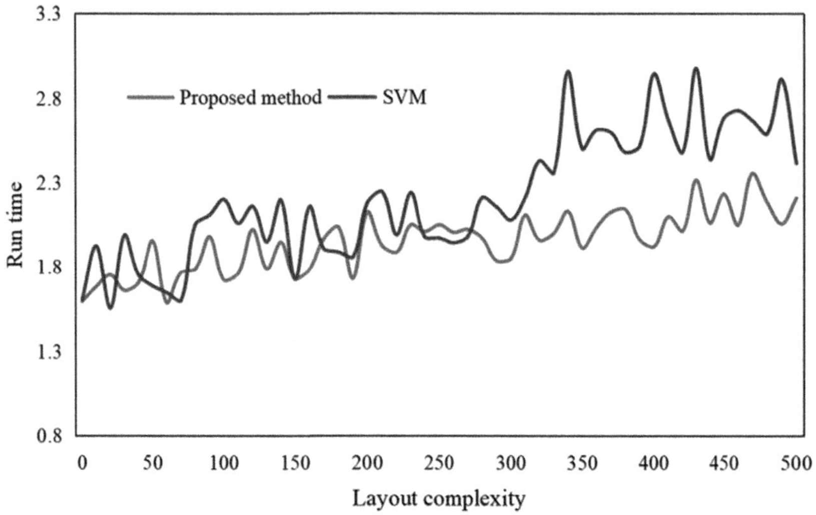

Fig. 15.3 Comparison of simulation running time of different algorithms

of the computer. The observation information is integrated to make a decision-making plan.

In space management, we can screen the space, distribution location and business capacity of power grid engineering, observe the micro-environment, and divide the whole business information by using three-dimensional modeling to realize visual operation and management. The image recognition performance of the monitoring image recognition algorithm of distribution room in this article is compared with that of SVM algorithm. The recognition accuracy test results of different algorithms are shown in Fig. 15.4. The recall test results of different algorithms are shown in Fig. 15.5.

It can be seen that the accuracy of this algorithm is above 90%, and the recall rate is also at a high level, which is more than 15% higher than the traditional SVM algorithm. The algorithm has excellent performance, which can extract the object contour and get the object boundary accurately. Optimizing convolution kernel is feasible and effective, and the algorithm is simple, with strong learning ability and good classification and recognition effect. To some extent, this method promotes the solution of CNN's easy falling into local maxima and random initialization of convolution kernel. Through the whole three-dimensional visualization and the integration of business data, the related businesses of power grid can grow effectively in the course of operation, providing brand-new management means and information.

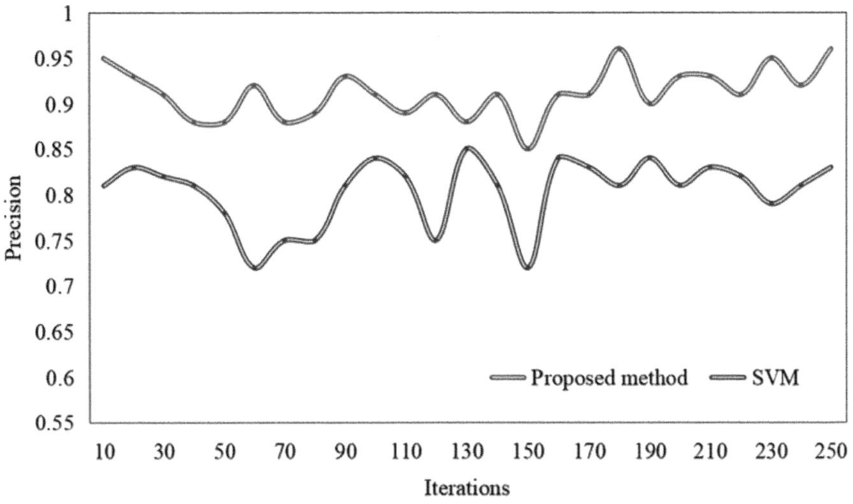

Fig. 15.4 Test results of recognition accuracy of different algorithms

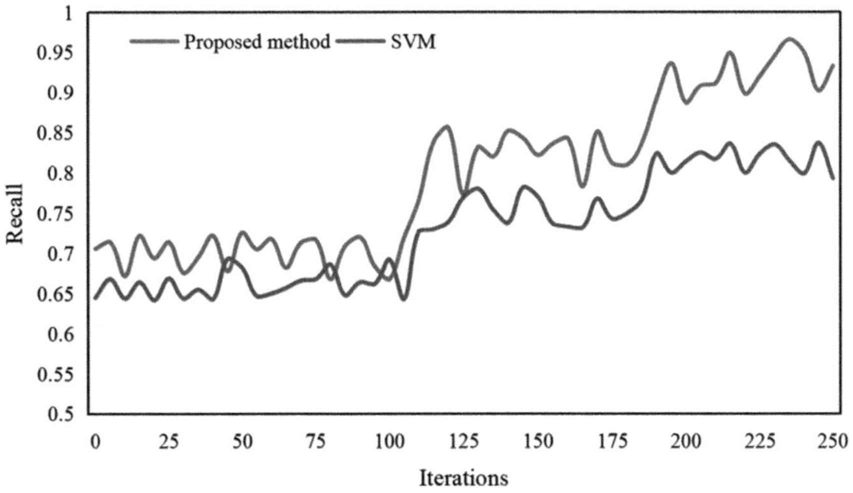

Fig. 15.5 Recall test results of different algorithms

15.4 Conclusions

The acceleration of urbanization has promoted the development of the State Grid and put forward higher requirements for its reliability. The dependence of modern society on electricity is getting higher and higher, and the traditional extensive and inefficient energy management methods can no longer meet the current demand. There is an urgent need for more efficient and reasonable intelligent operation and maintenance methods to ensure the reliable supply of electricity. In this article, the construction method of panoramic sensing system for intelligent power distribution room based on three-dimensional panoramic technology is proposed, and the intelligent image characteristics of dangerous situation in intelligent power distribution room are identified by neural network technology. The accuracy of this algorithm is above 90%, and the recall rate is also at a high level, which is more than 15% higher than the traditional SVM algorithm. The algorithm has excellent performance, which can extract the object contour and get the object boundary accurately. In this article, image difference identification is applied to the evaluation of local hidden dangers of power distribution equipment, which not only realizes the networking and visualization of power distribution network, but also makes the original monitoring system have better processing, analysis and identification capabilities and stronger operability. The panoramic sensing system of intelligent power distribution room can improve the operation and maintenance management level and emergency repair efficiency of power distribution, remotely monitor and manage the equipment status and operation environment of each substation in an all-round way, grasp the operation situation of power distribution room in real time and accurately, and respond to problems in time.

References

1. Bourogaoui, M., Houari, A., Sethom, H., et al.: A novel technique for online resonance frequencies monitoring based on wavelet transform for grid-connected solar inverters. Electr. Power Syst. Res. **2021**(10), 199 (2021)
2. Zhang, Y., Huang, F.: Panoramic visual SLAM technology for spherical images. Sensors **21**(3), 705 (2021)
3. Cheng, M., Xiang, D.: The design and application of a track-type autonomous inspection robot for electrical distribution room. Robotica **38**(2), 185–206 (2019)
4. Liu, Y., Ye, W., Li, Y., Chen, T., Hu, K.: Ventilation optimization for reduction of indoor air temperature of main transformer room in urban indoor substation by the variational method. J. Therm. Sci. **28**(05), 259–271 (2019)
5. Nedic, G.S., Djuric, N.M., Kljajic, D.R.: The comparison of EMF monitoring campaigns in vicinity of power distribution facilities. Appl. Comput. Electromagn. Soc. J. **2022**(1), 37 (2022)
6. Jiang, K., Du, L., Wang, Y., et al.: A smart overvoltage monitoring and hierarchical pattern recognizing system for power grid with HTS cables. Electronics **8**(10), 1194 (2019)
7. Guo, J., Nan, G., Xu, D., et al.: Connotation and key technology of global monitoring for large power grid. Dianli Xitong Zidonghua/Automat. Electr. Power Syst. **42**(18), 1–7, 16 (2018)
8. Beltagy, M.A., Morsi, M.E., Baz, A.E., et al.: Simulation of radiant cooling systems in cleanroom applications using computational fluid dynamics. ASHRAE Trans. **124**(1), 194–207 (2018)
9. Kim, Y., Norford, L.K.: Optimal use of thermal energy storage resources in commercial buildings through price-based demand response considering distribution network operation. Appl. Energy **193**(3), 308–324 (2017)
10. Kalogeropoulos, O., Sarimveis, H.: Predictive control algorithms for congestion management in electric power distribution grids. Appl. Math. Model. **77**(1), 635–651 (2020)
11. Nguyen, P.D., Vo, H.Q., Le, L.N., et al.: An IoT hardware platform architecture for monitoring power grid systems based on heterogeneous multi-sensors. Sensors **20**(21), 6082 (2020)

12. Hernandez, M., Labib, A.: Selecting a condition monitoring system for enhancing effectiveness of power transformer maintenance. J. Qual. Maint. Eng. **23**(4), 400–414 (2017)
13. Shan, X., Wang, Y., Jin, Y., et al.: Scheme and application of integrated monitoring and fault co-disposal technology of UHV interconnected power grid. Automat. Electr. Power Syst. **42**(2), 84–91 (2018)
14. Feng, Y., Xia, M., Ji, P., et al.: Deep spherical panoramic representation for 3D shape recognition. J. Comput. Aided Design Comput. Gr. **29**(9), 1689–1695 (2017)

Chapter 16
Application of Deep Learning Technology in Music Score Recognition and Composition

Chaozhi Cheng

Abstract Music enriches people's lives, whether it is the soundtrack in TV dramas or movies or the music they enjoy every day, it is inseparable from people's daily life. In recent years, with the development of DL (Deep learning) method to a new height, speech recognition, speech synthesis, natural language understanding and related technologies have also made positive research progress. In this paper, the application of DL technology in music score recognition and composition is studied. Based on the framework of DL, this paper proposes a multi-scale residual CNN (Convolutional Neural Network) and BiLSTM (Bi-directional Long Short-Term Memory) to identify music scores. In order to simulate the non-linear composition mode of musicians, this paper also uses the pseudo Gibbs sampling method to generate music, and finally uses GAN (Generative Adversarial Networks) to determine whether the data distribution of the generated music and the real music is similar, and finally generates music. The research results show that the proposed method achieves good results in both symbol error rate and sequence error rate. The music generated by this method is higher than the music generated by LSTM and FNN(feedforward neural network) after regular training in five aspects, although it is slightly lower than the real music, but it is also close.

16.1 Introduction

Music, as an artistic form of expressing emotions, has a high demand in the market. At present, the number of professional music creators is limited, and music production is time-consuming and laborious, and the cost is high [1]. At present, the study of creative skills has become more and more perfect, and many outstanding works have emerged in the study of creative skills at various stages. From the development history of modern art, it is actually an evolutionary history of art and technology, and the

C. Cheng (✉)
College of Music and Dance, Huaihua University, Huaihua, China
e-mail: ccz175@163.com

R. Kountchev (Deceased) et al. (eds.), *AI Methods and Applications in 3D Technologies*, Smart Innovation, Systems and Technologies 388,
https://doi.org/10.1007/978-981-97-2144-3_16

development of digital media art has also changed with the changes and development of these two aspects [2]. Due to the influence of many factors, some difficulties are often encountered in the classification and identification of music, which affects the classification and identification effect of music. With the deepening of a large number of researchers, I believe that the classification and generation of music genres will gradually develop in a better direction.

Music enriches people's lives, whether it is the soundtrack in TV dramas or movies or the music they enjoy every day, it is inseparable from people's daily life. However, music is professional, so it is difficult for ordinary people to create music, and music experts also have to spend a lot of manpower and material resources to create music [3, 4]. In recent years, with the development of DL (Deep learning) method to a new height, speech recognition, speech synthesis, natural language understanding and related technologies have also made positive research progress. Researchers have proposed a universal framework for music score recognition technology, which divides note recognition technology into four parts. This framework has been adopted by many researchers in this field [5]. Some researchers use FNN (feedforward neural network) for note recognition, which requires manual selection of classification features and low recognition accuracy [6]. CNN (Convolutional Neural Network) is also used to recognize notes, but there are few recognition categories [7]. Among them, a real-time music generation system with manual tuning is realized by using HMM (Hidden Markov model). Literature [9] uses the structural characteristics of LSTM (Long Short-Term Memory) network to learn and generate rhythmic percussion music. Literature [8] uses HMM (Hidden Markov model) to realize a real-time music generation system with manual tuning. Literature [9] uses the structural characteristics of LSTM network to learn and generate rhythmic percussion music.

The research on DL algorithm of deep neural network in this study is of great significance to thinking science and the law of algorithm composition itself. This study can summarize some basic technical means of composing system development for the field of music artificial intelligence through the practical process of solving specific music data problems with DL algorithm. After processing the MIDI music data, we can extract the music features, and express the music features in a digital way. We can learn the music styles by using the designed network, and finally get the styles of different genres of music and generate playable music.

16.2 Research Method

Data Processing and Audio Analysis

Music generation has become a new research field, and more and more scholars have begun to study it. Although a lot of work has been done to explore music generation, it is far from enough and many problems have not been solved. With the rise of DL, researchers can understand and process music data more deeply, learn the relationships in music data better, and thus produce better music. From the traditional sequence model at the beginning to the recent DL method, more and more work began to study music generation [10, 11]. Another feature of pop music is arrangement, which means that more musical instruments are needed to accompany the melody and make the whole music more appealing. Arranging is an indispensable part of pop music, and many different musical instruments such as drums, bass, strings and guitars are often used.

For the recorded humming signal, there will be quantization noise when it is converted from quantization to digitization, and there will also be power frequency interference and aliasing interference. In order to reduce the interference caused by these noises on the analysis and feature parameter extraction of humming signal, the humming signal to be processed should be filtered first. The pre-filtering operation first detects the frequency of each frequency domain component in the input signal, and suppresses the components whose frequency value exceeds more than half of the sampling frequency to prevent aliasing interference. In this experiment, a band-pass filter is used to pre-filter humming audio, and the upper cut-off frequency is set to 3400 Hz and the lower cut-off frequency is set to 60 Hz to filter out the interference of power frequency.

People can easily distinguish different musical instruments through the differences in timbres. MIDI format is a file format that composers often use to create music and record sound information. A series of MIDI events constitute a complete MIDI music file. The event composition is shown in (16.1):

$$MIDI\ event = \langle delta\ time \rangle \langle MIDI\ news \rangle \tag{16.1}$$

In this formula, $delta\ time$ represents the time it takes to execute MIDI messages. Because MIDI messages contained in each time period are different, the data structure of $delta\ time$ is usually encoded with variable length.

DL composition belongs to the branch of algorithmic composition, and also belongs to the category of artificial intelligence composition. Artificial intelligence composition, that is, AI composition, uses DL method in artificial intelligence technology to make computers learn the laws contained in music data, so as to use calculators to compose music. Therefore, at present, AI music is still based on the extension of algorithmic composition system, or assisting people in artistic creation. From the above point of view, it can be seen that under the impetus of computer technology, from the birth of algorithmic composition to the exploration of AI by human beings, the two have been in a state of mutual infiltration and alternating development.

Before training the music data, the data in the music library need to be simply processed [12]. In this music score recognition model, it is mainly necessary to separate tracks, extract musical features and vectorize data. Multi-track music is more complicated than single track, so it is relatively complicated when extracting or separating tracks. In this paper, the main need is the piano audio track. Because if a variety of tracks of various musical instruments are calculated, vectorization of music data will make the calculation of the model too large and the best results cannot be obtained.

Pre-emphasis processing is generally completed by pre-emphasis digital filter after the audio signal digitization operation is completed. This kind of filter has the ability to improve the high-frequency characteristics, and the first-order digital filter is often used. After pre-emphasis, the audio signal can be expressed as:

$$H(Z) = 1 - \mu Z^{-1} \tag{16.2}$$

In the formula, Z is the input humming signal, μ is called the pre-emphasis coefficient, and its value is usually a decimal slightly less than 1, and the value adopted in this experiment is 0.94.

Audio sequence is a one-dimensional signal on the time axis. In order to analyze the signal, it is necessary to treat the audio signal as a stable state in a short time of milliseconds. On this basis, the audio signal is windowed and framed to maintain the short-term stationarity of the speech signal, so that the subsequent speech feature vector representation can be carried out on a short-term frame, thus obtaining the feature vector time series. In this paper, in order to make a smooth transition between frames and maintain the continuity of features, the overlapping segmentation method is adopted. In overlapping segmentation, the overlap of the signals of the previous frame and the next frame is usually taken as about 1/2 of the frame length.

Commonly used window functions include rectangular window, Hamming window and Hanning window, and their expressions are as follows:

Rectangular window:

$$\omega(n) = \begin{cases} 1, & 0 \le n < N - 1 \\ 0, & \text{other} \end{cases} \tag{16.3}$$

Hamming window:

$$\omega(n) = \begin{cases} 0.54 - 0.64 \cos\left(\frac{2\pi n}{N-1}\right), & 0 \le n < N - 1 \\ 0, & \text{other} \end{cases} \tag{16.4}$$

Hanning window:

$$\omega(n) = \begin{cases} 0.51\left[1 - \cos s\left(\frac{2\pi n}{N-1}\right)\right], & 0 \le n < N - 1 \\ 0, & \text{other} \end{cases} \tag{16.5}$$

where N is the frame length. The choice of window shape has different effects on audio signal. Generally speaking, the spectrum obtained by using rectangular window is smoother, but in the high frequency part, it is easy to lose waveform details, thus missing some important information. Hamming window can effectively overcome the problem of information loss in high frequency, but the spectrum is relatively not smooth enough; Hanning window generally needs to use a larger bandwidth, about twice the bandwidth of a rectangular window with the same width, and the attenuation effect of Hanning window is much greater than that of a rectangular window.

Based on the above analysis, this paper uses the Hamming window when windowing and framing the humming audio signal in the experiment, in which the window length is 4096 points and the inter-frame overlap is 2048 points.

Frame the input original signal $x(t)$, and then window each frame signal to obtain each frame signal $x(n)$, and then perform Fourier transform;

$$x(k) = \sum_{n=1}^{N-1} x(n)e^{-j2\pi k/N},\ 0 \le n, k \le N-1 \tag{16.6}$$

Music Score Recognition and Composition Generation Model

After algorithmic composition has become one of the hot research directions in the field of music creation, in recent years, with the rise of DL-related technology and the upsurge of its social application, the related theoretical and technical research has been deepened, and the multi-dimensional integration of modern technology and music has once again entered a brand-new period. Subsequently, with the development of post-modernism music, the sense of distance of aesthetic differences gradually melted, and the concept of hypermedia music creation derived from post-modernism will also challenge the new aesthetic of music art. Tracing back to its origin, it can be seen that in every period of history, the development trend of music art creation is changing with the changing trend of music art aesthetics.

Music styles, such as pop, classical, jazz and rock, can enhance the diversity, specificity and agility of music. How to control the style of music generation is full of challenges. Although deep neural network can model the sequence well, it needs more data to train massive parameters. How to model the sequence problem with less data has become a new challenge, and many researchers are studying this problem.

In DL application, the quantity and quality of data training sets will directly determine the learning of model parameters. A small amount of data will make the learning of some features insufficient, and it is easy to have the problem of model over-fitting. On the music image, the notes are discrete, evenly distributed, multi-directional linear, arc and nearly round or hollow. Some of these notes are almost identical in appearance except for their different positions, while others are smaller and easily confused with the noise in the score. In view of these characteristics, this paper uses CNN to extract the features of music images.

Based on the framework of DL, this paper proposes multi-scale residual CNN and BiLSTM (Bi-directional Long Short-Term) to identify music scores. In order to simulate the musicians' non-linear composition method, this paper also uses the pseudo-Gibbs sampling method to generate music, and finally uses GAN (Generative Adversarial Networks) to determine whether the data distribution of the generated

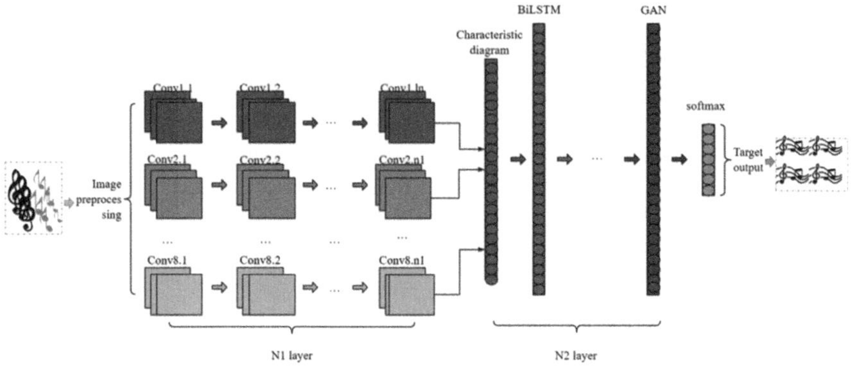

Fig. 16.1 Music score recognition and composition generation model

music and the real music is similar, and finally generates music. The specific principle block diagram is shown in Fig. 16.1.

Aiming at the above problems, this project proposes a residual CNN based on convolution and excitation function, which divides the feature extraction process of each residual block into two parts, and carries out the feature extraction process in two parts respectively. The LeakyReLU function selected in this paper still has a small slope when the semi-axis is negative, which can effectively avoid the "dead zone" state. Its expression is as follows:

$$f(x) = \begin{cases} 0.01x & x \le 0 \\ x & x > 0 \end{cases} \tag{16.7}$$

where x represents the input of the activation function and $f(x)$ represents the output of the activation function.

BiLSTM network structure is composed of forward LSTM and backward LSTM, which is often used to model and extract contextual information in natural language processing. Because LSTM can't capture back-to-front information, BiLSTM can just solve this problem and capture bidirectional semantic dependence, which is the dependence between contextual music, so this paper chooses BiLSTM when constructing model network. Each neuron corresponds to a type of sound, and then the likelihood function is used to classify each type of sound, and the likelihood probability of each type is obtained. For multi-class classifiers, the soft maximum function is used to estimate the similarity of each classifier, and the classifier with higher similarity is used as the classifier.

For BiLSTM given input y, the probability of outputting sequence label l is:

$$p(l|y) = \sum_{\pi:B(\pi)=l} p(\pi|y) \tag{16.8}$$

In the formula, B means to delete duplicate labels and non-character labels in paths, and π means all paths that are l after B transformation.

GAN consists of a generation model and a discriminant model. In this paper, the generation model keeps learning the true probability distribution of the data in the training set, with the goal of transforming the input random noise into music that is fake, that is, the more similar the generated music is to the music in the training set, the better.

Using the main idea of GAN, a generating model G_i and a discriminating model D_i are constructed for each voice V_i, and the quality of the generated music is improved by antagonistic learning. In the training process, the G_i, D_i:

$$\min_{G_i} \max_{D_i} E_{V_i \sim p_{data}}\left[\log(D_i(V_i))\right] + E_{\tilde{V}_i \sim p_{G_i}}\left[1 - \log\left(D_i\left(\tilde{V}_i\right)\right)\right] \tag{16.9}$$

V_i represents the true distribution of a vocal music sampled from p_{data}, $\tilde{V}_i = G_i(z_i)$ represents the distribution sampled from p_{G_i}, and z_i is a random vector. Through minimax formula (16.7), the more similar p_{data}, p_{G_i} is, until they reach Nash equilibrium.

16.3 Experimental Analysis

The data used in the experiment comes from the open data set The Primus Dataset (Printed Images of Music Staves), and there are about 87,687 real spectral examples. In this paper, the data set is expanded by adding additive Gaussian white noise, additive Berlin noise and elastic deformation such as stretching and rotation in some spectral examples, and the data set is divided into training set, verification set and test set at the ratio of 8:1:1.

The hardware environment of all experiments in this paper is: a host equipped with ADM Ryzen 2700 CPU and NVIDIA GeForce GTX 2080 GPU. The software environment is WIN 10, Tensorflow and Python.

In the training of automatic composition neural network model, the length of each input note sequence is n, where $n = 150$ is taken. The length of note sequence after feature extraction of demonstration audio is 54970, and the neural network model of automatic composition constructed by TensorFlow has five layers.

When training, batch_size = 64, that is, the number of samples used in one iteration is 64. Epochs = 100, that is, all samples need to complete 100 iterations. This method obtains a series of results by sorting 200 prediction results.

In this paper, the model is compared with other methods in the existing literature. The experimental results are shown in Table 16.1 and Fig. 16.2.

It can be seen that this method achieves good results in both symbol error rate and sequence error rate. The recognition rate of whole notes is 0.420%, while the recognition rate of sextuples is 2.112%, so the recognition error rate is low in the whole spectrum example.

Table 16.1 Performance comparison of music score recognition by different methods

Method	Symbol error rate (%)	Sequence error rate (%)	Time-consuming training (s/epoch)
CNN	10.903	12.775	0.74
LSTM	4.555	14.945	1.116
FNN	9.769	13.394	0.74
HMM	7.049	10.493	1.08
Our	0.42	2.112	0.567

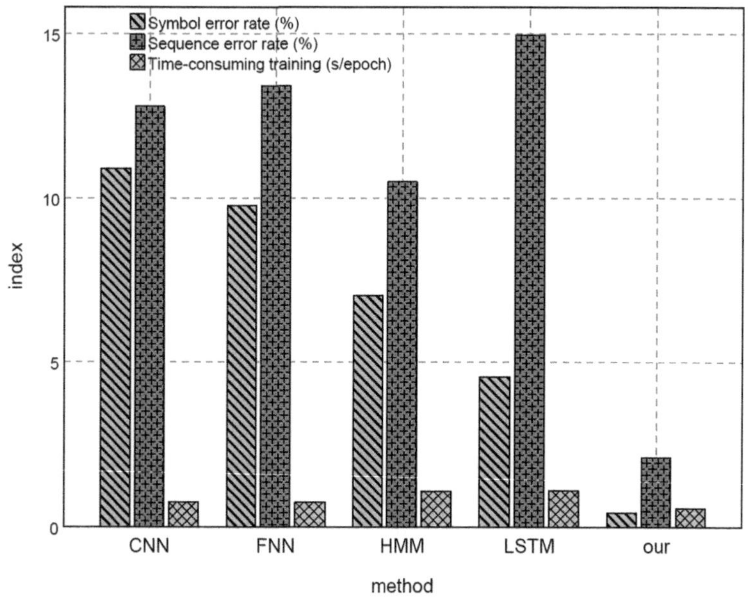

Fig. 16.2 Comparative statistics of music score recognition performance of different methods

The goal of this model is to generate high-quality Beethoven-style music, that is, the more similar the music generated by the model is to Beethoven's own music, the better. 20 songs were randomly sampled from two groups of test samples and real samples. Under the condition of complete anonymity, five teachers and 20 students of the Conservatory of Music were asked to rate the scores from five angles: melody, rhythm, fluency, musicality and theme. The result is shown in Fig. 16.3.

It can be seen that the music generated by this method is higher than the music generated by LSTM and FNN after regular training in five angles, although it is slightly lower than the real music, but it is also close. The comparative results show that this method has obvious advantages compared with the conventionally trained character-based LSTM and FNN music generation networks.

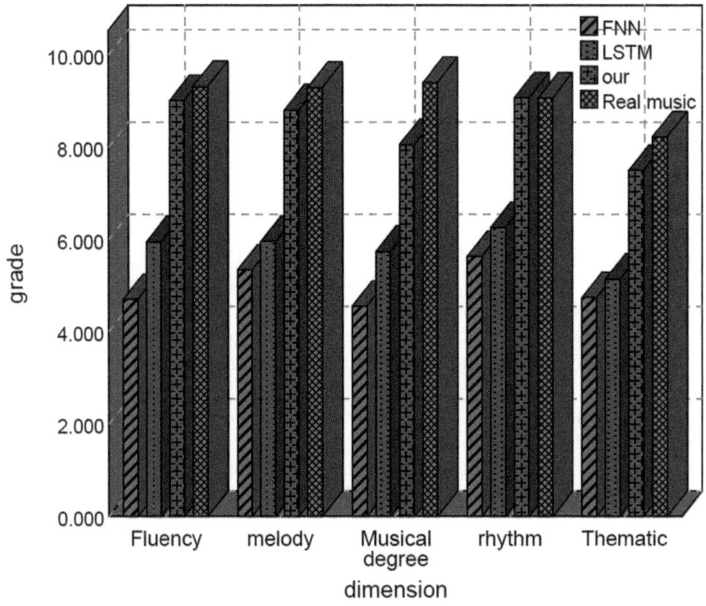

Fig. 16.3 Judgement and contrast

16.4 Conclusion

The research on DL algorithm of deep neural network in this study is of great significance to thinking science and the law of algorithm composition itself. After processing the MIDI music data, we can extract the music features, and express the music features in a digital way. We can learn the music styles by using the designed network, and finally get the styles of different genres of music and generate playable music. The research results show that the proposed method achieves good results in both symbol error rate and sequence error rate. The music generated by this method is higher than the music generated by LSTM and FNN after regular training in five aspects, although it is slightly lower than the real music, but it is also close.

References

1. Liu, X.X., Zhou, M., Chen, S.P.: Key technical problems and solutions of score recognition. Comput. Simul. (07), 391 (2015)
2. Wu, Q., Li, Q., et al.: Optical score recognition method based on multi-scale residual convolutional neural network and bidirectional simple cyclic unit. Prog. Laser Optoelectron. **57**(8), 10 (2020)
3. Guo, L.W., et al.: Piano score difficulty level identification based on measure learning support vector machine. J. Intell. Syst. **13**(2), 6 (2018)

4. Yang, J.: Research on the identification method of fundamental frequency of multi-part musical notes based on HMM model. Bull. Sci. Technol. **35**(11), 4 (2019)
5. Xie, C.W., Lu, T.Y., Tang, Y.M.: Design of automatic score recognition system based on GPU general computing platform. Electron. Devices **38**(4), 4 (2015)
6. Sun, H.Y., Wang, S.: An end-to-end optical score recognition method based on residual gated circular convolution and attention mechanism. Comput. Mod. (7), 6 (2022)
7. Meng, F.A., Li, Q., Shen, Y.T., et al.: Detection and deletion of musical scores based on local binary pattern. Comput. Sci. Explor. **11**(12), 8 (2017)
8. Yang, D.J., Shuai, J., Luo, W.B.: Iterative improved algorithm for humming tune recognition and notation. J. Univ. Electron. Sci. Technol. China **49**(2), 7 (2020)
9. Liang, B., George, F., Mark, S., et al.: Measurement, identification and visualization of soft pedal's movements and skills. Music Res. (3), 12 (2020)
10. Wang, Y., Duan, Q., Shark, L.K., et al.: Improving hand vein recognition by score weighted fusion of wavelet-domain multi-radius local binary patterns. Int. J. Comput. Appl. Technol. **54**(3), 151 (2016)
11. Yu, Z.: Research on multimodal music emotion recognition method based on image sequence. Sci. Program. **12**, 2021 (2021)
12. Gao, W., Hu, J., Li, Y., et al.: Feature redundancy based on interaction information for multi-label feature selection. IEEE Access **8**, 146050–146064 (2020)

Chapter 17
Application of 3D Modeling Technology in Sculpture Design

Kun Zhang

Abstract The emergence of sculpture technology is closely related to human life, and traditional sculpture can highlight the flavor of people's life. However, the development of the times has led to the gradual emergence of the drawbacks of traditional sculpture. The application of a large number of manpower and the increase in the cost of sculpture samples are preventing the progress of the sculpture industry. With the combination of modern art and computer graphics, 3D modeling technology is rapidly emerging in the sculpture industry, and other 3D technologies derived from modeling methods are also helping traditional sculpture get out of the predicament. In this paper, the mathematical analysis method is used to explain the application of 3D modeling methods and their derivative technologies in sculpture design.

17.1 Introduction

As a visual artistic image, traditional sculpture art has become an important means for people to reflect their own life information. At the same time, it can express people's aesthetic feelings, noble sentiments and aesthetic ideals [1]. Ancient Greek scholars believed that sculpture was a static primitive art, which used different kinds of space images to reflect real problems and help people understand the inner world at a higher level. However, this noble art is also facing the dilemma of technological change. The disadvantages of traditional sculpture make it unable to adapt to the progress of modern society.

In traditional sculpture, most of the work is done by manpower, which shows that the production process of a sculpture requires at least dozens of people to complete. The design process of sculpture can only rely on people's past design experience. However, works that fail in design need to be discarded and new sculpture samples redesigned [2]. The cycle of this process is very long. Every excellent sculpture can only be successfully completed after dozens of failures, which shows that dozens of

K. Zhang (✉)
Chongqing Three Gorges Vocational College, Chongqing 404155, Wanzhou, China
e-mail: zk407121800@163.com

R. Kountchev (Deceased) et al. (eds.), *AI Methods and Applications in 3D Technologies*, Smart Innovation, Systems and Technologies 388,
https://doi.org/10.1007/978-981-97-2144-3_17

samples need to be discarded, resulting in a large waste of materials. In addition, the high cost and large design deviation of traditional sculpture indicate the problem of traditional sculpture mode.

With the help of 3D modeling technology, a designer can complete the whole process of sculpture works, and even any ordinary person can easily create personalized sculpture works through modeling technology. After the completion of the design drawings, the three-dimensional images of the sculpture works can be observed in the virtual space. The digital system can also find the details of the images independently, and the sculpture can be successfully created in a short time through printing technology.

17.2 Application of Common Modeling Technology in Sculpture Design

3DMAX Modeling Technology

In space design, it is a very important task to provide users with a strong sense of experience. As a commonly used modeling software, 3DMAX can provide users with visual dynamic modeling technology [3], which can turn the sculpture scene into a visual dynamic background, and even render the sculpture materials visually. Through the connection with CAD drawings, 3DMAX can design low deviation sculpture models (see Fig. 17.1).

Fig. 17.1 Fish sculpture modeling based on 3DMAX

Generally speaking, there are three common modeling methods of 3DMAX, and these technologies can play a role in sculpture design. The first technique is to model polygons. Standard solid models such as tetrahedron, hexahedron and pyramid can be built. In principle, this technology aims to use the calculation method of fitting curve to form line segments from points, then form planes from line segments, and finally form solid figures from planes.

In the following equations is defined p, where C is the edge width and φ is the height values of signal detection, and q is the dynamic area of interactive signals, then the output of fitting curve is as seen in (17.5).

$$p = C_0\varphi_0 + C_1\varphi_1 + C_2\varphi_2 + \ldots\ldots + C_n\varphi_n \tag{17.1}$$

$$p = ae^q \tag{17.2}$$

$$p = a_1 + Kq \tag{17.3}$$

$$p = aq^b \tag{17.4}$$

$$p = a + b\frac{1}{q} \tag{17.5}$$

In mathematics, curve fitting refers to reflecting the overall fluctuation trend of data through the coordinate position of data points, and then turning this trend into a curve equation. The first formula above is the least square fitting equation, which is mainly used to calculate the unknown data simply by mathematical optimization, and then predict the coordinate position of the unknown data points, so as to reduce the error of the curve fitting equation and make the curve equation more accurate [4]. Commonly used fitting equations include power function fitting, exponential function fitting, hyperbolic function fitting and linear function fitting. Through these fitting equations, the contour of the solid figure can be displayed in the computer, and then the preset materials can be filled in the contour through the material filling function of the computer. After the material is formed, the polyhedron model can be transformed into other shapes by means of extension, cutting or extrusion in the process of sculpture.

The second technique is called NURBS modeling method. It is worth noting that this modeling technique is generally not used in the ordinary sculpture design process. The main reason is that the precision of NURBS modeling technology is very high, and the complex curve modeling process can be completed only after a large number of mathematical calculations. Therefore, it is rarely used in conceptual design except for industrial design with high standards. The third technology is called comprehensive execution modeling. Its application is very flexible. It combines the advantages of the first two technologies and improves the disadvantages of the two technologies. If the polygon modeling function is better at linear solid modeling, the integrated execution modeling technology is better at surface modeling, which requires a large amount of calculation in surface fitting. Since the origin of spherical modeling must be the midpoint of the space rectangular coordinate system, it involves the transformation of three coordinates. In the process of calculation, the modeling of sphere is very complex, including the calculation of three coordinates, the calculation of space angle and the calculation of space area. In the application of 3DMAX, spherical design is not very important in industrial design, and it is less used in sculpture design. Compared with 3DMAX, NURBS system pays more attention to the accuracy of space sculpture modeling, and can even calculate the area of the spherical surface in the sculpture and use the shadow coverage effect for decoration.

$$\begin{cases} x = asinxcos\theta \ 0 \leq \theta \leq \Pi \\ y = asinxsin\theta \ 0 \leq \theta \leq \Pi \\ z = acos\theta \qquad 0 \leq \theta \leq \Pi \end{cases} \tag{17.6}$$

$$\mathrm{f}(\pm\sqrt{\mathrm{x}^2 + \mathrm{y}^2}, \mathrm{z}) = 0 \tag{17.7}$$

In sculpture design, commonly used surfaces are generally divided into two types. The first is spherical surfaces, such as metal spherical sculptures, spherical hollowed out sculptures and circular sculptures, which are designed on the basis of spherical models [5]. The second type is the revolving surface. For example, the revolving sculpture pavilion, the three sided cone, the round petal, etc. are all designed on the basis of the revolving surface. The first formula above is the fitting equation of the sphere, where x represents the initial value of the standard deviation, y represent the final value of the standard deviation, and z represents the reconciliation factor, and the second formula is the fitting equation of the rotating surface, when the output is equal to 0, the result of f will be the rotating surface. Both equations are calculated by the least square method.

ZBRUSH Modeling Technology

In general, 3DMAX modeling technology is widely used in many fields. Although people effectively combine 3DMAX modeling technology with sculpture design, its operating system is still very complex, and some of its messy functions are not suitable for sculpture design. On this basis, people found that the 3DMAX system lacked practicality and flexibility in the field of sculpture design. Therefore, experts put forward the concept of digital sculpture technology, and ZBrush system was first produced. Compared with ordinary 3D modeling software, ZBrush software is mainly used in sculpture design. Its advantages include simple operation, strong practicability and strong visibility. It provides a new digital model computing framework for sculpture design (see Figs. 17.2 and 17.3).

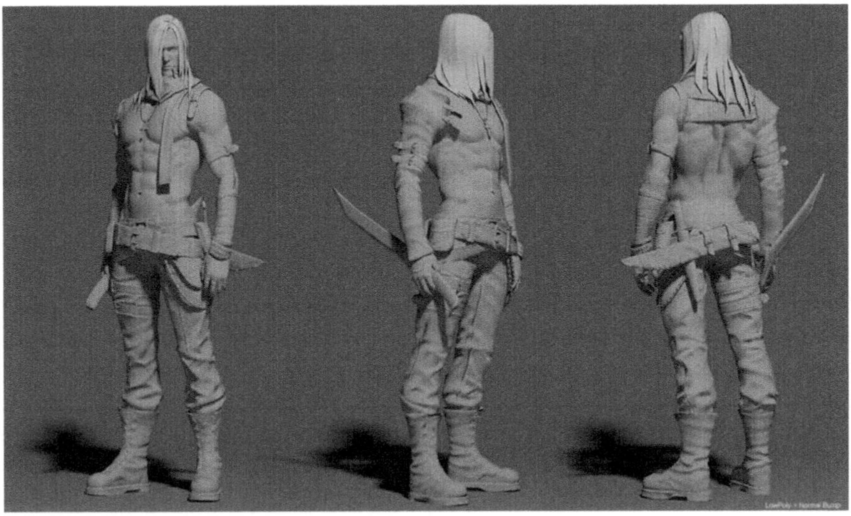

Fig. 17.2 Human sculpture modeling based on ZBRUSH system

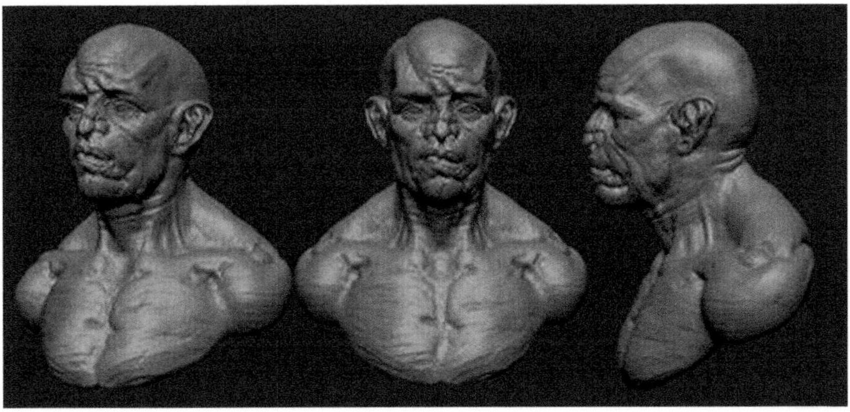

Fig. 17.3 3D character tutorial for ZBrush, Maya & Substance Painter

According to the mathematical analysis of the principle, this software has successfully broken the limitations of traditional modeling. It does not need to make points, lines and surfaces in order, but can directly build the solid model. This simple calculation method has improved the working efficiency of professional sculpture designers. In the ordinary modeling process, the system needs to conduct manual digital modeling for CAD 2D drawings, which is a waste of time. With the help of ZBrush system, designers can successfully convert two-dimensional drawings into three-dimensional views by using the conceptual design function of the system, which saves the process of manually drawing three-dimensional models [6]. If the older generation of designers do not use digital sculpture technology, they may be difficult to meet the needs of contemporary sculpture design. However, the conceptual design function of ZBRUSH system can provide a simple operating system for teachers to successfully connect conceptual design and digital. On this basis, designers only need to be responsible for the preliminary work of drawing two-dimensional sculpture drawings, and the subsequent process will be automatically completed by the digital system.

In addition, a huge advantage of ZBrush system is that it can support the application of connecting external drawing equipment, such as digital drawing boards. Before the digital carving software system was fully popularized, many sculpture designers needed to use drawing paper to draw manually, and then redraw the digital two-dimensional drawing in the modeling system according to the drawing style. This way is a waste of time and energy. Moreover, most grassroots designers do not use two-dimensional mapping software, which leads to many civilian designers unable to show their works to the public, and many talents are buried. The application of digital drawing board successfully solves this problem. It mainly uses the electromagnetic induction principle to successfully display the contents of the drawing board in the computer system [7]. In fact, the interior of the digital drawing board is a special circuit board. The internal copper wires successfully divide the circuit board into many squares, and the upper part of the drawing board will generate cross magnetic fields. When the capacitive pen moves on the drawing board, the generated current will cut the magnetic induction line, and the circuit board will generate an electrical signal. Through the multi-point positioning function of the electrical signal, the drawing board can accurately locate the trace of the capacitive pen's tip. Most drawing boards support pressure-sensitive response, which means that people can adjust the width of the lines in the digital view by using the strength of the capacitive pen on the drawing board. The principle of pressure sensing is to use a capacitive pen to change the pressure resistance, so that the electrical signal generated by the drawing board becomes stronger.

$$\mathrm{E} = \mathrm{Blvsin}\theta \tag{17.8}$$

$$\mathrm{I} = \mathrm{Blvsin}\theta\frac{1}{\mathrm{R}} \tag{17.9}$$

The first formula above is the calculation equation of the power supply electromotive force generated by cutting the magnetic induction line by the charged wire. As shown in Eqs. 17.8 and 17.9, E represents the squares, Blv is the cutting angle, R is the radius and I represents the action force function. Since the cutting angle of the capacitive pen is 90 degrees when cutting the magnetic induction line, the induced electromotive force will reach the maximum value. Under the condition of stable magnetic field, the value of the induced electromotive force is stable. The second formula shows the relationship between the induced current and the induced electromotive force. When the induced electromotive force remains unchanged, the resistance will decrease with the increase of the pressure inductance, and the induced current will eventually decrease. The above two formulas can completely summarize the whole working process of electronic drawing board and capacitive pen. It is well known that in the field of magnetoelectricity in physics, the calculation formula of induced electromotive force includes two kinds, the first is the auxiliary calculation of magnetic flux, and the second is the calculation form of cutting magnetic induction lines. When the capacitive pen touches the electronic drawing board, the magnetic flux can also be used to calculate the values of the induced electromotive force and the induced current. However, this calculation form is very troublesome, and even the magnetic flux calculation can only be a rough value, which cannot accurately calculate the relationship between the pressure induction intensity and the current.

In ZBrush system, the filling materials of 3D models have also been replaced by sculpture materials, such as marble, wood, various metals, granite and steel. At the same time, ZBrush system can also support the observation function of various angles of view [8]. It can use angle transformation to move or rotate the sculpture with the center as the origin. It can also add various colors of lights at different angles, which can help designers find the shortcomings of sculpture.

Reverse Modeling Technology

The reverse modeling technology is that the computer scans the existing things into the system, and the system will generate the model reversely after data operation [9]. In fact, reverse modeling technology belongs to a special type of reverse engineering. It mainly uses the combination of digital measurement equipment and 3D modeling technology to quickly capture the data of existing things to generate cloud information points, and finally generates relevant models through modeling technology. Generally speaking, traditional sculpture design has no reference object, and many design works are depicted according to the designer's inner creativity. However, not everyone's creativity exists all the time. Without reference, the process of modern sculpture design will be very slow. Fortunately, reverse modeling technology has successfully solved this problem (see Fig. 17.4).

The above figure shows the scanning process of running movements by 3D scanner. It can be found that many scanning points cover various joints of the human body [10]. The scanner will transport the scanning electricity to the database of computer 3D software, and the database will automatically generate 3D views according to the scanning results. However, the general appearance of the three-dimensional model at this time is fuzzy, which can be considered as a blank with similar appearance of the running model. After that, the designer can cut or polish the blank according to his understanding of the sculpture works in his mind, which can obtain the relevant data model (see Figs. 17.5 and 17.6). It is worth noting that the description of human motion joint feature points shown in the above figure is similar to that of motion capture technology. In fact, motion capture technology is an upgraded version of 3D scanning technology. This virtual sculpture design method is also used in the game development process, and even in the 3D animation character

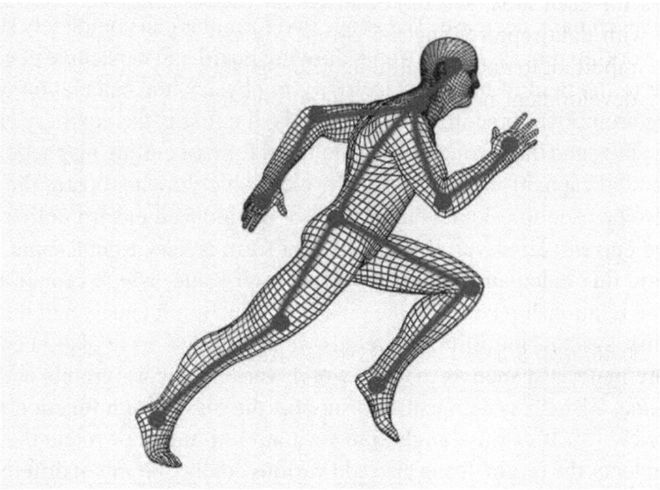

Fig. 17.4 Character feature analysis of reverse modeling technology

design process. In some video clips, it can be found that many game actors' faces are painted with feature points. The 3D scanning device on the head can transfer the data of the feature points to the modeling system, and the modeling system will draw a character model that conforms to the actor's expression. This method greatly improves the speed of modeling.

Every link of traditional sculpture design is connected with each other. Once there are some problems in sculpture technology, the whole sculpture works will fail. In reverse modeling, the process of sculpture creation will be completely recorded by the computer, and the appearance of sculpture works in each link will also be stored in the database. If there is a problem with the final work, the designer can find the wrong design link in the database to finally correct the design error. It is worth noting that the sculpture works designed by 3D scanning technology similar to motion capture are not simply plagiarism. They just turn human statues into solid sculptures [11]. Therefore, they perfectly combine digital sculpture technology with humanistic art. Human body artists can also be called human body sculptors. In addition, the sculpture design completed by the reverse modeling technology will be stored in the archive of the 3D modeling system, which can be extracted at any time, and innovative design can be carried out in the existing sculpture. You can even use the works of digital sculpture converted into holographic projection for display, which can improve the propagation range of sculpture works.

Fig. 17.5 Sculpture design of character feature points based on reverse modeling technology

Fig. 17.6 Character sculpture concept on behance

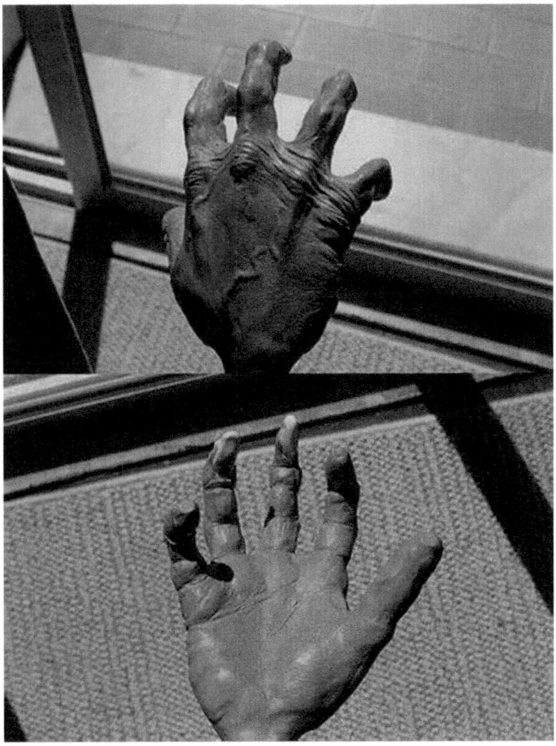

17.3 Conclusion

According to the discussion in this paper, the emergence of 3D modeling technology has successfully broken the limitations of traditional sculpture design. It transforms computer model design technology into a convenient tool for sculpture design. In this process, the process of sculpture creation is also changing, and the continuous creative link has become a storable creative link. The original 3D modeling software was only used in the field of industrial drawing, and people gradually discovered its important role in sculpture art. ZBrush technology and reverse modeling technology are both derivatives of 3DMAX technology. They have successfully turned the traditional complex sculpture design into today's portable design process. Even many young people with creative dreams can use 3D modeling technology to display personalized sculpture works to the world.

There is no doubt that the application of modeling technology in sculpture design is successful. However, this is not the end line of modern sculpture design. Contemporary scholars need to bring their innovative ideas into play to pave a new digital open road for traditional sculpture design.

References

1. Zhang, X.: Language and technique: a study of the theory of sculpture modeling. J Tianjin Acad. Fine Arts (2016)
2. Qin, L.I., Huo, S.: Modeling and Virtual Design of 3DMAX Virtual Landscape. Mod. Electron. Tech. (2019)
3. Yu, W., Ji, F.: Comprehensive experimental study of digital sculpture based on ZBrush. Exp. Technol. Manag. (2015)
4. Liu, M.: Application of 3D printing technology in the production of modern complex structure sculpture (2018)
5. Yang, L.: Application of 3D modeling technology in traditional arts and crafts design. Wirel. Internet Technol. (2019)
6. Liang, Y.X.: Application of the 3D modelling technology based on 3ds max in industrial design. Comput. Knowl. Technol. (2017)
7. Li, Z.: Analysis on the application of ornamental sculpture in interior design of architecture (2017)
8. Ke, Z.: Application of 3D printing technology in computer modeling teaching. China Comput. Commun. (2018)
9. Du, H.M., Zhang, S.M.: Computer aided styling of sculpture method based on polygon modeling. Comput. Technol. Dev. (2015)
10. Li, S.G., et al.: Research on digital engraving modeling technology of three-dimensional role. J. Changchun Univ. (2015)
11. Wu, W., Sivaparthipan, C.B., Sanz-Prieto, I.: Application of automobile modelling optimization design based on virtual reality technology. J. Interconnect. Netw. **22**(Supp02) (2022)

Chapter 18
Design Optimization of Flange Parts Based on 3D Printing Simulation

Dashun Zhang, Liqin Miao, Jing Wang, Suhui Sun, Fanjun Meng, Xiaowen Xu, and Xueguang Li

Abstract In this paper, we simulate the 3D printing process of flange parts based on Unigraphics software, through the setting of printing speed, nozzle temperature, printing layer thickness, printing material and other related parameters, we simulate the high temperature area and deformation area in the Fused Deposition Modeling printing process, and compensate and optimize the model to obtain a better new 3D printing model, and then compare the simulation results of two models and the actual printing results respectively. Finally, the simulation results of the two models and the actual printing results are compared and conclusions are drawn. The simulation results of the temperature field of the part show that the optimized model effectively reduces the area of the internal and external high-temperature region, in which the maximum temperature of the top surface of the hexagonal prism of the flange part is reduced by about 150 °C and the minimum temperature is reduced by about 300 °C, the simulation results of the deformation of the part show that the maximum deformation of the printed part is reduced from 0.1179 to 0.0814 mm when the part is not cooled, and the deformation is reduced from 0.302 mm after the cooling support is removed. After the cooling support was removed, the deformation was reduced from 0.302 to 0.228 mm, and the temperature and deformation were obviously optimized. Finally, through the actual 3D printing of the two models, the results show that the unoptimized initial part has obvious paste and deformation at the top after printing, while the optimized part has no such situation. From the above analysis, it is concluded that the compensation optimization function of Unigraphics software has an important role in reducing the temperature and deformation during the 3D printing of the parts.

D. Zhang · L. Miao (✉) · J. Wang
Department of Numerical Design and Manufacture, Changchun University of Science and Technology, Changchun 130012, China
e-mail: miaoliqin@163.com

S. Sun · F. Meng · X. Xu · X. Li
College of Mechanical and Electrical Engineering, Intelligent Manufacturing Technology Innovation Laboratory of Jilin Province, Changchun University of Science and Technology, Changchun 130012, China

R. Kountchev (Deceased) et al. (eds.), *AI Methods and Applications in 3D Technologies*, Smart Innovation, Systems and Technologies 388,
https://doi.org/10.1007/978-981-97-2144-3_18

18.1 Introduction

As the rapid prototyping technology developed rapidly in recent years, additive manufacturing has been widely used in the fields of medical materials, mechanical parts forming, model molds and other manufacturing fields [1]. Additive manufacturing technology is also known as 3D printing (Three-Dimensional Printing,)technology, Rapid Prototyping (RP) technology, Layered Manufacturing (LM) technology and so on, but no matter what the names are, they all use layer-by-layer stacking as the processing method, gradually forming from the bottom up. Additive manufacturing technology belongs to "accumulation" molding, the basic principle is a bit like the printing process of inkjet printers, laying the material on the working platform, only that the molding process needs to be laid layer by layer to form the workpiece, and its processing first generates a model with 3D software, slices the model, makes the model into multiple layers, then layer by layer The process starts by generating a model with 3D software, slicing the model into multiple layers, and then stacking them layer by layer to form the molded part, and then post-processing the finished product [2]. Fused deposition modeling (FDM) is one of the most used and widely used low-cost processes in 3D printing because it is easy to use, has inexpensive machines, and takes advantage of the hot-melt and adhesive properties of thermoplastic materials. Invented by Scott Crump in 1988, this printing technology is based on the principle of extruding a filamentary thermoplastic material from a heated nozzle and depositing the melt layer by layer according to a predetermined trajectory and rate to achieve three-dimensional molding [3, 4]. The most commonly used material for FDM is Acrylonitrile–Butadiene–Styrene (ABS), followed by Polycarbonate (PC), Polylactic acid (PLA), Polyphenylsulfone (PPSF), Ultem9085, and their blends [5]. After years of continuous development, this type of additive manufacturing technology currently occupies about 60% of the 3D printer market in the market, and is one of the most promising printing methods, FDM printers also have the advantages of high printing efficiency and no odor and no pollution, and small deformation after molding, which makes the users of FDM are increasing year by year, making it a broad commercial value [6]. However, at present, the actual application of 3D printing process, due to the technology itself is not mature enough, resulting in its practical application, it appears to be a complex process, low efficiency, in the actual application often can not play the expected effect [7]. With the popularity of printers and the gradual increase of users, how to print high-quality 3D printed parts to improve the printing efficiency has become one of the important issues considered by users and developers. In order to solve the problems of 3D printing process and achieve a seamless workflow from design to 3D printing, Siemens provides a solution to solve the possible problems of 3D printing process by using the rich features of the software and the integrated Materialize processor to achieve a seamless workflow from design to 3D printing, in addition to supporting powder layer fusion bed. The software also supports the optimization of 3D printing processes such as FDM [8], and the specific simulation optimization process is shown in Fig. 18.1, and this paper focuses on the optimization of the FDM process.

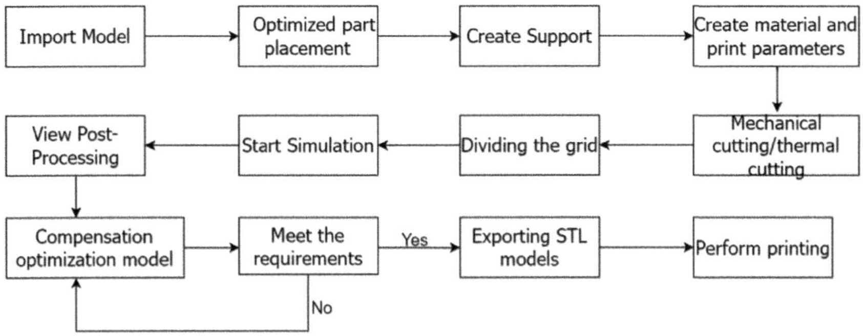

Fig. 18.1 Simulation optimization flow chart

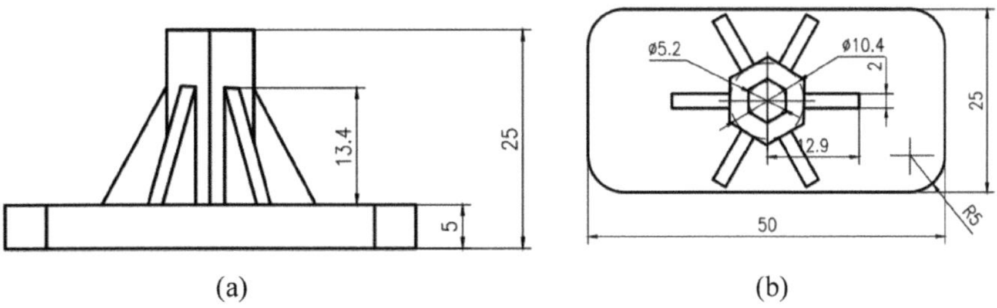

Fig. 18.2 Model size

18.2 Model Creation and Optimization of Print Orientation

3D Printing Model Creation

The specific shape and size of the flange parts required for the 3D printing simulation optimization process established by UG are shown in Fig. 18.2.

Optimization of Part Print Orientation

The model is imported into the additive manufacturing module, and the component optimization function is used to optimize the printing position of the part. select the fused deposition modeling (FDM) printer in the parts optimization interface, set the maximum extension angle to 45°. After the analysis of the surface area, support volume, print time and other related data comparison, the direction shown in Fig. 18.3 for the printing direction when the printing time and printing support materials spent to achieve the minimum value, can achieve the purpose of saving time and reducing costs, so this direction selected as the final printing direction.

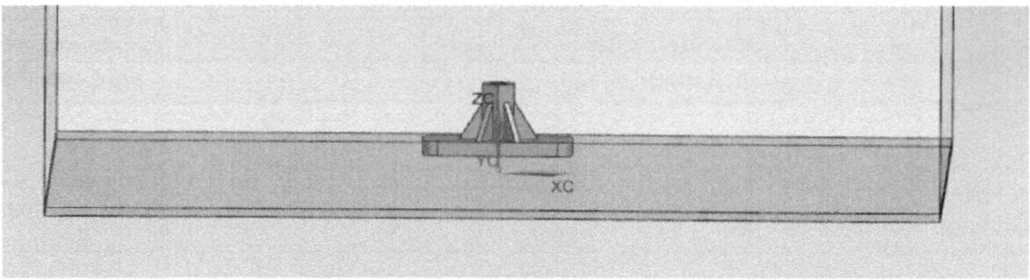

(a)

(b)

Fig. 18.3 Final position of the part and related parameters **a** Final position of the part **b** Related date

18.3 Pre-processing Related Parameter Settings

Printing Parameters and Material Settings

After determining the best print position, in order to ensure the accuracy of the simulation, parameters such as print speed, nozzle temperature, layer thickness, material need to be defined, various parameters configurations are shown in Table 18.1.

When setting the material, it is necessary to define various parameters of ABS. The mass density is set to 1.05e-06 kg/mm^3, Poisson's ratio is set to 0.38, specific heat is set to 1.8×10^9 μJ/(kg·K), and the thermal expansion coefficient is 7e-05 °C-^{1}. The thermal conductivity is set to 0.2256 W/mK [9].

Table 18.1 Table of parameters

Project	Parameter value
Simulation type	Thermodynamic simulation
Nozzle temperature (°C)	280
Print speed (mm/s)	100
Layer thickness (mm)	9.2
Material	ABS plastic

Fig. 18.4 Mechanical slicing and thermal slicing

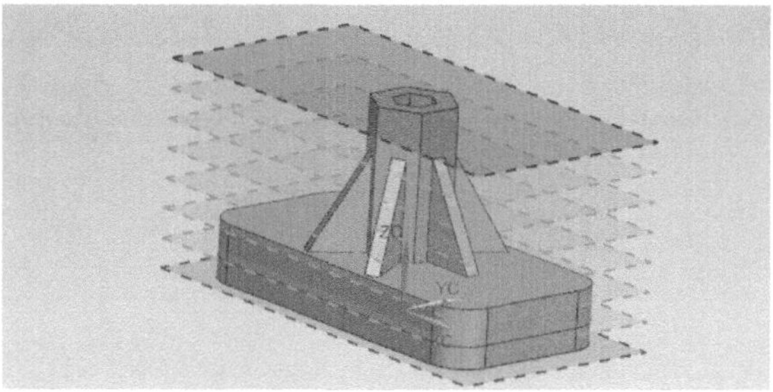

(a) Mechanical slicing

(b) Thermal slicing

Definition of Mechanical Section, Thermal Section and Grid

Before the simulation, in order to calculate the temperature field and possible deformation during the printing process, the mechanical slices and thermal slices shown in Fig. 18.4 were first created for the model.

After defining the slices, divide the grid and set the grid size to 0.5 mm. The 3D grid model is shown in Fig. 18.5. After meshing, determine the required calculation options, and check the calculation of thermal, mechanical and stiffness, Then the simulation is started.

18.4 3D Printing Process Simulation Results and Analysis

Temperature Field Distribution

The fused deposition molding process is mainly a process in which the material goes from the solid state to the molten state and back again and is accompanied by the spatial transfer of the material, and thermal analysis is one of the main research problems in FDM because of the phase change involved [9]. Figure 18.6 shows the temperature field during the 3D printing process, from which it can be seen that the

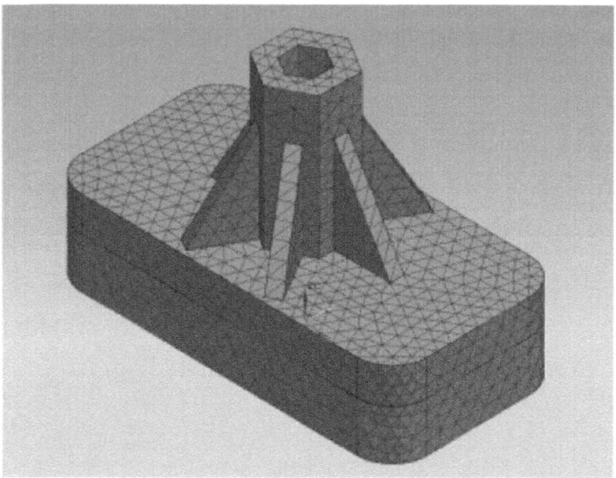

Fig. 18.5. 3D meshing

high temperature is concentrated at the top position of the hexagonal prism, where the highest temperature is about 774 °C. From the top surface to take a point to view the general temperature change of the surface can be concluded that the surface temperature changes between about 753 °C to 766 °C, and then respectively along the X axis and Y axis for the profile can be seen that the internal temperature distribution of the part is not optimistic, high temperature is concentrated in the upper part of the flange parts hexagonal, which can be inferred that too high temperature is likely to have an impact on the finished 3D printing.

Model Deformation Situation

The quality problems that tend to arise in the molded part during fused deposition molding 3D printing process include problems or defects such as too much surface roughness, low precision, and low strength. When these faults or defects arise during the process, the shrinkage deformation process is caused by the change of the structural form inside the material, especially the temperature field and stress inhomogeneity caused by the molten phase change of the material [9]. The deformation of the model after simulation is shown in Fig. 18.7, where the left side of the figure shows the deformation after the end of printing and the right side shows the deformation after the removal of the cooling support, from which it can be seen that the deformation after printing is distributed on both sides of the bottom side of the model with a maximum deformation of 0.1179 mm, and the maximum deformation after the removal of the cooling support occurs at the top of the hexagonal prism with a maximum deformation of 0.302 mm.

Through the analysis of the temperature field and deformation, we can see that the upper part of the flange model has high temperature and deformation, so we need to compensate for the deformation to improve the printing accuracy.

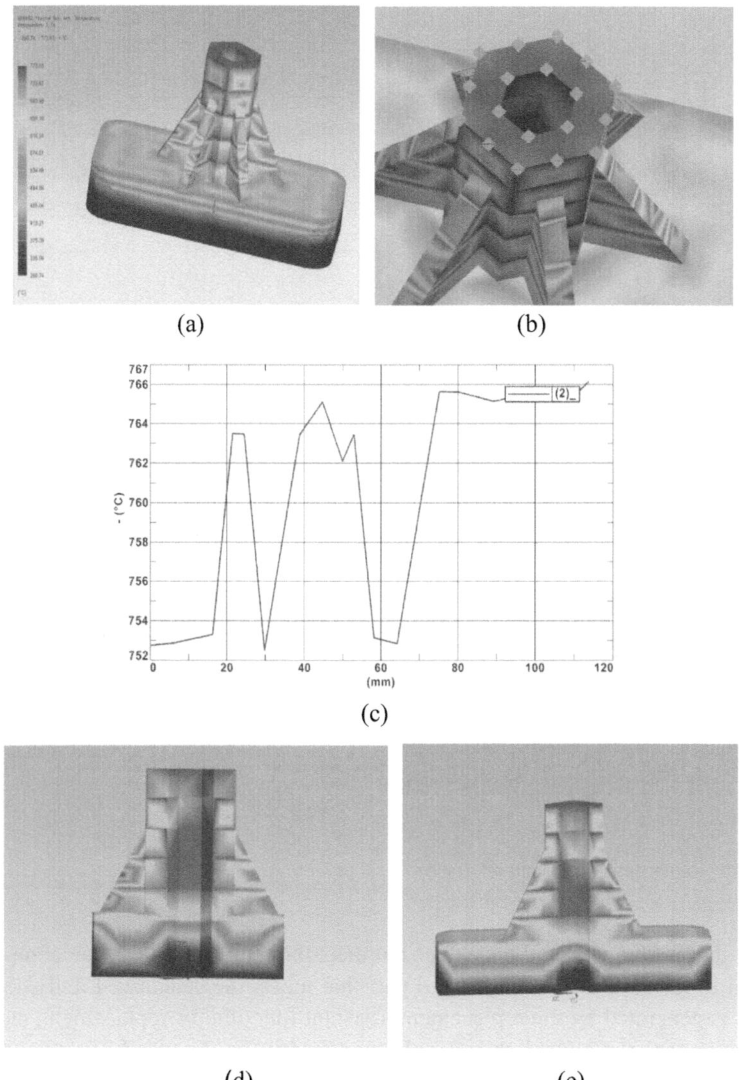

Fig. 18.6 Model simulation temperature distribution. **a** Overall temperature field distribution **b** Selection of surface **c** Graph of temperature variation along the path **d** Internal temperature distribution along the X-axis cut **e** Internal temperature distribution along the Y-axis cut

Compensation Optimization of the Model

The UG compensate model function can compensate the high temperature and deformation area of the model based on the results of the simulation of the original model to reduce the actual printing error and get a more suitable 3D printed part. If the

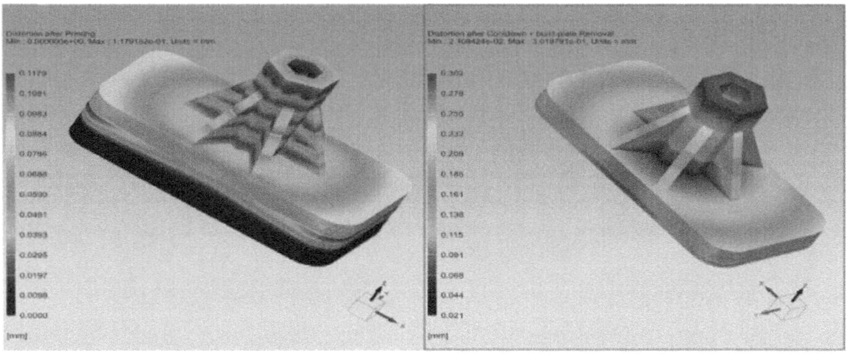

Fig. 18.7 Distribution of deformation cases

Fig. 18.8 Replacement assistant interface

model that is compensated once does not meet the requirements, it can compensate several times until find a 3D printed part that meets the standard. Firstly, the new model is generated by the replacement assistant function for replacement, and then the new model is matched with the old model entity, surface and line respectively, and all 114 edges, 41 faces and 1 entity of the part are matched to generate the new compensation model as shown in Fig. 18.8.

Optimized Recalculation

The model after compensation optimization is re-imported to the simulation module for simulation, and the simulation is performed again after inputting the same parameters, the simulation results are shown in Fig. 18.9.

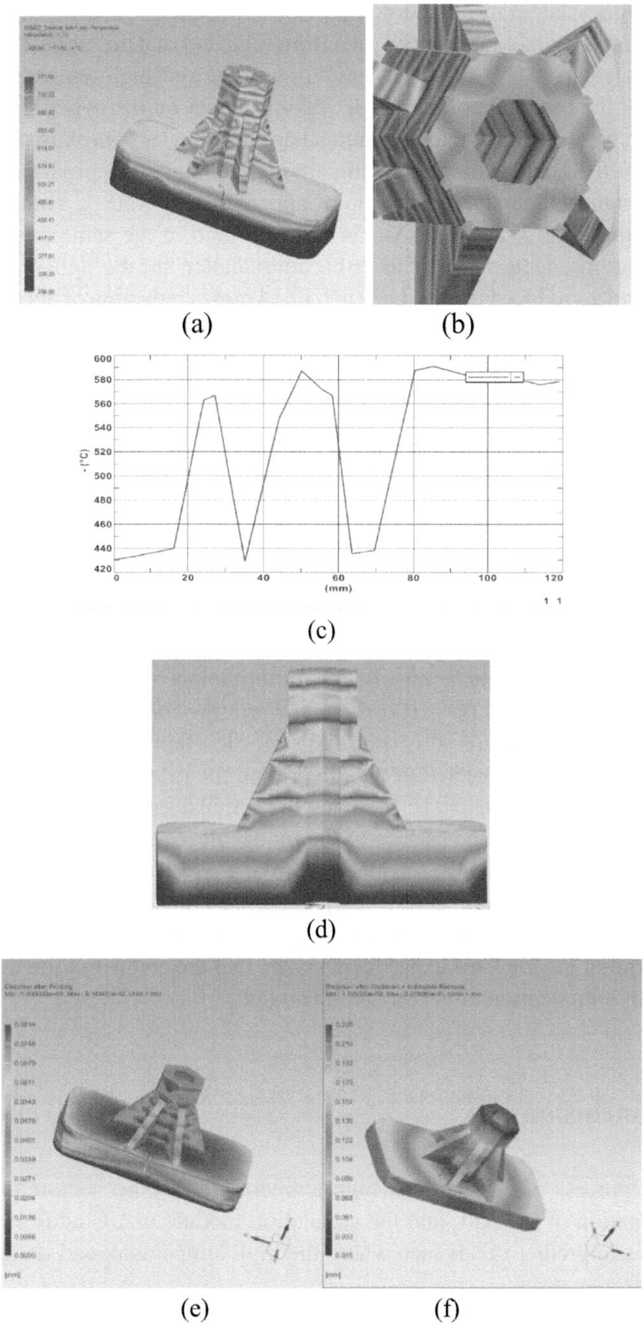

Fig. 18.9 Optimized model post-processing situation **a** Temperature field distribution **b** Surface selection **c** Graph of temperature variation along the path **d** Internal temperature distribution along the X-axis cut **e** Internal temperature distribution along the Y-axis cut **f** Distribution of deformation cases

According to the analysis and comparison with the original model before the compensation deformation, it can be seen from Fig. 18.9 that the highest temperature field is reduced from 774 °C to 771 °C°C, and the red high temperature area is improved by the external color change. Taking points on the upper surface of the hexagon again shows that the temperature change fluctuates approximately between 430 °C and 590 °C, with a minimum temperature reduction of approximately 300 °C and a maximum temperature reduction of approximately 150 °C compared to the original model. The X-axis and Y-axis were profiled at the same position as the initial simulation, and the red area inside became smaller, and the highest deformation after printing was reduced from 0.1179 to 0.0814 mm, a reduction of about one-third. The deformation after the cooling support was removed was reduced from 0.302 to 0.228 mm, a reduction of about one-fifth. The effect of this compensation is to reduce the temperature and model deformation.

Part Actual Printing Comparison Verification

After getting the model after compensation and optimization, export the STL model file for printing, at this time the exported STL model file is divided into the original model and the model after compensation deformation, print and compare the original model and the model after compensation deformation, the printer is selected as shown in Fig. 18.10 UP BOX + 3D printer, the printing software is selected with the printer UP Studio, the material is used ABS plastic Then the model was imported into the software and connected to the 3D printer, the printing results are shown in Fig. 18.11, from the figure can be seen that the original model printed on the upper part of the obvious paste, the surface of the model after compensation optimization is not obvious paste, it can also see that the top of the two models are deformed, but the initial model deformation is obviously larger than the model after compensation optimization, the left side of the figure is the model after compensation, the right side is the initial model. This experiment shows that the compensation deformation has a certain optimization effect on the 3D model.

18.5 Conclusion

(1) In the process of 3D printing, high temperature or other factors may lead to deformation of the part, and the simulation module of UG additive manufacturing can predict in advance where the high temperature and deformation of the part may occur.
(2) By comparing the simulation results of the initial model and the model after compensating for the deformation, it was found that after the optimization of the original model, the area of the high temperature area was significantly reduced, and the cooling of the internal area and the top surface was especially obvious,

Fig. 18.10 UP BOX + 3D printer

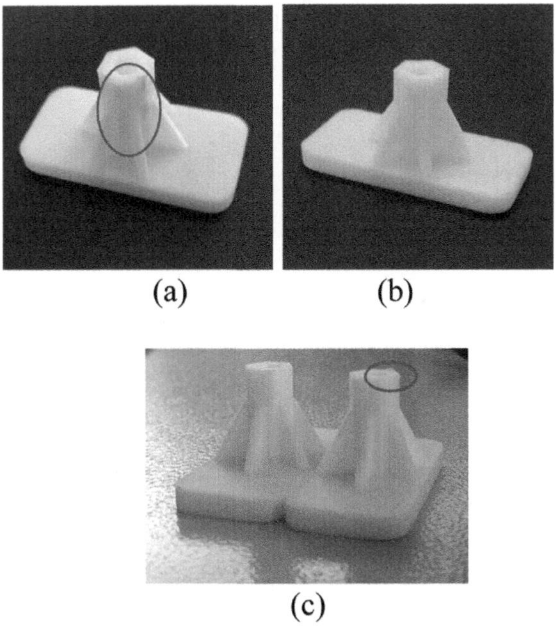

Fig. 18.11 Comparison of printing results **a** Original part **b** Optimized part **c** Top deformation comparison

in which the area of the high temperature in the internal area was reduced by more than one-half, and the temperature change range of the top surface was reduced from 753 °C to 766 °C to 430 °C to 590 °C,The highest deformation after printing without cooling was reduced from 0.1179 to 0.0814 mm, a reduction of about one-third, and the deformation after cooling support removal was reduced from 0.302 to 0.228 mm, a reduction of about one-fifth, which shows that the model after compensating for deformation has a certain effect on improving the quality of 3D printed products.

(3) By comparing the actual 3D printing of the original part and the compensated optimized part, the results show that the top of the original part has obvious paste and deformation, while the optimized part does not have obvious such condition on the outside. It can be concluded that the temperature field can be reduced and the surface quality of the component can be improved with compensating optimization of the part. The top distortion of the model without optimization is noticeably higher than that of the optimized model after printing, as can be shown by analysis of the deformation results at the top of the model before and after optimization compensation. and it can be seen that the part after compensating for the deformation is more suitable for 3D printing.

(4) The method used in this paper places more emphasis on the compensation optimization of the structure of the three-dimensional model than most traditional methods used to improve the production quality of the product during the 3D printing molding process. The technology can be taken into consideration in the ensuing development process combined with the simulation of process parameters to simulate and optimize the 3D printing modeling process, from more levels to consider the possible problems in the 3D printing modeling process and analyze and solve

References

1. Lu, N., Zheng, J., Zhao, X.: Temperature field analysis and optimization of the heating platform of FDM rapid prototyping machine. China Mech. Eng. **30**(01), 90–95 (2019)
2. Wang, Y.N.: Study on the influence of hot bed temperature field on the accuracy of forming parts based on FDM. Inner Mongolia University of Science & Technology (2020)
3. Liu, Y.Z.J., Xia, C.L., Zhang, J., Jiang, Z.Z.: Application progress and prospect of fused deposition molding 3D–printing. Eng. Plast. Appl. **45**(3), 130–133 (2017)
4. Zhang, X., Fan, W., Liu, T.: Fused deposition modeling 3D printing of polyamide-based composites and its applications. Compos. Commun. **21**, 100413 (2020)
5. Dudek, P.: FDM 3D printing technology in manufacturing composite elements. Arch. Metall. Mater. **58**(4), 1415–1418 (2013)
6. Tianmin, G., Li, Y., Zhang, Y.: Study on mechanical properties of material formed with fused deposition molding technology. China Plast. **35**(6), 68 (2021)
7. Tian, Y., et al.: 3D printing remanufacturing current problems and countermeasures analysis. Technol. Innovat. Appl. (15), 51 (2015)
8. NX Additive Manufacturing Solutions. https://file.bagevent.com/reso-urce/20190220/162036622827473.pdf
9. Zhang, B.Q., Meng, F.Y., Pan, J.C., Wang, R.D., Yu, S.Y.: Simulation and research on FDM molding process based on thermal-force coupling. Manuf. Automat. **39**(07), 39–42 (2017)

Chapter 19
The Application of 3D Laser Scanning Technology in Land Surface Subsidence Monitoring

Chao Tang, Jinhong Rong, Yinggang Guo, Wenjiang Cai, Alimusu Liang, Jiaze Xu, WeiWei Zhu, Xirong Tao, and Junying Feng

Abstract Due to certain human and natural factors, land surface subsidence and collapse phenomena have emerged endlessly, and the demand for periodic monitoring of the land surface is becoming increasingly urgent. This article focuses on the needs and characteristics of land surface subsidence monitoring, and uses three-dimensional (3D) laser scanning technology to obtain point cloud data of different periods on the land surface of a certain area. The M3C2 (Multi scale Model to Model Cloud Comparison) algorithm is used to compare and detect point cloud data to extract land surface subsidence data, and the precision of this method is evaluated. Experiments show that the measurement precision of this method is within 4 mm, which provides a useful reference for the application of 3D laser scanning technology in deformation monitoring.

19.1 Introduction

Due to various factors such as mineral development, excessive groundwater consumption, and changes in human or natural factors, land surface subsidence and even collapse have occurred frequently in recent years [1], resulting in significant property losses and posing a threat to the safety of people's lives. Therefore, land surface subsidence monitoring has become an important task in current social life, and the workload of this task is increasing year by year, with increasingly high precision requirements [2].

Land surface subsidence monitoring belongs to the category of deformation monitoring. The traditional deformation monitoring scheme mostly uses conventional surveying equipment such as total station and level, which is simple and effective and has high precision of monitoring results, but it is inefficient, time-consuming

C. Tang (✉) · J. Rong · Y. Guo · W. Cai · A. Liang · J. Xu · W. Zhu · X. Tao · J. Feng
Northwest Institute of Nuclear Technology, Xi'an, China
e-mail: tangchao10@163.com

R. Kountchev (Deceased) et al. (eds.), *AI Methods and Applications in 3D Technologies*, Smart Innovation, Systems and Technologies 388,
https://doi.org/10.1007/978-981-97-2144-3_19

and laborious to obtain the surface information in time [3]. In order to make up for the shortcomings of traditional monitoring methods, modern monitoring equipment and methods are constantly emerging, such as three-dimensional (3D) laser scanning technology, measurement robots, InSAR, etc. [4]. The emergence of such new technologies has effectively changed the limitations of traditional monitoring methods, improved monitoring efficiency, and reduced human and material resources.

In order to study the applicability and accuracy of 3D laser scanning technology in land surface subsidence monitoring, we analyzed the precision of 3D laser scanning system in the monitoring process based on the error propagation law [5], and designed relevant experiments to fully verify it, so that 3D laser scanning technology can be used to timely obtain the information of land surface subsidence, thus providing a reference for deformation monitoring.

19.2 Basic Principles and Methods

3D laser scanning technology mainly relies on 3D laser scanners for measurement. Using the principle of laser ranging, by recording a large number of dense points on the surface of the measured object, including 3D coordinates, reflectivity, and texture information, the 3D model of the measured object and various graphic data such as lines, surfaces, and volumes can be quickly reconstructed [6]. The monitoring surface of the research area is scanned by using this technology to build a multi-period model, which reflects the trend of overall settlement and deformation and displacement of the research area [7]. The detailed process is shown in Fig. 19.1.

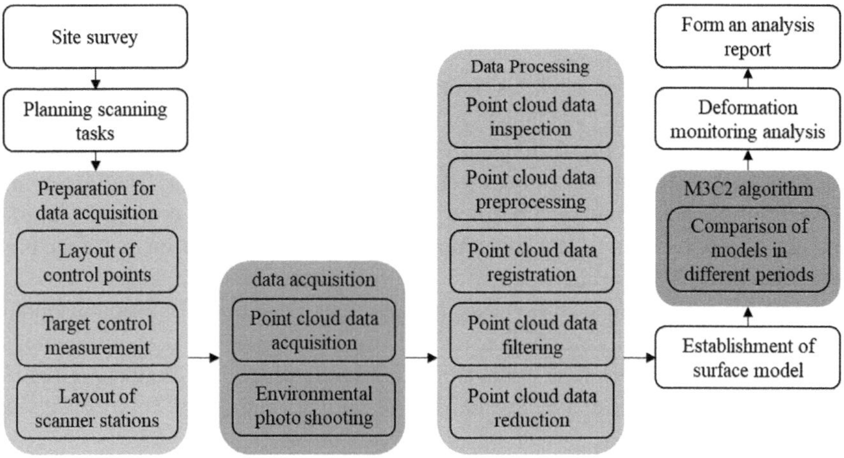

Fig. 19.1 The process of surface subsidence monitoring using 3D laser scanning technology

As shown in Fig. 19.1, the first step is to conduct a site survey of the survey area to plan the scanning task, clarify the specific scanning method, scanning resolution, etc. The second step is data acquisition preparation, including the layout of scanning control points, measurement of target control points, and layout of scanner stations. The third step is data acquisition, mainly including the acquisition of point cloud data and environmental photos. The fourth step is data processing, which includes operations such as such as inspection, preprocessing, registration, filtering, and reduction of scanned data. The fifth step is to establish a 3D model to compare the data in different periods, obtain the deformation monitoring data, evaluate the data precision, and finally form an analysis report.

In this study, we used the M3C2 (Multi scale Model to Model Cloud Comparison) algorithm for comparative detection of two-period point clouds [8]. This method can avoid mesh or grid processing of point cloud data, and the algorithm has good robustness against data loss or uneven point cloud density. The core idea of the algorithm is as follows:

(1) Select the reference point cloud data to calculate core points. Core points are essentially a subset of reference point cloud data, calculated by setting the minimum interval of a point cloud. Core points have lower spatial resolution and more uniform distribution compared to the original reference point cloud data. Its advantage is that it can accelerate calculation speed and make the calculation results easier to transform into grid form.
(2) Calculate the surface normal of 3D space. For a given core point i, first search for neighborhood points within the radius of D/2, name these points set N, and then fit all neighborhood points within set N into an optimal plane, obtaining the normal of that plane. Among them, the standard deviation of the distance from each point in set N to the fitting plane is used as an indicator to measure the roughness near the core point i.
(3) Calculation of the distance between two-period point clouds. After determining the normal at core point i, define a cylinder with a radius of d/2, whose axis passes through core point i and whose direction is consistent with the normal at core point i. The defined cylinder will intersect with two-period point clouds, generating two subsets of point cloud numbers N_1 and N_2, respectively. By projecting the points in each subset onto the axis of the defined cylinder, the distribution of two projection points on the axis of the cylinder can be obtained. The mean positions of the two projection point distributions are denoted as the average positions i_1 and i_2 of the normal directions of the two-period point clouds at the core point i, with their standard deviations $\sigma_1(d)$ and $\sigma_2(d)$ will be used as a local estimate of the normal direction roughness of two-period point clouds at the core point i. The distance between i_1 and i_2 is denoted as the local value L(i) of the distance between the two-period point clouds.

Among them, D can be denoted as the normal scale, and d can be denoted as the projection scale. The value of D will affect the normal direction and also affect the roughness of the reference point cloud at the core point i. The value of d will affect the number of point clouds participating in the calculation of the average position, thereby affecting the local value L(i) of the distance between the two-period point clouds, and is related to the roughness of the comparison point cloud in the normal direction at the core point i.

The steps (1) and (2) are shown in Fig. 19.2, where $\overrightarrow{N}$ is the normal vector of the fitting plane at the core point i, σ_1(D) represents the roughness of the core point i with a radius of D.

The step (3) is shown in Fig. 19.3, where i_1 and i_2 represent the average positions of the two-period point clouds at the core point i, σ_1(d) and σ_2(d) represents the roughness estimation of two-period point clouds with a radius of d at the core point i, while L_{M3C2}(i) represents the distance between the two-period point clouds at the core point i.

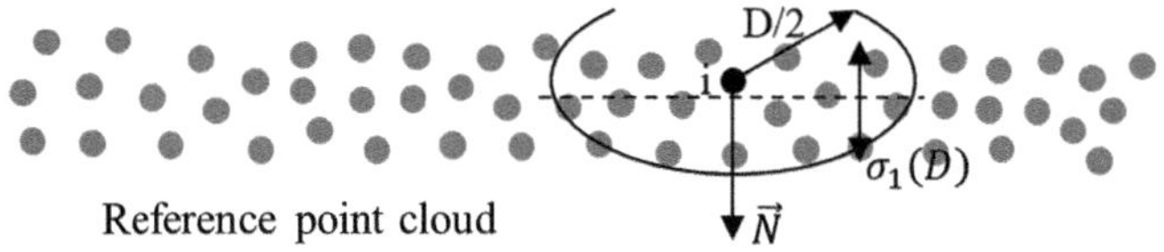

Fig. 19.2 The schematic diagram of M3C2 algorithm core steps 1 and 2

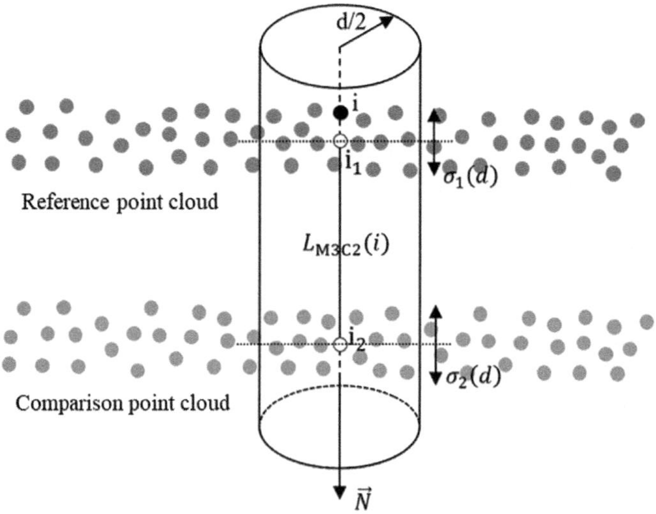

Fig. 19.3 The schematic diagram of M3C2 algorithm core step 3

Fig. 19.4 Establishment of control network and scanning implementation

19.3 Analysis of Surface Subsidence in a Certain Area

Establishment of Surface Model

In order to achieve the prediction research of surface subsidence rate in the target area, deformation monitoring is carried out by periodically collecting surface point cloud data, with a sampling interval of 1 year. The target area has a range of 20 m × 30 m, the monitoring object is the Yardang landform. Establish a high-precision control network through an industrial measurement system, set up a mandatory observation pier, as shown in Fig. 19.4, and conduct periodic scanning of the target area. The working site, point cloud data, and model as shown in Fig. 19.5.

Surface Subsidence Analysis

By placing the two-period models in the same coordinate system, the surface morphology of the two-period models can be compared and the elevation difference between the two-period data can be calculated to obtain the settlement value. Thus, it is possible to visually observe and analyze the deformation situation. In this study, we compared and analyzed the scanning data of the first period (data obtained in March 2019) and the second period (data obtained in April 2020) of the target area based on the M3C2 algorithm, as shown in Figs. 19.6 and 19.7. In our result, the first period is used as a reference object, and compared with the point cloud data of the second period, the blue area represents surface subsidence, the red area represents surface uplift, and the white area represents insignificant surface changes. We conducted statistical calculations on the data of the difference histogram and found that the main range of surface changes in the target area between the first and second periods was −0.06 to 0.02 m. Among them, the average annual settlement rate of the settlement part is about 1.27 cm.

a. Target area

b. Target area scanning point cloud

c. Target area model

Fig. 19.5 Research on surface subsidence monitoring model

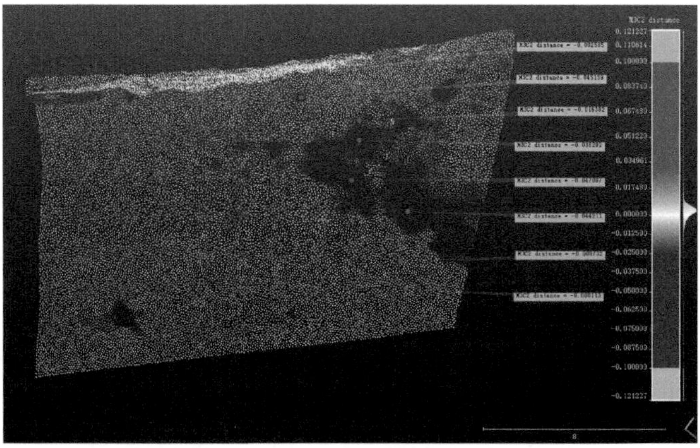

Fig. 19.6 The comparison results of two-period point clouds in the target area

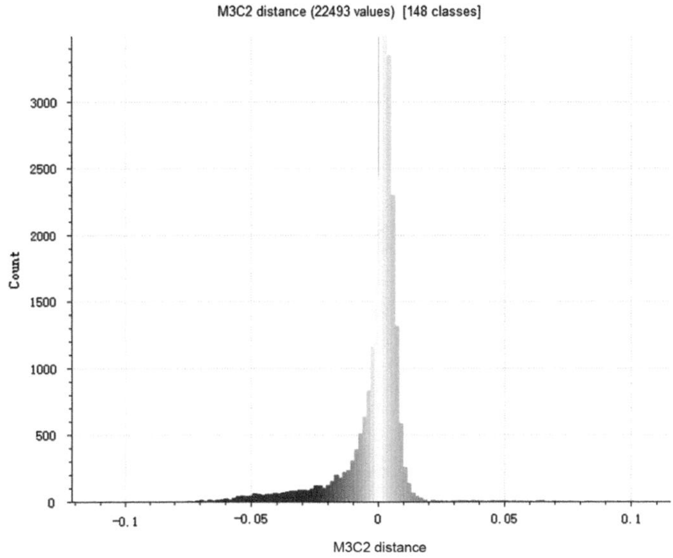

Fig. 19.7 The difference histogram of two-period point clouds in the target area (unit: m)

Precision Evaluation

In this study, we analyzed the precision of the 3D laser scanning method by analyzing the error sources and related error analysis in the process of surface subsidence monitoring. The error sources of 3D laser scanning method mainly include five items: instrument system error m_i, control network measurement error m_n, point cloud acquisition error m_a, point cloud registration error m_r, and point cloud comparison error m_c. Among them, the instrument system error m_i is mainly determined as 2 mm by the equipment calibration error, and the control network measurement error m_n is determined as 0.1 mm by the plane control precision limit.

The point cloud acquisition error is measured by the laboratory industrial measurement system [9]. In this study, we established a high-precision control network in the laboratory using an industrial measurement system, with a precision of 0.1 mm, as shown in Fig. 19.8. We measured 5 black-and-white targets on the wall using the established control network, and obtained the center coordinates of the 5 black-and-white targets using point cloud data obtained from the Trimble TX8 laser scanner. The error range of point cloud acquisition was obtained by comparing the pre and post data, as shown in Table 19.1. According to Table 19.1, the point cloud acquisition error m_a is within 2 mm.

The point cloud registration error is provided by Trimble RealWorks software [10]. The point clouds scanned by two stations are registered by extracting black-and-white targets, and the registration error report is shown in Table 19.2.

The comparison error of point clouds is measured through experiments. In this study, based on the control variable method, we scanned a wall twice in the laboratory

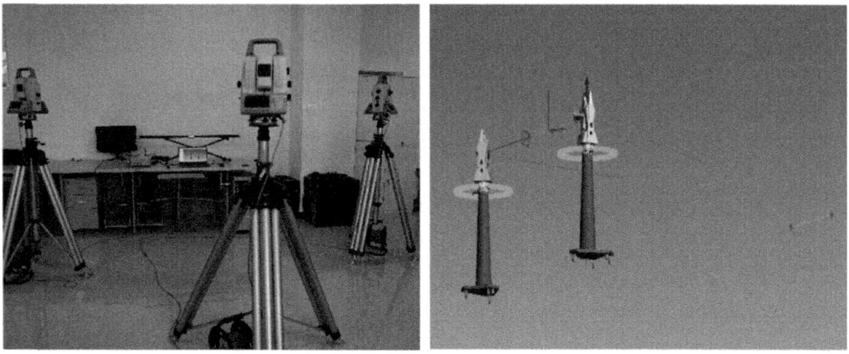

Fig. 19.8 Establishment of high-precision control network in the laboratory

Table 19.1 Precision evaluation of black-and-white target coordinates (unit: mm)

Name	Coordinates of control network measurement			Coordinates obtained from scanning point cloud data			Spatial distance
	X	Y	Z	X	Y	Z	
P1	−4636.07	−1598.74	−203.28	−4637.52	−1599.40	−203.75	1.66
P2	−3600.30	1977.64	996.04	−3600.37	1978.57	996.35	0.98
P3	−1365.92	2545.90	−170.56	−1365.36	2546.14	−170.61	0.61
P4	5697.99	−3832.15	864.91	5699.53	−3832.18	864.90	1.54
P5	−1634.35	−3896.70	1247.71	−1634.94	−3897.16	1247.93	0.78

Table 19.2 Registration error report for extracting black-and-white targets (unit: mm)

Target	Registration residual	Average error
P1	0.53	0.75
P2	0.60	
P3	0.69	
P4	0.76	
P5	1.16	

environment to obtain the comparison point clouds, and compared two point clouds through the M3C2 algorithm to obtain the algorithm detection precision. The selected random point comparison precision is shown in Table 19.3.

According to the law of error propagation, the error of the 3D laser scanning method is determined as m_s:

$$m_s = \pm\sqrt{m_i^2 + m_n^2 + m_a^2 + m_r^2 + m_c^2} \approx \pm 3.49\,\text{mm}$$

Table 19.3 Comparison precision of randomly selected points (unit: mm)

Serial number	Comparison error	Average error
1	2.02	1.90
2	1.81	
3	1.80	
4	1.97	
5	2.00	
6	1.75	
7	1.96	

where m_i denotes the instrument system error, m_n denotes the control network measurement error, m_a denotes the point cloud acquisition error, m_r denotes the point cloud registration error, and m_c denotes the point cloud comparison error.

19.4 Summary

The 3D laser scanning technology can achieve a comprehensive, efficient and accurate point cloud data acquisition in the target area. Compared with traditional monitoring methods, the data acquisition speed is faster and the data is more complete. By comparing the scanned data in different periods, this method can reflect the subsidence of the target area, and the measurement error is controlled within a reasonable range, which has strong practicability and reliability. In practical engineering applications, we should pay attention to the data errors introduced by data acquisition, point cloud registration and point cloud comparison processes to meet the needs of high-precision deformation monitoring.

References

1. Li, X.B., Lin, Z.J., Yang, X., et al.: Research on 3D real scene fusion modeling based on laser scanning and oblique photography (in Chinese). Laser J. **42**(8), 166–170 (2021)
2. Wang, J., Cai, Y.S., Peng, Z., et al.: Real time monitoring and early warning of land subsidence based on 3D laser scanning technology (in Chinese). Nonferrous Metals (Mining Section) **70**(6), 12–15 (2018)
3. Wang, L., Zhou, J.W., Wang, H.S., et al.: Deformation monitoring of ancient towers based on 3D laser scanning technology (in Chinese). Bull. Surv. Mapp. **7**, 120–124,129 (2020)
4. Zhu, S.G., Nie, S.G.: Application research of 3D Laser scanning technology in building deformation monitoring (in Chinese). Value Eng. **37**(31), 259–261 (2018)
5. Wang, H.Y., Wang, R.H.: Application of error theory in 3D laser scanning of point cloud data processing (in Chinese). Geosp. Inf. **17**(10), 45–48 (2019)
6. Xu, H., Zhang, W.C., Lin, N.: Research on the application of 3 D laser scanning technology in the construction of 3D modeling of underground building (in Chinese). Urban Geotech. Invest. Surv. **6**, 82–85 (2015)

7. Chen, Q.M., Huang, F.M., Song, P.: Application of UAV and 3D laser scanner technology in beach profile dynamic monitoring and analysis of measuring accuracy (in Chinese). J. Appl. Oceanogr. **40**(3), 529–539 (2021)
8. Hou, T.C., Liu, Y.W., Su, Y.M.: Geometric identification and damage detection of structural elements by terrestrial laser scanner. International Society for Optics and Photonics (2016)
9. Li, G.Y., Li, Z.C.: Principles and Applications of Industrial Measurement Systems (in Chinese). Surveying and Mapping Publishing House (2011)
10. Peng, L., Huang, Z.H., Cheng, P., et al.: The application of trimble realworks in True 3D modeling of point cloud data. Bull. Surv. Mapp. **6**, 155–156 (2017)

Chapter 20
Gesture Recognition in Complex Background Based on Fusion of Hog and Faster R-CNN

Yumin Yang, Shuobo Xu, Kefeng Li, Keming Li, Jindi Li, and Jiaqi Wang

Abstract Gesture recognition in simple backgrounds performs well, but it is still challenging in complex background. Because the gesture is indistinguishable from the surrounding background, gesture recognition becomes difficult. To reduce the impact of complex backgrounds on gesture recognition, we decided to adopt target detection to extract gestures area following by gesture recognition. So, we propose a gesture recognition network based on fusion of Hog and Faster R-CNN. By introducing HOG-SVM module, we fine-tune the loss value of the network to enhance the accuracy of gestures. Then, SE (Squeeze and Excitation) spatial attention mechanism is introduced to improve the accuracy of prediction bounding box. The results of the experiment show that the accuracy can reach 95.8% on PostHand dataset, which is 2.4% higher than Faster R-CNN.

20.1 Introduction

Human gesture recognition is considered to be a way for computer to understand human behavior. For gesture recognition in simple background, the traditional methods can achieve high accuracy, but it don't perform well when the gestures in complex background. Deep learning has experienced rapid growth, and it has been introduced in gesture recognition to get good performance. Recognizing gestures accurately in complex backgrounds is still a difficult task, as it involves dealing with diverse gestures and complex scenes. Despite the progress made in gesture recognition technology, there is still room for improvement in this area.

Gesture techniques based on computer vision [1–14] can be classified into two ways: traditional and deep learning methods. In some traditional methods [1, 2], features of the hand, like skin color or texture are extracted and sent into a classifier for recognition. As other non- targets of gesture images, it is not conducive to extracting

Y. Yang · S. Xu · K. Li (✉) · K. Li · J. Li · J. Wang
School of Information Science and Electric Engineering, Shandong Jiaotong University, Jinan, China
e-mail: seafrog1984@hotmail.com

R. Kountchev (Deceased) et al. (eds.), *AI Methods and Applications in 3D Technologies*, Smart Innovation, Systems and Technologies 388,
https://doi.org/10.1007/978-981-97-2144-3_20

hand contour features, so traditional gesture detection methods entail the problems of low accuracy. The method utilizes convolutional neural networks to extract image features, that has stronger robustness and can better extract hand features. Therefore, the current mainstream gesture recognition method is deep learning method.

Many methods have been applied to gesture recognition in various computer vision tasks, including image classification and object detection based on deep learning. In image classification, there are MobileNet [5], EfficientNet [6] and other models. For gesture recognition of simple background, the traditional way is beyond the deep learning method with very little data. For gesture recognition under complex background, the deep learning approach has proven to be more effective than traditional methods. However, the background of the gesture can significantly impact the ability of gesture recognition, and the complex background is not conducive to the feature extraction of the classification model, and the accuracy is lower than the gesture recognition under simple background. Therefore, we decided to adopt the object detection method, firstly extract the gesture, and then recognize it, to mitigate the impact of complex background on recognition. In object detection tasks, such models as RCNN [7], Fast RCNN [8], Faster R-CNN [9] and so on, however, these models have the problem of inaccurate gesture detection and recognition. As illustrated in Fig. 20.1 shows the detection results of Faster R-CNN. It is evident that the model recognizes "one" as "thumb Up".

To address this challenge, we designed a fusion method based on HOG and Faster R-CNN to detect and recognize gestures in a complex background more accurately. Our contributions can be classified into two categories:

- We propose a fusion way of HOG and Faster R-CNN to reduce impact of complex background on gesture recognition, thereby enhancing the accuracy of network recognition.
- By adding attention mechanism, the accuracy of gesture detection is improved.

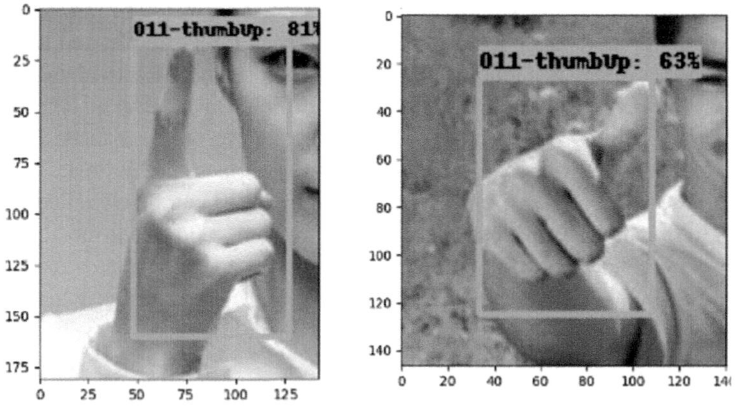

Fig. 20.1 Recognition result based on Faster R-CNN

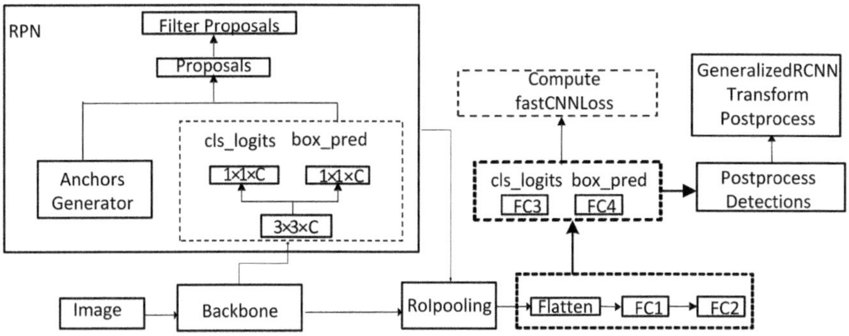

Fig. 20.2 Faster R-CNN

20.2 Related Work

Faster R-CNN

As illustrated in Fig. 20.2, the Faster R-CNN target detection model consists of three main steps. (1) The first step of the Faster R-CNN target detection model involves sending the input image to a feature extraction network to obtain a feature map. (2) Next, the Region Proposal Network (RPN) is utilized to generate candidate boxes, which are then placed into the feature map to obtain the corresponding feature matrix. (3) Finally, the feature matrix is passed through the RoI pooling layer to obtain a 7*7 feature map, which is then fed into the fully connected layer to obtain the prediction results.

Attention Mechanism

SE Attention mechanism [10] is a significant approach to improve network capabilities. SE attention, as shown in Fig. 20.3, designs a channel attention mechanism, effectively establishes the channel correlation between features, and improves the performance of convolution neural network.

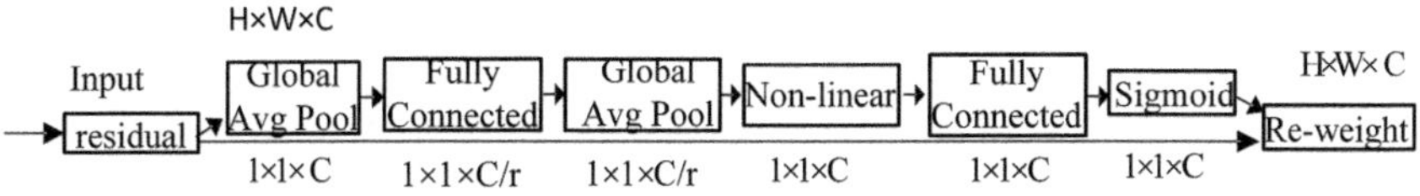

Fig. 20.3 SE attention

Hog

HOG (Histogram of Oriented Gradients) feature detection algorithm [11] is a descriptor used to characterize images' local gradient direction and gradient intensity distribution characteristics. The fundamental concept is that the local objects' shape and representation in an image can be accurately depicted by the distribution of edges' gradient or directional density.

20.3 Fusion of Hog and Faster R-CNN

Hog-Faster R-CNN Gesture Recognition Network

Figure 20.4 depicts the overall architecture of the adopted network which includes Faster R-CNN framework, HOG-SVM, and SE attention. First, generate a feature map. The feature map goes through the SE attention module and then enters the RPN part of Faster R-CNN. RPN divides the generated proposals into foreground and background. After that, the foreground is fed into RoI pooling layer and HOG-SVM module. In RoI Pooling layer, the foreground classified prediction result is generated as pre_1. Meanwhile, to acquire the foreground's location, we leverage the location regression function and feed the location parameters into the Hog-SVM module. The gesture regions can be extracted by HOG module an d then sent to the SVM classification network for prediction, with the prediction result as pre_2. Then pre_1 and pre_2 are fused according to $P = (1\text{-c}) \times \text{pre\_1} + c \times \text{pre\_2}$. The parameter c controls the ratio between the two prediction results. The fed fused results become a loss function and participate in the propagation of the network.

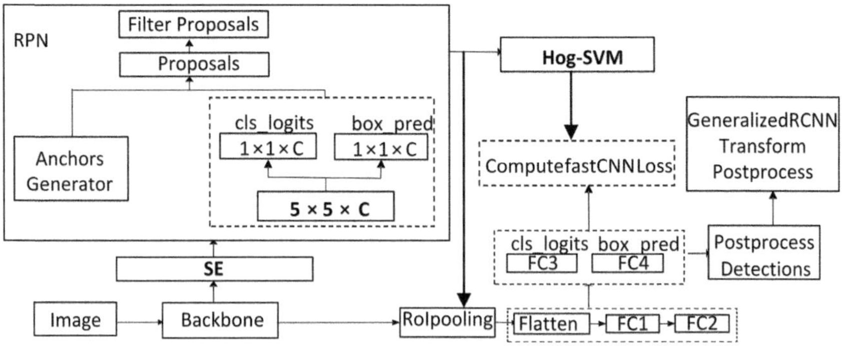

Fig. 20.4 Hog-Faster R-CNN(EF)

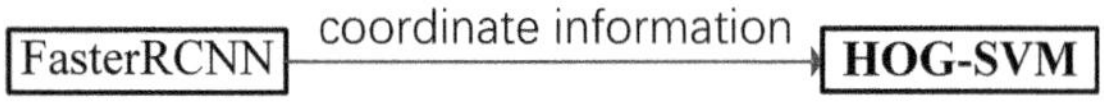

Fig. 20.5 Hog-faster R-CNN(SF)

Fusion of HOG-SVM and Faster R-CNN

Serial Fusion of HOG-SVM Model and Fsatrt R-CNN

The overall architecture of our appro ach is shown in Fig. 20.5. Firstly, the purpose of using Faster R-CNN is to retrieve the coordinates of the target box from the output image. Then, the RoI is extracted from the output results, and the extracted image content is fed into the Hog-SVM model for image recognition.

Embed Fusion of Faster R-CNN and HOG-SVM Module

The embed fusion method of HOG-SVM and Faster R-CNN is shown in Fig. 20.4. We feed coordinate information obtained from the RPN part into HOG-SVM module. The module uses coordinate information to extract RoI from images and sends RoI into the SVM module for the result prediction. Then, we get a new prediction, which is obtained by fusing pre_1 with pre_2 of the Faster R-CNN. The new prediction is returned to the model's loss function. To train for both regression (classification and bounding-box), each training RoI is assigned a gt class label u and a gt bounding boxs regression target v. We employ a multi-task loss function L on each labeled RoI:

$$L(p, u, t^u, v) = L_{cls}(p, u) + \lambda[u \geq 1]L_{loc}(t^u, v) \quad (20.1)$$

The hyper-parameter λ determines the trade-off between the two losses, and all experiments in this study use $\lambda = 1$. The Iverson bracket indicator function $[u \geq 1]$ evaluates to 0 when $u < 1$ and 1 otherwise. The log loss for the true class U is denoted by L_cls(), and the bounding-box regression loss for the ground-truth bounding-box regression target V is denoted by L_loc(). The classification loss can be calculated using the following Eq. (20.2):

$$\mathrm{L_{cls}}(p, u) = -\log(p_u) \quad (20.2)$$

where p represents the Softmax probability distribution predicted by the classifier, $\mathrm{p} = \mathrm{p}_{0,\ldots,}\mathrm{p}_{\mathrm{k}}$ $(p_0 \ldots\ldots p_k)$, t^u corresponds to the regression parameter $(t_x^u, t_y^u, t_w^u, t_h^u)$ for the predicted class u by the bounding box regressor, and v corresponds to the regression parameter (v_x, v_y, v_w, v_h) of the actual target bounding box.

In the Softmax probability distribution, we combine Faster R-CNN with the output results of the HOG-SVM module to get a new Softmax probability distribution.

$$p = (1 - \mathrm{c})(p_0 \ldots\ldots p_k) + c(p_0^* \ldots\ldots p_k^*) \quad (20.3)$$

where $(p_0 \ldots\ldots p_k)$ represents the output probability of Faster R-CNN model; $(p_0^* \ldots\ldots p_k^*)$ represents the output probability of HOG-SVM module; parameter c controls the ratio of the two predictions, and c = 0.1 is used in all experiments.

SE Attention Mechanism

The SE attention mechanism belongs to category of channel attention, and its purpose is to assign more weight to crucial channels. By doing so, the network can concentrate on the channels that require attention. To reduce disruption caused by non-target objects in gesture images, the SE attention mechanism is integrated into Faster R-CNN to precisely locate gesture objects.

We embed SE attention between Backbone and RPN. The SE attention mechanism assigns different weights to each channel, and important channels of the feature map are given higher weights. So, network can focus more on gestures, and the influence of complex background on the network is reduced. The bounding box regressor adjusts the anchors to fit the object more efficiently, thus network can predict the boundary frame of the recognized object more accurately.

As the large proportion of gestures in the image, the 3 × 3 convolution kernel of RPN part was changed to 5 × 5 convolution kernel. This can improve the receptive field, and the network can better classify foreground and background.

20.4 Experimental Results and Analysis

The operating system used in the experiment is windows10, the GPU is 3080, the processor is Intel corei9-10,900@ 3.7 GHz, and the deep learning framework is pytorch1.2.0.

Data Sets and Evaluation Indicators

In the experiment, the public gesture recognition data set Post Hand [12–15] is used for the training and prediction. In this data set, there are 2740 gesture pictures in 14 categories with complex backgrounds, as shown in Table 20.1.

To test the accuracy of the HOG-FasterR-CNN, Average Precision (AP) and Mis Rate (MR) are used to evaluate the detection results, and Mean AP (mAP) is used to measure the performance of detection network.

Table 20.1 Gesture recognition data set

Post hand dataset							
Gesture type	One	Three	Four	Five	Six	Nine	Ok
Number of pictures	364	122	121	262	130	133	198
Gesture type	Heart	Yearh	Love	Gun	Thumb	Pick	Fist
Number of pictures	130	309	174	130	191	155	372

Table 20.2 Method contrast

Model	HOG-SVM	Faster R-CNN	Hog-faster R-CNN(SF)	Hog-faster R-CNN(EF)
Accuracy	0.910	0.934	0.924	0.942

Experimental Results and Analysis

Fusion of HOG-SVM and Faster R-CNN

To verify the effectiveness of our method, HOG-SVM, Faster R-CNN, and Hog-Faster R-CNN are carried out on Pose Hand dataset that display on Table 20.2.

The experimental results are shown in Table 20.2 indicate that the EF has best accuracy. HOG fearture with SVM classifier performs worse than deep learning method in gesture recognition under complex backgrounds. Hog-Faster R-CNN(SF) directly usses the results of backbone to extract gesture area that are sent to HOG-SVM module to do recogntion. The extraction of the gesture region can reduce the effect of the complex background to some extent, so this method performs better than HOG-SVM. But the Faster R-CNN detection ability is not enough to get accurate bounding boxes, this inaccuracy leads more errors to HOG-SVM than Faster R-CNN for calssification, so the SF result is a little lower than Faster R-CNN. In the Hog-Faster R-CNN(EF) network, the prediction results of HOG-SVM moudel and Faster R-CNN network are combined to calculate the loss. It helps the network to better propagate back and adjust parameters, so the network in embed fusion mode has the highest accuracy.

SE Attention Mechanism

See Figs. 20.6 and 20.7 for the experimental results of adding attention and changing the receptive field:

As in Fig. 20.6, we network with SE attention mechanism has the smallest bounding box regression loss. The SE attention mechanism adds a different weight to each channel, which can network more focused on useful channels and locate gesture objects more precisely. So, the bounding box regressor can more precisely adjust the anchors to fit the target, and the network can predict the target bounding

Fig. 20.6 Comparison of attention

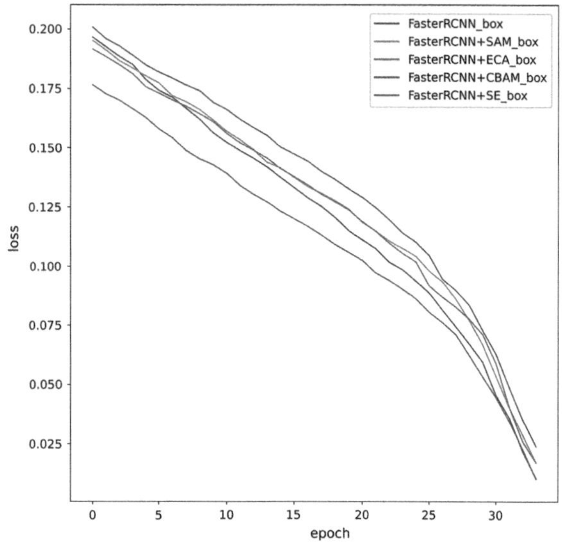

Fig. 20.7 Change receptive field

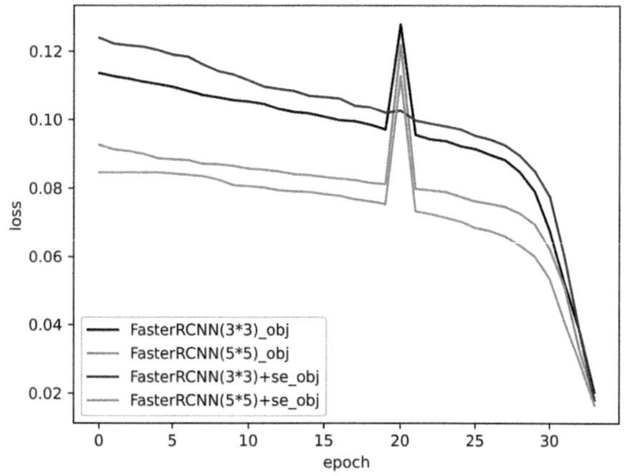

box more accurately. In the RPN of original network, the receptive field of 3×3 convolution kernel is 491, but the maximum of the given target box in network is 512. The receptive field value is less than the size of the given target frame, which is not conducive to target detection. So, 3×3 convolution kernel is changed to 5×5. The receptive field of the convolution kernel becomes 555, greater than the maximum value of the given target box. As shown in Fig. 20.7, this method improves the network for gesture recognition rate.

Table 20.3 SE attention and receptive field

Faster R-CNN	HOG-SVM	SE (3*3)	SE (5*5)	Accuracy
√				0.934
√	√			0.942
√		√		0.938
√	√	√		0.954
√	√		√	**0.958**

Table 20.4 Comparison

Model	Accuracy
MobileNet	0.893
Swin-transformers	0.924
YoloV3	0.928
CNN + Bangla sign language [16]	0.926
Skin and wrist algorithms [17]	0.930
Proposed method	**0.958**

Ablation Experiments

To effectively compare the efficacy of each module, we did an ablation experiment, and the findings are presented in Table 20.3.

The results of the experiment show that the addition of the HOG-SVM module and SE attention mechanism in the Faster R-CNN network can effectively improve the gesture recognition accuracy under complex backgrounds, and mAP increases by 2.4% compared with the original network.

Table 20.4 presents a comparison between the proposed work and several published studies on hand detection, highlighting the different methodologies employed.

Table 20.4 demonstrates that the proposed algorithm can provide a valuable knowledge contribution with a novel methodology and a satisfactory recognition rate.

20.5 Conclusion

The presented research introduces a gesture recognition network that combines Faster R-CNN and HOG. To enhance the recognition accuracy of Faster R-CNN, HOG-SVM module is incorporated. Additionally, the RPN part of Faster R-CNN is equipped with the attention of SE to improve the accuracy of bounding box prediction, thereby improving detection and recognition capabilities. However, the proposed method's real-time detection speed lags behind one-stage object detection methods significantly. Therefore, future research will reduce the size of the algorithm model, make it better deployed on Edge device, improve the response speed of the model, and enable it to process the input gesture pictures in real time.

References

1. Nielsen, E., Canalis, L.A., Tejera, M.H.: Hand gesture recognition for human-machine interaction. UNION Agency (2004)
2. Wang, W., Pan, J.: Hand segmentation using skin color and background information. In: 2012 International Conference on Machine Learning and Cybernetics, IEEE, vol. 4, pp. 1487–1492 (2012)
3. Al Farid, F., Hashim, N., Abdullah, J.: Vision-based hand gesture recognition from rgb video data using svm. International Workshop on Advanced Image Technology (IWAIT) 2019, SPIE, vol. 11049, pp. 265–268 (2019)
4. Liu, S., Liu, Y., Yu, J., Wang, Z.: Hierarchical static hand gesture recognition by combining finger detection and hog features. J. Gr. **20**(6), 781–788 (2015)
5. Howard, A., Sandler, M., Chu, G., Chen, L.-C., Chen, B., Tan, M., Wang, W., Zhu, Y., Pang, R., Vasudevan, V., et al.: Searching for mobilenetv3. Proceedings of the IEEE/CVF International Conference on Computer Vision, pp. 1314–1324 (2019)
6. Tan, M., Le, Q.: Efficientnet: rethinking model scaling for convolutional neural networks. International Conference on Machine Learning, pp. 6105–6114. PMLR (2019)
7. Girshick, R., Donahue, J., Darrell, T., Malik, J.: Rich feature hierarchies for accurate object detection and semantic segmentation. Proceedings of the IEEE Conference on Computer Vision and Pattern Recognition, pp. 580–587 (2014)
8. Girshick, R.: Fast R-CNN. 2015 IEEE International Conference on Computer Vision (ICCV), pp. 1440–1448. Santiago, Chile (2015). https://doi.org/10.1109/ICCV.2015.169
9. Ren, S., He, K., Girshick, R., Sun, J.: Faster R-CNN: Towards real-time object detection with region proposal networks. Adv. Neural Inf. Process. Syst. **2** (2015)
10. Hu, J., Shen, L., Sun, G.: Squeeze-and-excitation networks. In: Proceedings of the IEEE Conference on Computer Vision and Pattern Recognition, pp. 7132–7141 (2018)
11. Dalal, N., Triggs, B.: Histograms of oriented gradients for human detection. In: 2005 IEEE Computer Society Conference on Computer Vision and Pattern Recognition (CVPR'05), vol. 1, pp. 886–893. Ieee (2005)
12. Al Farid, F., Hashim, N., Abdullah, J., Bhuiyan, M.R., et al.: A structured and methodological review on vision-based hand gesture recognition system. J. Imaging **8**(6), 153 (2022)
13. Chistyakov, I., Chepin, E.: Gesture recognition system based on convolutional neural networks. IOP Conference Series. Materials Science and Engineering, vol. 498, no. 1. IOP Publishing (2019)
14. Köpüklü, O., Gunduz, A., Kose, N., Rigoll, G.: Online dynamic hand gesture recognition including efficiency analysis. IEEE Trans. Biom. Behav. Identity Sci. **2**(2), 85–97 (2020). https://doi.org/10.1109/TBIOM.2020.2968216
15. Marin, G., Dominio, F., Zanuttigh, P.: Hand gesture recognition with leap motion and kinect devices. In: 2014 IEEE International Conference on Image Processing (ICIP), pp. 1565–1569 (2014)
16. Tasmere, D., Ahmed, B.: Hand gesture recognition for bangla sign language using deep convolution neural network. In: 2020 2nd International Conference on Sustainable Technologies for Industry 4.0 (STI) (2020)
17. Yadav, K., et al.: Hand gesture recognition using improved skin and wrist detection algorithms for Indian sign. J. Netw. Commun. Emerg. Technol. (JNCET) **9**(2) (2019)

Part III
Intelligent Methods and Applications

Chapter 21
Inspiration from the Fuzziness of Traditional Garden Architecture Space to Modern Architectural Space Design

Songlin Wu

Abstract In order to explore the inspiration of traditional landscape architecture space fuzziness to modern architectural space design, this paper analyzes modern architectural space design by combining the fuzziness of traditional landscape architecture space. In order to extract the fuzzy features in traditional Garden design, this paper uses intelligent fuzzy control algorithm as the feature extraction algorithm. From the perspective of intelligent feature analysis, the visual features of Garden design images are analyzed through space construction methods, and the fuzziness in Garden design is quantitatively processed with fuzzy control methods to extract the corresponding fuzzy features, and the intelligent algorithm is used for design information fusion, Intelligent display of reconstructed design images. From the simulation experiment, we can see that the application effect of fuzziness in traditional architectural space in modern architectural space is very good, which can be used for reference in current architectural space design.

21.1 Introduction

Chinese traditional architectural space aesthetics has the characteristics of fuzziness, and fuzziness is uncertainty, which is not to tell the whole thing, but to make it exist between likeness and non-likeness, and it is a form of "leaving the square and escaping from the circle". Moreover, fuzziness is by no means vague, but an artistic choice and arrangement of objective things according to subjective emotions, excluding insignificant factors and making them vague. Because of its uncertainty and polysemy, this processing method has a great dynamic effect on thinking. Leaving room and thought-provoking can often make a small world and a local treatment meaningful and attractive.

S. Wu (✉)
Wuhan Qingchuan University, Wuhan, China
e-mail: wsl326390739@163.com

R. Kountchev (Deceased) et al. (eds.), *AI Methods and Applications in 3D Technologies*, Smart Innovation, Systems and Technologies 388,
https://doi.org/10.1007/978-981-97-2144-3_21

In architecture, it belongs to a space that is specious, ambiguous, and indescribable. It is often located between two different types of spaces, such as outdoor, open, closed, etc., resulting in the ambiguity, uncertainty, ambiguity, grayness of the space Thus rich in implicitness and thought-provoking, more than spatial connections, transitions, extensions, etc. [1].

Virtual space, also known as symbolic space, does not have a clear form of spatial segmentation. Within a defined space, the space is redefined through local changes in the interface, relying entirely on the audience's associations and psychological feelings to achieve. For example, a discussion space surrounded by seats does not have a clear interface segmentation. There are many methods for handling virtual spaces, such as dividing columns, arranging interlayers, and staggered ceilings and floors. It is also possible to create virtual functional spaces with different characteristics through variations in low walls, partitions, railings, screens, rolling curtains, curtains, greenery, and lighting [2].

A type of virtual space refers to the virtual image reflected by indoor mirrors, which brings people's attention to the virtual space behind the mirror, resulting in a visual effect of expanding the space. Sometimes, through the refraction of several mirrors, it can create an illusion of three-dimensional space for originally flat objects [3]. Objects close to the mirror can also create the illusion of complete objects (circular) by transforming incomplete objects (arch bridges on water in gardens). In particularly narrow spaces, mirrors are often used to expand the sense of space, and the illusion of mirror decoration is used to enrich the indoor landscape. In addition to the mirror, another type is a large image with a certain depth of field, which leads people's gaze towards the distance and creates a profound intention of space. Another type is to pursue mysterious, novel, and unpredictable effects in specific spaces, using colorful, abstract patterns, and dynamic linear exaggerations. In terms of shape, distortion, breakage, inversion, and dislocation are often used to create endless fantasies [4].

Grey space is a type of space, and its application in landscape architecture is influenced by the characteristics of the landscape itself, making its design distinctive compared to other types, with various composition methods from individual to group, from group to surrounding environment. How to coordinate the design of garden landscape gray space with the surrounding environment is a very important content, which requires designers to have a certain understanding and understanding of the properties of garden landscape gray space [5].

The gray space of garden landscape includes the evolution and contrast of space, with variable and uncertain factors, enhancing the sense of ambiguity and bringing vitality to life. The gray space of landscape architecture is characterized by horizontal displacement, intersection, and strong dispersion. However, for the vertical enclosure interface of the gray space, its spatial volume and boundaries are not complete. In addition, the infiltration of other multi-level spaces and the influence of adjacent spaces make the uncertainty of the enclosure form of the gray space more obvious [6]. In the actual design process, in order to maximize the coordination between spaces, this form of gray space will be connected with adjacent spaces through methods such as enclosure, infiltration, connection, and fusion. Therefore, its freedom and ambiguity are more evident [7].

In order to ensure the natural connection between the gray space and adjacent spaces to the maximum extent, the enclosing interface of the garden landscape will intersect, penetrate, and transform with each other to a certain extent. Therefore, the richness of the boundary form of the gray space in this situation is clearly reflected, and most of it is in a free, interlaced, and discontinuous form, making it difficult to find a specific defining position. Overall, the form of grey space in garden landscapes is greatly influenced by the form of adjacent spaces and has a strong dependence on the surrounding environment. Therefore, the intersperse and infiltration of space is a very important attribute of grey space in garden landscapes [8].

The gray space of garden landscape still belongs to a typical spatial form in essence, and its main role in designing spatial structures is to transition, transform, contrast, and connect between different spatial forms. Strictly speaking, the form of grey space is mainly determined by two factors, namely the structure of adjacent spaces and their spatial layout. Influenced by these two factors, the structural organization of grey space has a certain degree of richness and variability, and in turn, due to these characteristics of grey space structure, adjacent spaces have been extended and expanded to a certain extent. There is a certain commonality between architectural spatial structure and garden landscape spatial structure, and relevant experience in architectural spatial design can be drawn and applied to the design of garden landscape gray space [9].

Garden landscape gray space is a dialogue and communication between heterogeneous spaces, mainly due to the richness and diversity of the structural organization of gray space, which is mainly influenced by the adjacent spatial structure. The mutual infiltration and interaction between the two make it difficult to describe the form of gray space in professional terms. In addition, due to the influence of its design theory, in the actual design process of the gray space, its specific location coordinates, scales, boundaries, etc. are in a relatively vague state, so from the overall analysis, the shape of the gray space has Connected sum mobility [10].

People have reliable requirements for the surrounding environment, and specific landscapes and structures are set up in the gray space, jointly forming a space with a sense of security and enclosure. In the garden landscape space, service-oriented facilities such as pavilions, corridors, railings, and benches are set up, which not only can be used by people, but also can bring a sense of enclosure and domain to people. The natural, friendly, comfortable, and convenient conditions make people feel that the space can be relied on and have a sense of psychological security [11].

This paper combines the fuzziness of traditional garden architecture space to analyze the modern architectural space design, from which we can get inspiration and improve the effect of modern architectural space design.

21.2 Fuzzy Model of Garden Space

Acquisition of Space Environment Design Image

We construct a pixel space fusion model of landscape spatial environment design image, Template matching of visual reconstruction of landscape space environment design image as shown in Fig. 21.1.

We assume that the gray pixel set of landscape spatial environment design image is (i, j), which is the pixel center, and construct the feature visual reconstruction model of landscape spatial environment design image by sharpening template block combination method. In the gray pixel feature distribution space, the gradient feature components of the landscape spatial environment design image are obtained from the gray value I_{swk} of the landscape spatial environment design image collected in the *k-th* sub-band as follows:

$$P_{rk} = \left(\frac{\sum_{j=1}^{c} I_{swk}(1, j)}{c}, \frac{\sum_{j=1}^{c} I_{swk}(2, j)}{c}, \ldots, \frac{\sum_{j=1}^{c} I_{swk}(i, j)}{c}, \ldots., \frac{\sum_{j=1}^{c} I_{swk}(r, j)}{c} \right) \tag{21.1}$$

$$P_{ck} = \left(\frac{\sum_{j=1}^{r} I_{swk}(i, 1)}{r}, \frac{\sum_{j=1}^{r} I_{swk}(i, 2)}{r}, \ldots, \frac{\sum_{j=1}^{r} I_{swk}(i, j)}{r}, \ldots., \frac{\sum_{j=1}^{r} I_{swk}(i, c)}{r} \right) \tag{21.2}$$

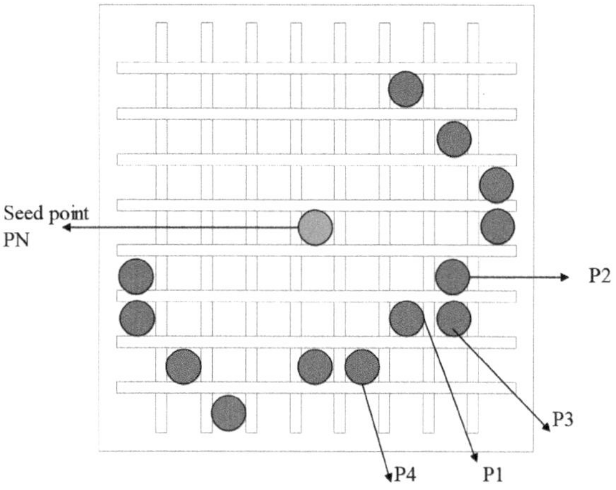

Fig. 21.1 Template matching of visual reconstruction of landscape space environment design image

We assume that the gray pixel set of landscape spatial environment design image is (i, j), P_{rk} represents the LGB Vector quantization matrix feature component of the landscape space environment design image, and P_{ck} represents the motion blur feature component.

Among them, C is the number of LGB vector quantization matrix columns of landscape space environment design image, and r is the feature quantity of motion blur. Combined with the fuzzy pixel regional feature fusion reconstruction method, the distribution pixel set of landscape spatial environment artistic features is obtained.

Image Feature Fusion

The visual feature distribution of landscape spatial environment design image is as follows [12]:

$$G(\vec{x}) = \sum_{j=1}^{p} G_{\mathrm{j}}(\vec{x}) \tag{21.3}$$

$G(\vec{x})$ is the quantified value of local visual features. The adaptive fusion method is used to reconstruct the image vision of landscape design, and the edge vision reconstruction model of landscape space environment design is constructed, and the fuzzy paste of landscape space environment image is obtained.

The proximity function is:

$$fitness(\vec{x}) = f(\vec{x}) + (C\mathrm{t})^{\alpha} \sum_{j=1}^{p} G_{\mathrm{j}}^{\beta}(\vec{x}) \tag{21.4}$$

$f(\vec{x})$ is a continuous two-dimensional function, and $C\mathrm{t}$ represents the boundary fitness value, if we assume that the coordinates of landscape space environment design P_N are (X_{PN}, Y_{PN}), then we compare the edge point coordinates (X_k, Y_k) of all landscape space environment design on L with those of P_N.

When $\mathrm{x_k} > X_{PN}, i_L = i_L + 1$ i;

When $\mathrm{x_k} < X_{PN}, i_L = i_L - 1$;

When $\mathrm{x_k} = X_{PN}, i_L = i_L + 0$;

The fitness function of landscape information fusion of landscape art:

$$fitness(\vec{x}) = \begin{cases} f(\vec{x}), \text{If feasible} \\ 1 + \mathrm{r}G(\vec{x}), \text{otherwise} \end{cases} \tag{21.5}$$

$$W_u(a,b)=e^{i2\rho k\ln a}\times\frac{K}{\sqrt{a}}\left\{\begin{array}{l}\left[\dfrac{ae^{\frac{j2\pi f_{\min}}{a}(b-b_a)}}{f_{\min}}-\dfrac{e^{\frac{j2\pi f_{\min}}{a}(b-b_a)}}{f_{\max}}\right]+\\ j2\pi(b-b_a)\left[\begin{array}{l}Ei(j2\pi f_{\max}(b-b_a))-\\ Ei\left(\dfrac{j2\pi f_{\min}}{a}(b-b_a)\right)\end{array}\right]\end{array}\right\} \tag{21.6}$$

$W_u(a,b)$ is a represents the reconstruction model function. Among them, $b_a=(1-a)\left(\frac{1}{af_{\max}}-\frac{T}{2}\right)$, and $Ei(\cdot)$ represent landscape space environment visual information feature recombination output, combined with model recognition method, landscape design [13].

Design Optimization

After extracting and traversing all edge points on L, we set σ_1^2 as the local variance of landscape spatial environment design image, σ_η^2 as the optimization coefficient of landscape spatial environment design image, $\beta=\max\left[\frac{\sigma_1^2-\sigma_\eta^2}{\sigma_1^2},0\right]$. The gradient descent method is used to reconstruct the landscape spatial environment vision in different regions, so that the sparse eigenvalue of landscape spatial environment design image satisfies $C\in S$. According to the sparse prior representation results, the optimal visual reconstruction threshold at the m-th frame (x,y) of the landscape space environment design image $F_m(x,y)$ is obtained. Based on the approximate sparse representation method, the template matching of landscape spatial environment design images is carried out, and the matching coefficients are as follows:

$$g_i^*=\begin{cases}Rs_j, z\le i\le x-y\\ g_i, Otherwise\end{cases} \tag{21.7}$$

In the formula, R represents a standard constant, which is combined with the template matching method of the block area to design the landscape space environment and distributed detection, and the contour point matching model is used to extract the edge features of the landscape space environment design.

The maximum grayscale value for the image analysis part of landscape spatial environment design is:

$$n_{pg}=\frac{\mu_{pb}}{(\mu_{00})^{\gamma}} \tag{21.8}$$

The information reconstruction model of landscape space environment design is as follows:

$$g(x, y) = f(x, y) + \varepsilon(x, y) \tag{21.9}$$

Among them, $f(x, y)$, $g(x, y)$ and $\varepsilon(x, y)$ represent the original landscape spatial environment image, reconstructed image and gray image respectively. To sum up, the interactive genetic optimization design of landscape design is carried out.

In this paper, an interactive genetic algorithm based visual reconstruction algorithm for landscape spatial environment design images is proposed. Combined with block area template matching method, the feature visual reconstruction of landscape space environment design image is carried out. Based on the method of adaptive feature matching of local features, the information fusion model of landscape spatial environment vision is constructed, and the expression model of landscape spatial environment artistic features is constructed. Under the optimization of genetic evolution, the information fusion expression of landscape art is obtained as follows [14]:

$$g = \mathrm{k} \otimes f + n \tag{21.10}$$

Among them, $\otimes$ represents the convolution operator. Vector set fusion processing is carried out on the collected landscape spatial environment design images, and the decomposition model of landscape spatial environment visual features is constructed, and the best distinguishing feature values of landscape spatial environment visual features are obtained as follows:

$$s_{PPM}(t) = \sum_{\mathrm{i}=-\infty}^{\infty} \sum_{j=0}^{N_{\mathrm{p}}-1} p\big(t - iT_{\mathrm{s}} - jT_{\mathrm{p}} - c_j T_{\mathrm{c}} - a_i \varepsilon\big) \tag{21.11}$$

$$s_{PPM}(t) = \sum_{\mathrm{j}=-\infty}^{\infty} d_j p(t - jT_{\mathrm{s}}) \tag{21.12}$$

T_{s} is the iterative width of genetic evolution. Landscape space environment design as shown in Fig. 21.2.

21.3 Research on the Fuzziness of Garden Space

People's real experience in space is composed of multiple dimensions. In the landscape space design of emotional sustainability, experience space is the origin and function is the condition. Through the design of landscape elements, artistic conception space is created to guide cognitive behavior, that is, function, design elements, artistic conception and cognitive behavior (Fig. 21.3). In addition, the functional space design is guided by behavioral needs and cognition, and the design symbols are conveyed through landscape elements, which condenses the hidden life experience, culture and yearning for beauty in everyone's heart into the artistic conception of

Fig. 21.2 Landscape space environment design

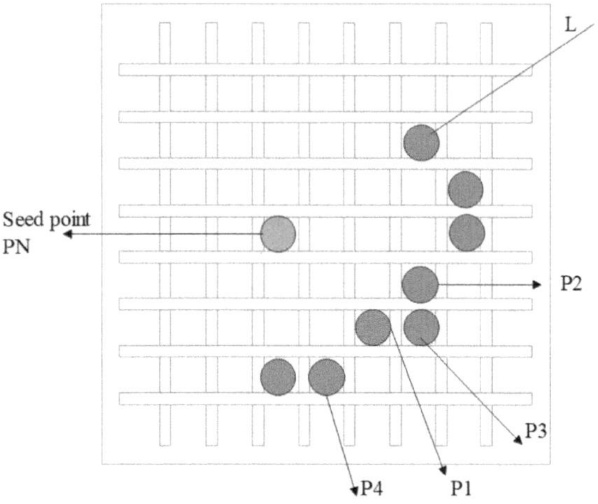

(a) Left seed point block matching

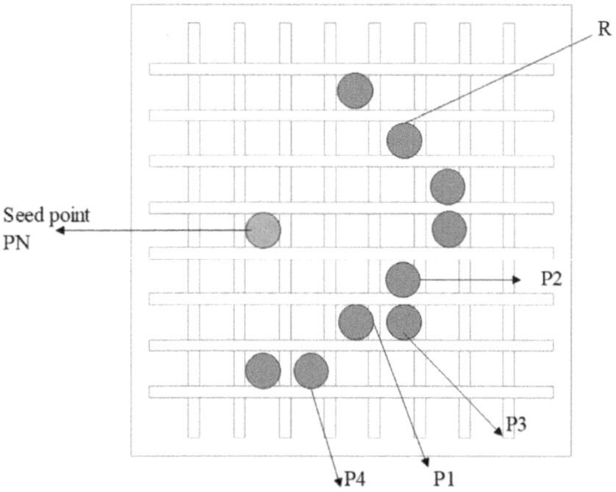

(b) Right seed point block matching

garden space, and completes the complete experience from space to emotion, which is the infinite extension of design itself. Therefore, only by paying full attention to experience in design can we study the appropriate design strategy more pertinently and effectively.

Emotional perpetuation in landscape is actually a process of asking how the "feeling" in psychological thinking is integrated with the design concept, that is, a process of asking how the landscape space performance can get a satisfactory spatial feeling

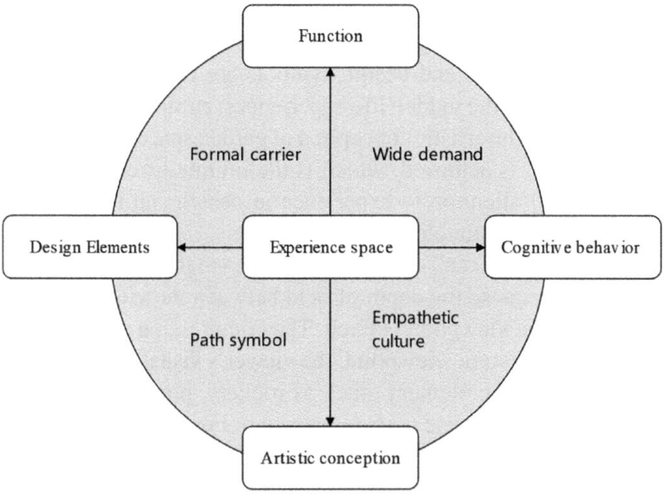

Fig. 21.3 Fuzzy experience space dimension

and obtain a complete meaning in people's minds, which requires designers to understand space from multiple angles. The fuzzy experience perception model is shown in Fig. 21.4.

The real experience of a person in space is composed of multiple dimensions. In the design of an emotionally sustainable landscape space, taking the experience space as the starting point and satisfying functions as the conditions, through the design of landscape elements, an artistic conception space is created to guide cognitive

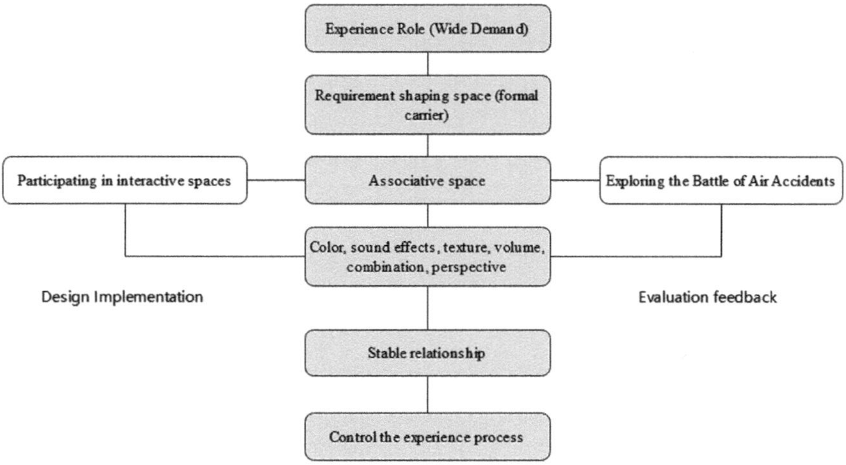

Fig. 21.4 Fuzzy experience perception model

behavior, namely the four dimensions of function, design elements, artistic conception, and cognitive behavior (Fig. 21.4). Behavioral needs and cognition guide the design of functional spaces, and design symbols are conveyed through landscape elements, By condensing the hidden life experiences, culture, and longing for beauty in everyone's heart into the artistic conception of garden space, a complete experience from space to emotion is achieved, which is the infinite extension of design itself. Therefore, paying full attention to experience in design can lead to more targeted and effective research on suitable design strategies.

In this paper, "scene surface" is extended: it refers to the picture with spatial depth that the viewer sees because the depth of field between the viewpoint and the scene is compressed when the viewpoint is fixed. The compression of depth of field is due to the fact that under the static viewpoint, the viewer's visual perception of the scene is presented by each scene element (such as rockery, pine tree, shrub, etc.) in the form of one frame and two scene units (as shown in Fig. 21.5). Therefore, the falling point of the line of sight constantly switches and jumps among the scene units, which leads to the visual illusion that the depth of field is compressed.

As shown in Fig. 21.6, it is an example diagram of spatial fuzziness in traditional gardens. This diagram is processed by the model of this paper, and the spatial fuzziness in garden architectural design can be found by close-up, which has a certain promotion effect on the spatial design of modern buildings. On this basis, through several groups of experimental simulation, the application effect of fuzziness in traditional architectural space in modern architectural space is verified. Moreover, this paper evaluates the application effect, and finally obtains the evaluation results shown in Table 21.1.

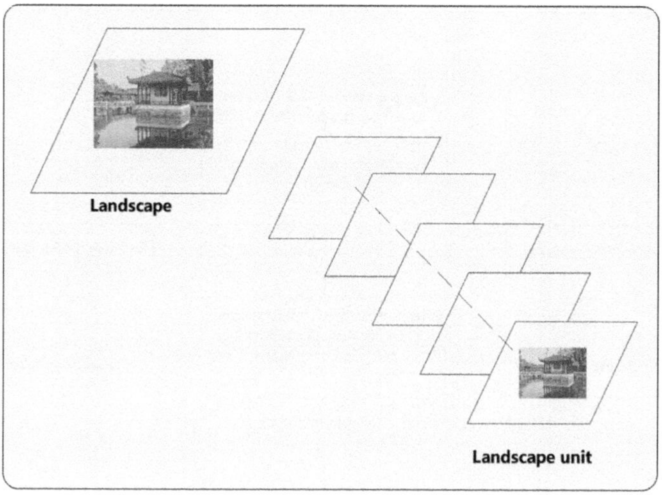

Fig. 21.5 Scene generation diagram

Fig. 21.6 Example of spatial fuzziness of traditional garden architecture

Table 21.1 Verification of the application effect of fuzziness of traditional architectural space in modern architectural space

Number	Evaluate	Number	Evaluate	Number	Evaluate
1	83.327	15	79.764	29	88.624
2	78.722	16	79.410	30	83.253
3	86.139	17	84.705	31	86.539
4	79.649	18	81.872	32	81.430
5	83.861	19	82.075	33	80.705
6	81.672	20	78.455	34	87.506
7	87.442	21	79.286	35	83.933
8	84.980	22	86.711	36	79.878
9	82.891	23	78.520	37	84.790
10	87.261	24	82.938	38	79.679
11	87.619	25	88.222	39	80.172
12	80.991	26	85.197	40	82.919
13	87.200	27	82.238	41	82.356
14	82.881	28	85.325	42	80.089

From the above simulation experiments, we can see that the application effect of fuzziness in traditional architectural space in modern architectural space is very good, which can be used for reference in current architectural space design.

21.4 Conclusion

In this paper, "scene surface" is defined as the picture with spatial depth seen by the viewer when the depth of field between the viewpoint and the scene is compressed when the viewpoint is fixed. Because the scene surface has the fuzzy attribute between two-dimensional and three-dimensional, the viewer's perception of space is often blurred under the static viewpoint. This visual phenomenon is defined as "spatial blurring" in this paper. The viewer can subjectively reorganize and assemble the landscape from the static viewpoint by following the guidance of the garden path from the dynamic viewpoint, thus forming a holistic cognition of space. This transformation mode between landscape surfaces is defined as "spatial connection" in this paper. Therefore, this paper analyzes the typical cases with infinite sense of space in classical gardens from two aspects of "spatial fuzziness" and "spatial convergence", and verifies that the application effect of traditional architectural spatial fuzziness in modern architectural space is very good by combining intelligent algorithm (Fuzzy Space Algorithm).

In future research, high-resolution remote sensing images, land use status maps, traffic maps, soil maps, and other data will be comprehensively considered to extract more accurate remote sensing information, in order to improve the accuracy of landscape classification and simulation.

References

1. Pinto-Correia, T., Almeida, M., Gonzalez, C.: Transition from production to lifestyle farming: new management arrangements in Portuguese small farms. Int. J. Biodivers. Sci. Ecosyst. Serv. Manag. **13**(2), 136–146 (2017)
2. Yi, Z., Fei, Z., Ruiqin, L.: The path of rural revitalization in rapidly urbanizing areas: taking Southern Jiangsu as an example. China City Plan. Rev. **27**(4), 24–33 (2018)
3. Antic, M., Santic, D., Kašanin-Grubin, M., Malic, A.: Sustainable rural development in Serbia-relationship between population dynamicss and environment. J. Environ. Prot. Ecol. **18**(1), 323–331 (2017)
4. Cecchini, M., Zambon, I., Pontrandolfi, A., Turco, R., Colantoni, A., Mavrakis, A., Salvati, L.: Urban sprawl and the 'olive' landscape: sustainable land management for 'crisis' cities. GeoJournal **84**(1), 237–255 (2019)
5. Kaindoa, E.W., Finda, M., Kiplagat, J., Mkandawile, G., Nyoni, A., Coetzee, M., Okumu, F.O.: Housing gaps, mosquitoes and public viewpoints: a mixed methods assessment of relationships between house characteristics, malaria vector biting risk and community perspectives in rural Tanzania. Malar. J. **17**(1), 1–16 (2018)
6. Hulko, W., Hovanes, J.: Intersectionality in the lives of LGBTQ youth: Identifying as LGBTQ and finding community in small cities and rural towns. J. Homosex. **65**(4), 427–455 (2018)

7. Herrero-Jáuregui, C., Arnaiz-Schmitz, C., Herrera, L., Smart, S.M., Montes, C., Pineda, F.D., Schmitz, M.F.: Aligning landscape structure with ecosystem services along an urban–rural gradient. Trade-offs and transitions towards cultural services. Landsc. Ecol. **34**(7), 1525–1545 (2019)
8. Ferretti, V., Gandino, E.: Co-designing the solution space for rural regeneration in a new world heritage site: a choice experiments approach. Eur. J. Oper. Res. **268**(3), 1077–1091 (2018)
9. Le, P. d. W.O., Cohuet, S., Ndazima, D., Kucharski, A.J., Juan-Giner, A., Flasche, S., Edmunds, W.J.: Characteristics of human encounters and social mixing patterns relevant to infectious diseases spread by close contact: a survey in Southwest Uganda. BMC Infect. Dis. **18**(1), 1–12 (2018)
10. Keane, M., Chen, Y.: Entrepreneurial solutionism, characteristic cultural industries and the Chinese dream. Int. J. Cultural Policy **25**(6), 743–755 (2019)
11. Chen, C., LeGates, R., Fang, C.: From coordinated to integrated urban and rural development in China's megacity regions. J. Urban Aff. **41**(2), 150–169 (2019)
12. Ngugi, H.N., Mutuku, F.M., Ndenga, B.A., Musunzaji, P.S., Mbakaya, J.O., Aswani, P., LaBeaud, A.D.: Characterization and productivity profiles of Aedes aegypti (L.) breeding habitats across rural and urban landscapes in western and coastal Kenya. Parasit. Vectors **10**(1), 1–12 (2017)
13. Beyene, A.D., Mekonnen, A., Randall, B., Deribe, R.: Household level determinants of agroforestry practices adoption in rural Ethiopia. For. Trees Livelihoods **28**(3), 194–213 (2019)
14. Stojanović, J., Kokotović Kanazir, V., Stojanović, M.: Does small town with touristic function have demographic potential?. J. Geogr. Inst. Jovan Cvijic SASA **67**(2), 145–162 (2017)

Chapter 22
The CAA Captioner–Enhancing Image Captioning with Contrastive Learning and Attention on Attention Mechanism

Zhao Cui

Abstract Image Captioning is a complex interdisciplinary task, situated at the intersection of CV (computer vision) and NLP (natural language processing). In recent years, with the development of encoder decoder frameworks and attention mechanisms, although significant progress has been made in this task, there is still a need to optimize the accuracy and completeness of text generation to improve the generalization ability of the model. We propose an Image Captioning method that combines a contrastive learning-enhanced ResNet101 encoder with an Attention on Attention (AoA) improved Transformer decoder to achieve better performance in generating accurate and natural descriptions. We utilize contrastive learning to enhance the model's understanding of image content. Additionally, we apply data augmentation techniques to enhance the model's generalization ability, and optimize the attention mechanism of the Transformer decoder using the AoA module. Our experimental results demonstrate the effectiveness of contrastive learning in handling complex scenes, multiple objects, and detailed descriptions, as our approach achieves superior performance on widely adopted Image Captioning datasets.

22.1 Introduction

Image Captioning is an interdisciplinary task between the computer vision (CV) and natural language processing (NLP) fields in deep learning. The primary goal is to enable machines to analyze image content and generate textual descriptions corresponding to the image content, making it a typical task in the multi-modal domain. Compared to common CV tasks, such as image classification, object localization, image segmentation, and object detection, Image Captioning is a more complex and heavy task.

Z. Cui (✉)
Ocean University of China, Qingdao, China
e-mail: cuizhao@stu.ouc.edu.cn

R. Kountchev (Deceased) et al. (eds.), *AI Methods and Applications in 3D Technologies*, Smart Innovation, Systems and Technologies 388,
https://doi.org/10.1007/978-981-97-2144-3_22

This task requires taking an image as input, analyzing its features to recognize the entities within, and further extracting the relationships among these features to obtain higher-level information on the connections among entities in the image. Then, by combining relevant NLP methods, the "image language" is translated into a human-readable text.

The main purpose of Image Captioning is to enable computers to learn to describe images, bringing the task's outcomes closer to humans' multisensory perception of the world. As a result, it holds significant importance in the fields of CV and NLP and takes a crucial part in advancing AI technology. It can be applied in medical imaging for auxiliary diagnosis, improving accessibility technology applications, and enhancing human–computer interaction, optimizing search engines, and intelligent robot design, etc.

The development of Image Captioning tasks has been relatively short. Traditional methods can be traced back to 2010 [1, 2], while deep learning-based Image Captioning emerged in 2014. Google's Neural Image Caption (NIC) [3] model, proposed in 2015, is the most representative and practical achievement. This paper first introduced the encoder-decoder framework for solving the Image Caption task, using CNN as the image feature extractor and encoder, and RNN as the extractor and decoder from image information to text language. In 2016 [4], the attention mechanism was introduced, and the encoder's structure was improved. In 2017, the transformer [5], a decoder based on Multi-head attention, was introduced for NLP tasks. With appropriate modifications, the transformer structure can be well applied to Image Captioning tasks and has achieved significant improvements.

Despite the remarkable progress in Image Captioning tasks achieved by deep learning in recent years, there are still many challenges: generating accurate, complete, and natural descriptions, and improving the model's generalization ability are key research breakthroughs needed.

We propose an improvement to the Image Captioning model that combines contrastive learning loss with traditional cross-entropy loss, while also enhancing the attention mechanism to Attention on Attention (AoA). Under these conditions, our model can achieve relatively better accuracy.

We proposes an Image Caption method based on a contrastive learning enhanced ResNet101 encoder and an Attention on Attention (AoA) Transformer decoder. The main contributions are as follows:

- The integration of a contrastive learning strategy to enhance the encoder's ability to represent image content. By employing contrastive learning, our model can distinguish images with similar visual features and better capture local information within the image.
- The utilization of data augmentation techniques to improve the model's generalization ability. We apply various data augmentation methods, such as random masking, cropping, flipping, and rotation, to increase the model's adaptability to images with different perspectives and scales.
- The adoption of the AoA module to optimize the Transformer decoder's attention mechanism, resulting in more accurate and natural descriptions. The AoA module

can better balance global and local context information, improving the model's performance when generating longer sentence descriptions.

22.2 Related Works

Traditional Image Caption Methods

Before the rise of deep learning methods, Image Caption tasks mainly relied on template-based [2] or retrieval-based [1] methods. Template-based methods generate descriptions using predefined grammatical structures, while retrieval-based methods retrieve the most matching description from a pre-built description library. The limitations of these methods mainly lie in the lack of diversity and flexibility in the generated descriptions and insufficient model generalization capability.

Application of Deep Learning in Image Caption

With the rapid development of deep learning, methods based on convolutional neural networks (CNNs) and recurrent neural networks (RNNs) have become the mainstream in Image Caption (Show and Tell [3]). Among them, attention-based methods, such as Show, Attend and Tell [4], have achieved remarkable results. Such methods typically employ a CNN as an encoder to extract image features and a RNN as a decoder to generate descriptions. By introducing the attention mechanism, the model can adaptively focus on key areas within the image to generate a more accurate description.

Contrastive-Based Visual Representation Learning

Contrastive learning methods were initially proposed to discover which data points are similar or dissimilar. They learn representations from unlabeled datasets (hence called self-supervised learning). In recent years, contrastive learning has made significant progress in visual representation learning. By contrasting the differences between positive and negative samples, models can learn more effective image representations. Representative works like SimCLR [6] and MoCo [7–9] demonstrate the effectiveness of contrastive learning in unsupervised and semi-supervised learning tasks. These methods inspired us to apply contrastive learning to Image Captioning tasks to improve the model's understanding of image content.

Transformer and Its Application in Image Caption

Since the introduction of [5], the Transformer architecture has achieved breakthrough results in the field of natural language processing. Compared with traditional RNNs, Transformers have stronger long-distance dependency modeling capabilities and higher parallelism (Fig. 22.1).

Transformer was initially introduced in machine translation tasks and quickly applied to various types of natural language processing and computer vision tasks, such as text generation, text classification, image classification, object detection, and object segmentation. In recent years, Transformer has been successfully applied to Image Caption tasks, such as Meshed-Memory Transformer [10]. These efforts confirm the potential of Transformer in generating image descriptions. At the same time, AoA optimizes the output information of the attention mechanism through a gated form to enhance its generalization ability.

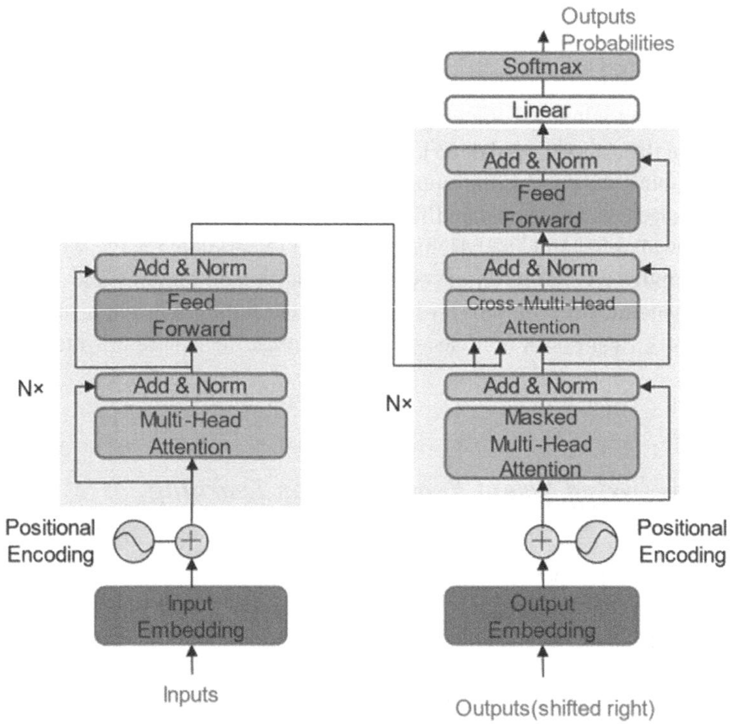

Fig. 22.1 Traditional transformer architecture

22.3 Methods

In conventional model training, the number of learning samples for the same feature in the commonly used datasets is usually very limited. In this case, when there are many types of samples and the feature similarity is close, our model will have difficulty in capturing the essential features accurately, resulting in unfavorable situations such as underfitting or overfitting. By using image processing methods such as masking, cropping, rotating, introducing noise, and adjusting chroma, we can increase the number of image sample learning and generate positive and negative examples. Then, based on this, contrastive learning can help the model better capture the key points of features and the essential differences between features.

In Image Captioning tasks, we need to encode the input image and decode the output text. This section introduces two new methods for image encoding and text decoding: an improved ResNet101 based on contrastive learning as the image input encoder, and a Transformer model based on attention mechanism and adaptive gating mechanism as the text output decoder.

ResNet Encoder Structure Using Contrastive Learning

Application of Contrastive Learning in Our Model

We use the contrastive learning method to improve the ResNet101 [11] model to enhance the quality of image encoding. In the training process, we use the contrastive learning method to learn the feature representation of the ResNet101 model. Specifically, we use a pair of noisy images as input, with one being a positive sample and the other a negative sample. We feed these two images into the ResNet101 model to extract their respective feature representations. Subsequently, we employ a contrastive loss function during the model training process. The objective of this loss function is to minimize the feature representation distance between positive samples, while maximizing the distance between negative samples.

The loss of contrastive learning uses Noise Contrastive Estimation (NCE) loss function, that is:

$$L_{NCE}(x_i, x_j) = -\log\frac{\exp(\mathrm{sim}(x_i, x_j)/\tau)}{\sum_{k=1,k\neq i}^{2N}\exp(\mathrm{sim}(x_i, x_k)/\tau)} \tag{22.1}$$

The output of the MLP layer is where the objective function of contrastive learning works. Assuming we have a batch of N samples, the total number of samples in the batch is 2N, with N pairs of positive samples obtained through data augmentation. In the SimCLR [6] framework, for each positive sample pair, the remaining 2(N−1) samples serve as negative samples. It is important to note that these negative samples are generated from the data within the same batch. The $\mathrm{sim}(x_i, x_j)$ is the cosine

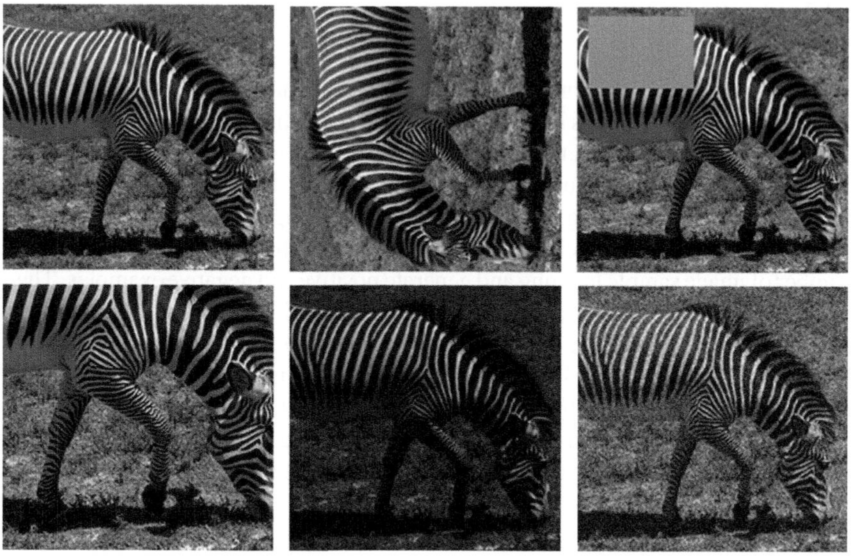

Fig. 22.2 Examples of data augmentation

similarity calculation formula in the matrix space, $\mathrm{sim}(u, v) = \frac{u^T v}{||u|| \cdot ||v||}$, τ is the temperature coefficient.

In the proposed approach, the focus is on calculating the distance between positive sample pairs, while the negative samples only contribute to the denominator of the contrastive loss function. Through training, the loss function gradually decreases, leading to a smaller distance between positive sample pairs. Meanwhile, a larger distance is encouraged between negative sample pairs, resulting in a smaller overall loss.

To augment the training data and enhance the model's generalization ability, we utilize the following data augmentation techniques in contrastive learning:

These include: random masking, random cropping, random affine transformations (rotation, reflection, and shear mapping), adding Gaussian noise, and random color jittering (brightness, contrast, and saturation). We show examples of data augmentation processing for an image in Fig. 22.2.

Our overall model structure is showed in Fig. 22.3.

ResNet101 Model's Framework

We use the ResNet101 model based on contrastive learning to extract the feature representation of the image. The ResNet101 model is a deep convolutional neural network composed of numerous residual blocks. Its main feature is to solve the gradient vanishing problem in deep neural networks through shortcut connections,

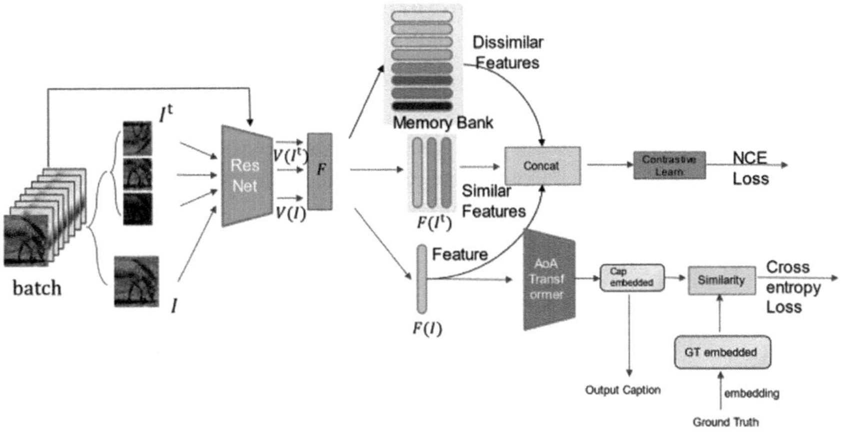

Fig. 22.3 Application of contrastive learning and overall model structure

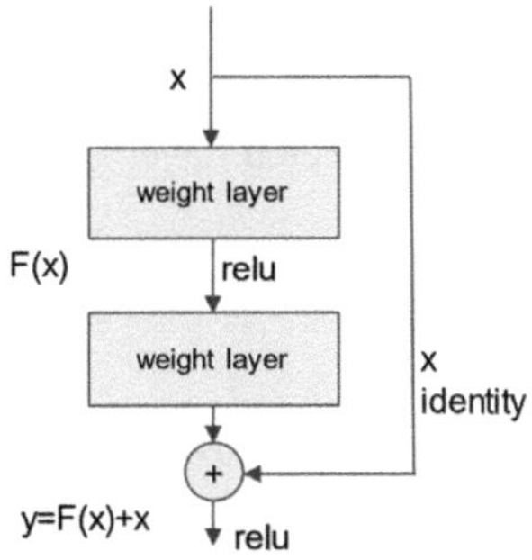

Fig. 22.4 One of the residual blocks

allowing the network to perform deeper training more effectively. The residual learning unit is shown in the following Fig. 22.4.

Since the residual will not actually be zero, this also allows deep stacked layers to generally learn new features based on the input features, giving the model better performance.

The ResNet101 model consists of 5 residual block groups, each containing multiple residual blocks. In each residual block group, the size and depth of the feature maps change, with the depth increasing by 4 times. In each residual block, we use two convolutional layers, each with a 3×3 kernel size, followed by a BN (batch normalization) layer and a ReLU activation function. The input and output of each residual block are connected through shortcut connections to avoid the gradient vanishing problem.

Transformer Text Decoder Structure Improved by AoA

The Transformer model, which utilizes the self-attention mechanism, is an encoder-decoder model widely recognized for its remarkable performance in machine translation tasks. In our approach, we use the Transformer model as the text output decoder, incorporating attention and adaptive gate control mechanisms. This choice allows us to enhance the quality and efficiency of text generation.

Optimization of Attention Mechanism

Based on the traditional self-attention mechanism (Fig. 22.5), we introduced the AoA (Attention on Attention) module [12], depicted in Fig. 22.6. The AoA module assigns a weight value to each attention weight to optimize attention allocation.

Attention mechanism: Multi-head attention consists of multiple one-head attention. A multi-head attention has n heads, and the weights of the i head are ${W_i}^Q$, ${W_i}^K$, ${W_i}^V$. The input q, k and v matrices input each one-head attention respectively. Each head output matrix is stitched according to the characteristic (dim) dimension to get a new matrix, multiplying W^O to get the output (in fact, it can also be a fully connected Layer Linear), and the output shape is still the same.

$$f_{Att}(Q, K, V) = Softmax(\frac{QK^T}{\sqrt{d_k}})V \tag{22.2}$$

$$MultiHead(Q, K, V) = Concat(head_1, ..., head_n)W^O \tag{22.3}$$

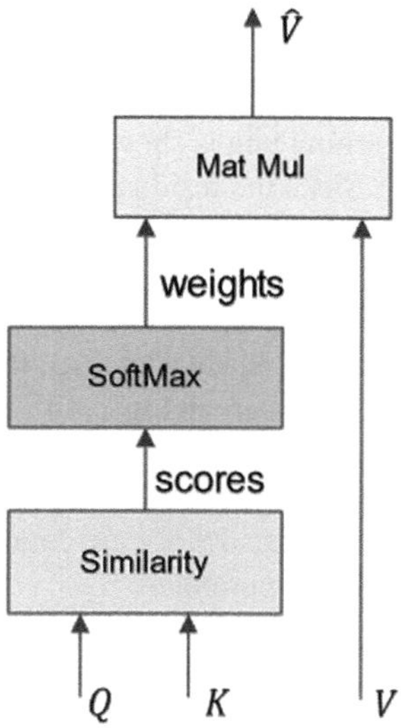

Fig. 22.5 Traditional attention mechanism

Fig. 22.6 Attention on attention module

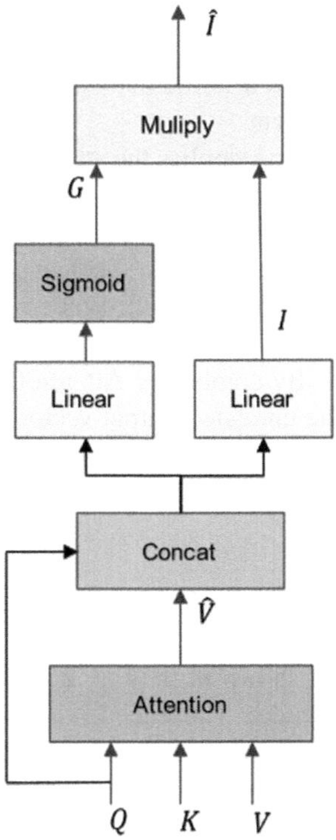

Among this,

$$head_i = f_{Att}(qW_i{}^Q, kW_i{}^K, vW_i{}^V) \tag{22.4}$$

The f_{Att} (attention mechanism) produces weighted averages for each query, irrespective of the relationship between Q (query) and K/V (key/value). Consequently, the attention module may generate weighted average vectors that contain irrelevant or misleading information, even in the absence of a correlated vector.

Therefore, the AoA module (Fig. 22.6) is introduced to measure the correlation between attention outcome and the query. The AoA can produces a "text information vector" I and a "note gate vector" G through separate linear transformations. These transformations are conditioned on the attention outcome and the current contextual query (referred to as Q):

$$G(Q, K, V) = W_q^g Q + W_v^g f_{Att}(Q, K, V) + b^g \tag{22.5}$$

$$I(Q, K, V) = W_q^i Q + W_v^i f_{Att}(Q, K, V) + b^i \tag{22.6}$$

where W_q^i, W_v^i, W_q^g, $W_v^g \in R^{D\times D}$, b^i, $b^g \in R^D$, and D is the dimension of Q and V.

$$AoA(f_{Att}, Q, K, V) = \sigma(G(Q, K, V)) \odot I(Q, K, V) \tag{22.7}$$

where σ is the activation function of sigmoid, where ⊙ means multiplication by element.

AoA applies the attention gate by performing element-wise multiplication on the information vector. By incorporating a gate-controlled mechanism, it enhances the accuracy of attention allocation. The AoA module can be embedded in each self-attention layer of the Transformer decoder. Specifically, we insert the AoA module between the multi-headed self-attention mechanism and the feed-forward neural network (FFN). This enables the model to optimize attention at multiple levels.

By combining Attention on Attention with transformer, we can avoid querying the unrelated output vectors of the attention layer in the vector, so that the decoder model is clearer when learning how to combine visual and language characteristics, so that the model can generate better captions.

AoA-Transformer's Architecture

As shown in the Fig. 22.7, the model includes an N-head encoder and an N-head decoder. The encoder is based on the traditional Transformer model, which can convert the input image feature representation into semantic encoding feature representations. In each decoder layer, we use the self-attention mechanism to capture semantic information in the annotations and the dual-layer attention mechanism to interact and fuse information in images and annotations. In the adaptive gating mechanism, we use the gating vector G and the gate control ratio to adjust the parameters of the gate control mechanism to better meet the requirements of Image Caption tasks.

Training and Optimization Strategy

Training Goal and Loss Function

Training goal is to minimize the negative log-likelihood of the description, that is,

$$L = -\sum_{i=1}^{N} log P(y_i|x_i) \tag{22.8}$$

where x_i is the input image and y_i is the correct description of the output. In order to introduce a comparative learning strategy, we combine NCE loss with negative logarithmic likelihood loss to form a total loss function:

$$L_{total} = L + \lambda L_{NCE} \tag{22.9}$$

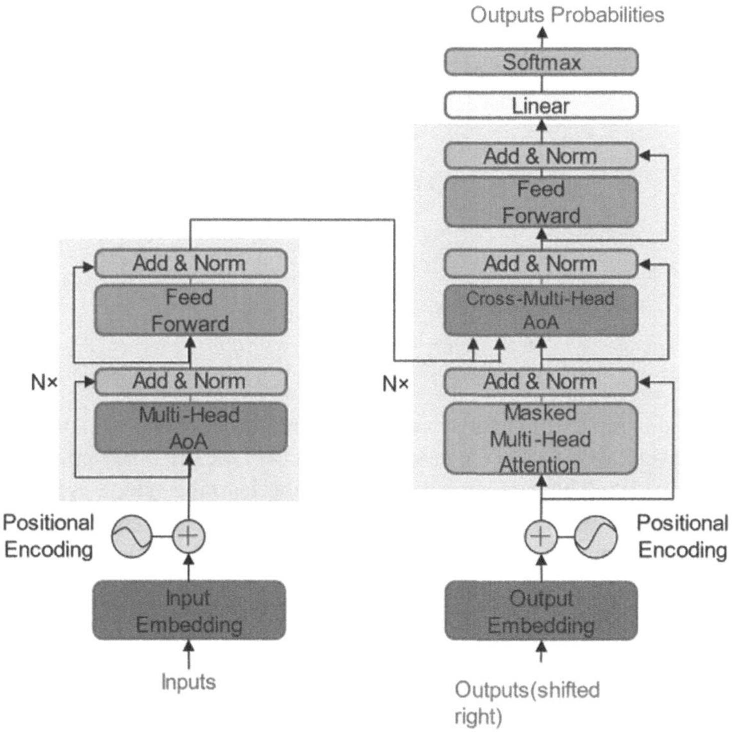

Fig. 22.7 Transformer text decoder structure improved by AoA

where λ is used to control the weight of the two loss functions.

Optimization Algorithm and Learning Rate Adjustment Strategy

During the training process, we use Cosine Annealing strategy [13] to adjust the learning rate, that is:

$$\eta_t = \eta_{min} + \frac{1}{2}(\eta_{max} - \eta_{min})(1 + \cos(\frac{\pi t}{T})) \tag{22.10}$$

where η_t is the learning rate of the tepoch, η_{min} and η_{max} are the minimum and maximum learning rates respectively, and T is the total number of training epoch.

We used AdamW [13] optimizer to optimize the loss function, which is a gradient descent algorithm with adaptive learning rate. In each training cycle, we randomly select a batch of data pairs, calculate Loss function and update the model parameters. The relevant formulas of AdamW optimizer are as follows:

$$m_t = \beta_1 m_{t-1} + (1 - \beta_1) g_t \tag{22.11}$$

$$v_t = \beta_2 v_{t-1} + (1 - \beta_2) {g_t}^2 \tag{22.12}$$

$$\widehat{m}_t = \frac{m_t}{1 - \beta_1^t} \tag{22.13}$$

$$\widehat{v}_t = \frac{v_t}{1 - \beta_2^t} \tag{22.14}$$

$$\theta_t = \theta_{t-1} - \frac{\eta}{\sqrt{\widehat{v}_t} + \xi} \widehat{m}_t - \lambda \eta \theta_{t-1} \tag{22.15}$$

where m_t and v_t represent the first-order and second-order momentum of the tth time step respectively, θ_t represent the model parameters of the t-th time step, g_t represent the gradient of the t-th time step, η indicates the learning rate, ξ is a very little number to prevent dividing by 0, λ is the attenuation coefficient in L2 regularization, $\widehat{m}_t$ and v_t are the modified estimates of deviation. In the AdamW optimizer, β_1 and β_2 represent the attenuation coefficients of first-order momentum and second-order momentum respectively, usually set to 0.9 and 0.999. The initial learning rate is set to 1e-4.

Regularization and Overfitting Control

To prevent overfitting, we adopt the following strategies:

(1) Weight decay: Add an L2 regularization term to the loss function to penalize excessively large weight values. The L2 regularization term formula is as follows.

$$L_{reg} = \lambda \cdot \sum_{i=1}^{n} {w_i}^2 \tag{22.16}$$

where λ is the attenuation coefficient, set to 1e-4, n is the number of weights, and w_i is the weight value.

After adding a regular term, the total loss function is:

$$L_{total} = L + \lambda L_{NCE} + L_{reg} \tag{22.17}$$

(2) Dropout: In the Transformer decoder, we add the Dropout layer between the AoA layer and the FFN layer to randomly discard some neurons. Dropout probability is set to 0.1.
(3) Early Stopping: Monitor the performance of the model on the validation set. Stop training when the performance of the validation set has not improved for

several consecutive periods and roll back the model parameters to the state with the best performance.

In summary, we propose an Image Captioning method based on a contrastive learning ResNet101 encoder and an AoA-enhanced Transformer decoder. The method introduces contrastive learning strategies, data augmentation, AoA module, and training and optimization strategies to improve the model's performance in Image Captioning tasks, achieving high-quality Image Caption generation. The experiment displays that the execution of this model performs really good in image subtitle generation tasks.

22.4 Experiments

Datasets and Preprocessing

This study uses MS COCO [14], Flickr8k [15] and Flickr30k [16] datasets for training and evaluation. The MS COCO dataset contains approximately 120 k training images, 5 k validation images, and 40 k test images. These datasets contain numerous images along with their corresponding descriptive texts, providing rich resources for training and evaluating the model. For preprocessing, we first resize the images to a fixed size (300 $\times$ 300) and normalize them. By adjusting the image to a smaller and identical size, the computational complexity of the model can be reduced, and the training, reasoning, and convergence of the model can be accelerated. For the text data, we tokenize the sentences and use a word embedding matrix to map words to fixed-dimension vectors.

We implement the proposed model using PyTorch. During training process, we utilize AdamW to optimize the learning rate through a Cosine Annealing adjustment strategy. For contrastive learning and data augmentation, we adopt methods such as random cropping, horizontal flipping, and color jittering.

Evaluation Metrics

To comprehensively evaluate the proposed method, we use four commonly used evaluation metrics: BLEU, ROUGE-L, CIDEr, and METEOR [17]. BLEU is calculated by comparing the difference between machine translation output and one or more reference translations. The metric is based on the degree of matching between a set of predicted N-grams (from 1 to N) and N-grams in the reference translation. ROUGE-L is a calculation method based on the longest common subsequence, used to measure the similarity between the generated summary and reference summary. It not only considers the matching of words or phrases but also their positions in the text, thus better reflecting the coherence and consistency of the generated text.

CIDEr computes the similarity score between the description generated by the model and the reference description, then uses some algorithms (e.g., averaging or maximizing) to aggregate the scores into a single metric, which is more in line with human subjective evaluation. METEOR calculates similarity scores by converting candidate translations and reference translations into sets of feature vectors and then measuring the similarity between them. These metrics have their own advantages and measure the similarity between the generated captions and reference captions from different perspectives.

Experimental Results

We evaluated the performance of our proposed model using the aforementioned metrics and compared it with the AoA baseline model and its previous state-of-the-art (SOTA) counterparts. This comparison allowed us to assess the practical improvements brought by our model from a scoring perspective (Table 22.1).

Upon comparison, we observed that our model achieved improvements across all metrics. Compared to AoANet, our model showed a 0.8 and 0.3 percentage point increase in BLEU@1 and BLEU@4, respectively, and a 0.4, 0.3, and 2.6 percentage point increase in METEOR, ROUGE-L, and CIDEr, respectively. While our focus was primarily on optimizing and comparing a single model scenario, it is apparent that there is potential for further improvement in the performance of our model when applied in Ensemble/Fusion settings. It should be noted that our model has demonstrated a notable enhancement in the accuracy of generated descriptions. However, it is important to acknowledge that additional enhancements can be achieved by incorporating optimization training in Ensemble/Fusion situations or utilizing other Score Optimization techniques.

Table 22.1 Erformance evaluation on COCO datasets, with B@N, M, R, and C representing BLEU@N, METEOR, ROUGE-L, and CIDEr scores respectively. All values are presented as percentages (%)

	B@1	B@2	B@3	B@4	M	R	C
NIC [3]	66.6	46.1	32.9	24.6	–	–	–
NIC + Att [4]	71.8	50.4	35.7	25.0	23.0	–	–
LSTM-A [18]	75.4	–	–	35.2	26.9	55.8	108.8
RFNet [19]	76.4	60.4	46.6	35.8	27.4	56.8	112.5
Up-Down [20]	77.2	–	–	36.2	27.0	56.4	113.5
GCN-LSTM [21]	77.3	–	–	36.8	27.9	57.0	116.3
AoANet [12]	77.4	–	–	37.2	28.4	57.5	119.8
Ours	**78.2**	**61.7**	**48.1**	**37.5**	**28.8**	**57.8**	**122.4**

Experimental Analysis

Simply observing the metrics does not fully reflect the value of the improvements made to the model. By examining the following image captioning examples, we can further analyze the actual performance of our model and assess whether its semantic understanding aligns with human habits. By comparing the output results of the Baseline model (AoA-LSTM), we can further analyze whether our attention mechanism has been improved.

As shown in Fig. 22.8, it can be observed that our model can capture the main information with higher relevance to the image's theme, and the generated textual descriptions are more accurate. Therefore, it can be said that the attention allocation of our model is more precise and efficient.

22.5 Conclusions and Future Work

We propose an Image Captioning method based on a contrastive learning ResNet101 encoder and an Attention on Attention (AoA) Transformer decoder. The experimental results indicate that our method has achieved significant performance on commonly used Image Captioning datasets. Moreover, our approach exhibits clear advantages in handling complex scenes, multiple objects, and detailed descriptions.

Although our method has achieved good performance in Image Captioning tasks, there are still some aspects that can be improved. Future research directions include:

- Exploring more efficient and stable contrastive learning strategies to further improve the quality of image representations.
- Research more advanced attention mechanisms to better capture the complex relationship between images and text.
- Seeking better encoder-decoder structures, such as using Swin Transformer [22] for feature extraction optimization in the encoder and employing diffusion models [23] for language generation optimization.
- Apply the methods of this article to other multi-modal tasks, such as video subtitles, video grounding, visual Q&A, etc., to explore its generalizability and adaptability in different tasks.
- Combine other natural language processing technologies, such as pre-training language models, to further improve the accuracy and semantic coherence of generated captions.

By addressing these future research directions, we aim to enhance the performance and broaden the applicability of our proposed Image Captioning approach.

	Ours: A large jetliner sitting on an airport runway with mountains behind it. Baseline: A large jetliner sitting on the grass. GT1:A large jetliner sitting on top of an airport runway. GT2:A passenger plane sits on a runway with mountains behind it. GT3:An airplane landing on a runway in an airport.
	Ours: A man and a young girl are riding a skateboard. Baseline: A group of people riding skateboards down a street. GT1: A man riding a skateboard next to a woman. GT2: The skateboarder has a passenger on his skateboard. GT3: A girl hitching a ride with her dad on a skateboard.
	Ours: A book sitting on top of a wooden bench. Baseline: A wooden bench sitting next to a wooden bench. GT1: A book on a wooden bench on a street. GT2: A book sitting on top of a wooden desk. GT3: A california travel guide book on a park bench.
	Ours: A man in a hat holding a frisbee in his hand. Baseline: A man is standing next to a car on a water. GT1: A man holding a white object in his hand. GT2: A man wearing a hat and holding an object in his hand. GT3: A man holding a frisbee in a parking lot near water.
	Ours: A cat is laying on a bed with a blanket. Baseline: A cat laying on top of a bed. GT1: A cat under a blanket looking at something. GT2: A cat is peeping out from under a blanket. GT3: A gray tiger cat sleeping on a bed under a blanket.

Fig. 22.8 Examples of captions produced by CAA (Ours method) and AoA-LSTM (A baseline model), along with the corresponding ground truth captions

References

1. Torralba, A., Russell, B.C., Yuen, J.: LabelMe: online image annotation and applications. P Ieee **98**(8), 1467–1484 (2010). https://doi.org/10.1109/Jproc.2010.2050290
2. Aydin, Z., Murray, J.I., Waterston, R.H., Noble, W.S.: Using machine learning to speed up manual image annotation: application to a 3D imaging protocol for measuring single cell gene expression in the developing C. elegans embryo. Bmc Bioinformatics, vol. 11 (2010). Artn 8410.1186/1471–2105–11–84
3. Vinyals, O., Toshev, A., Bengio, S., Erhan, D.: Show and tell: a neural image caption generator. 2015 IEEE Conference on Computer Vision and Pattern Recognition (Cvpr), pp. 3156–3164 (2015). https://doi.org/10.1109/cvpr.2015.7298935
4. Xu, K. et al.: Show, attend and tell: neural image caption generation with visual attention. Pr. Mach. Learn. Res. **37**, 2048–2057 (2015)
5. Vaswani, A. et al.: Attention is all you need. Adv. Neur. In. **30** (2017). https://doi.org/10.48550/arXiv.1706.03762
6. Chen, T., Kornblith, S., Norouzi, M., Hinton, G.: A Simple Framework for Contrastive Learning of Visual Representations. arXiv (2002). https://doi.org/10.48550/arXiv.2002.05709.
7. Chen, X., Xie, S., He, K.: An Empirical Study of Training Self-Supervised Vision Transformers. arXiv (2104). https://doi.org/10.48550/arXiv.2104.02057
8. Chen, X., Fan, H., Girshick, R., He, K.: Improved baselines with momentum contrastive learning. arXiv (2003). https://doi.org/10.48550/arXiv.2003.04297
9. He, K., Fan, H., Wu, Y., Xie, S., Girshick, R.: Momentum Contrast for Unsupervised Visual Representation Learning. arXiv (2020). https://doi.org/10.48550/arXiv.1911.05722
10. Cornia, M., Stefanini, M., Baraldi, L., Cucchiara, R.: Meshed-Memory Transformer for Image Captioning. arXiv (2020). https://doi.org/10.48550/arXiv.1912.08226
11. He, K.M., Zhang, X.Y., Ren, S.Q., Sun, J.: Deep residual learning for image recognition. In: Proceedings of the IEEE Conference on Computer Vision and Pattern Recognition, pp. 770–778 (2016). https://doi.org/10.1109/Cvpr.2016.90
12. Huang, L., Wang, W.M., Chen, J., Wei, X.Y.: Attention on attention for image captioning. In: Proceedings of the IEEE/CVF International Conference on Computer Vision, pp. 4633–4642 (2019). https://doi.org/10.1109/Iccv.2019.00473
13. Loshchilov, I., Hutter, F.: Decoupled Weight Decay Regularization. p. arXiv:1711.05101. https://doi.org/10.48550/arXiv.1711.05101
14. Lin, T.Y., et al.: Microsoft COCO: common objects in context. Lect. Notes Comput. Sc. 8693, 740–755 (2014). https://doi.org/10.1007/978-3-319-10602-1_48
15. Hodosh, M., Young, P., Hockenmaier, J.: Framing image description as a ranking task: data, models and evaluation metrics. J. Artif. Intell. Res. 853–899 (2013)
16. Young, P., Lai, A., Hodosh, M., Hockenmaier, J.: From image descriptions to visual denotations: new similarity metrics for semantic inference over event descriptions. TACL **2**, 67–78 (2014)
17. Chen, X., et al.: Microsoft COCO Captions: Data Collection and Evaluation Server. arXiv (2015). https://doi.org/10.48550/arXiv.1504.00325
18. Yao, T., Pan, Y., Li, Y., Qiu, Z., Mei, T.: Boosting Image Captioning with Attributes. arXiv (2016). https://doi.org/10.48550/arXiv.1611.01646
19. Jiang, W., Ma, L., Jiang, Y.G., Liu, W., Zhang, T.: Recurrent Fusion Network for Image Captioning. arXiv (2018). https://doi.org/10.48550/arXiv.1807.09986
20. Anderson, P., et al.: Bottom-Up and Top-Down Attention for Image Captioning and Visual Question Answering. arXiv (2017). https://doi.org/10.48550/arXiv.1707.07998
21. Yao, T., Pan, Y., Li, Y., Mei, T.: Exploring Visual Relationship for Image Captioning. arXiv (2018). https://doi.org/10.48550/arXiv.1809.07041
22. Liu, Z., et al.: Swin transformer: hierarchical vision transformer using shifted windows. In: 2021 Ieee/Cvf International Conference on Computer Vision (Iccv 2021), pp. 9992–10002 (2021). https://doi.org/10.1109/Iccv48922.2021.00986
23. Smith, S.J., et al.: Continual Diffusion: Continual Customization of Text-to-Image Diffusion with C-LoRA. arXiv (2023). https://doi.org/10.48550/arXiv.2304.06027

Chapter 23
Vehicle Scheduling Problem of Multiple Distribution Centers Based on Improved Ant Colony Algorithm

Caolu Wang

Abstract Optimizing Vehicle Routing Problem (VRP) has become an important element to reduce the pressure in the logistics industry. VRP problems with more constraints are usually solved with heuristic algorithms, so this article adopts the Ant Colony Optimization (ACO) algorithm to address the university meal distribution problems under the epidemic situation. For the problem that the ant colony algorithm easily falls into the local optimum, the optimal path obtained in each iteration is optimized by adding 2-optimization (2-opt). The improved algorithm is applied to the vehicle scheduling problem of multiple distribution centers, and it's indicated that the algorithm has practical significance compared with dynamic programming algorithm and ant colony algorithm without 2-opt.

23.1 Introduction

Although China's logistics industry is developing rapidly, under the situation that the new coronavirus pneumonia epidemic is spreading, it is facing huge pressure in transportation, personnel control and other aspects [1], so optimizing the vehicle path has become an important element to reduce the pressure of the logistics industry. At the same time, in order to effectively prevent the spread of virus, major universities across the country have been implementing online teaching, and some schools have even implemented an online teaching mode of closed accommodation [2], which has led to the problem of student meal delivery. As a part of the logistics industry chain, distribution of meals needs being optimized to effectively address the demands of college students.

Vehicle routing problem (VRP) aims to achieve the goal of the shortest vehicle path with the minimum total costs, the shortest consumed time and so on under the given constraints by means of which the distribution center arranges the route and

C. Wang (✉)
Department of Applied Chemistry, Faculty of Science, Northeastern University, Shenyang 110000, Liaoning Province, China
e-mail: 20202018@stu.neu.edu.cn

R. Kountchev (Deceased) et al. (eds.), *AI Methods and Applications in 3D Technologies*, Smart Innovation, Systems and Technologies 388,
https://doi.org/10.1007/978-981-97-2144-3_23

time of the vehicles reasonably to deliver different quantities of goods requested by each of certain number of customers from the distribution center to the customer in a system which exists a relationship of supply and demand [3]. The main methods for solving such problems are exact algorithms, heuristic algorithms and meta-heuristic algorithms, etc. Moreover, the optimal solution cannot be obtained when the problem size increases so that the meta-heuristic algorithm is usually used to solve complex practical problems [4]. The heuristics commonly used to solve VRP consist of genetic algorithm [5], simulated annealing algorithm [6], forbidden search algorithm [7], particle swarm algorithm [8], ant colony algorithm [9], etc. Le Thi Diem Chau et al. proposed three algorithms such as guided local search, forbidden search and simulated annealing for the tractor truck routing problem to optimize the total logistics cost and the amount of internal tractor trucks used [10]. Zaghdoud R et al. proposed an improved particle swarm method to solve the automatic guided vehicle scheduling so that total travel distance of all the automatic guided vehicles is minimized [11].

In addition to other heuristic algorithms, Chinese and foreign scientists have made some progress in ant colony algorithm. Wang Xiufan et al. improved the 2-opt algorithm according to the ant colony algorithm for the multi-objective optimization problem in logistics delivery, which has faster pace and reduces the cost of logistics delivery with effect [12]. Wang et al. used a new heuristic function and pheromone update strategy to propose an optimized colony algorithm [13]. Pala et al. simplified the service by service vehicle between hotels and airports as a multi-objective vehicle path problem with capacity constraints and proposed an ant colony optimization algorithm to reduce the visit time [14]. Yanqin et al. studied the constraints of emergency logistics and a mathematical model of the emergency logistics vehicle routing problem was established by using an ant colony algorithm [15]. So in view of the problem that the ant colony algorithm easily falls into the local optimum, the optimal path in this article obtained in each iteration is optimized by adding 2-optimization (2-optimization). The optimized algorithm is used for the vehicle scheduling problem of multiple distribution centers, and it's indicated by the experimental results that the algorithm has practical significance compared with dynamic programming algorithm and ant colony algorithm without 2-opt.

23.2 Question Formulation

Preliminary Research

In the campus under the serious situation of the epidemic, school cafeterias accomplish the task of delivering boxed lunches to students to effectively mitigate the spread of the virus through delivery vehicles. The location of student dormitories and cafeterias, the demand of each student, the capacity of the vehicles, and the weight of the students' boxed lunches are known. The vehicles have the following constraints when making boxed lunch deliveries:

(1) The cafeteria's box lunches can meet the demand of all students and there are enough vehicles to make the delivery;
(2) The boxed meals in demand can be mixed;
(3) The boxed meals demanded by each student cannot exceed the load limit of each vehicle;
(4) Only one vehicle can be used to transport the boxed meals required by each student;
(5) The vehicle has a transportation volume limit;
(6) The vehicle leaves the cafeteria and eventually goes back to the cafeteria.

Building the Model

The purpose of the problem is to discover a route with the lowest cost while satisfying the above constraints. In the vehicle distribution model of multiple distribution centers established based on the above analysis, considering the actual situation of Northeastern University campus, the distribution center of canteens and customer points of dormitories can be considered as a connected network G = (V, E, A), where the vertex set V denotes the set of N canteens and P dormitories. $\{v_1, v_2, \cdots, v_N\}$ indicates N canteens, $\{v_N+1, v_N+2, \cdots, v_P\}$ indicates P dormitories. E is an undirected set of edges indicating the time cost between each point, and A is a directed set of arcs indicating the existence of feasible paths between each point. Edges of E and arcs of A are assigned values (indicating distance, time cost, etc.), and 'V', 'A', and 'E' are subsets of V, A, and E respectively. The objective function is made optimal under the satisfaction of constraints (customer's demand, vehicle's load, vehicle's volume, etc.). The objective function can be taken as the shortest total vehicle travel, the lowest cost of delivery, the shortest delivery time, the least number of vehicles to be delivered, etc. Assume that $d_i(n)$ denotes the distance from the ith dormitory to the nth canteen, where N denotes the total amounts of canteens. Assume that the set is $D_i(n) = \{d_i(n)\}$, $n = 1, 2, \cdots, N$, $d_{im} = \min D_i$, then the ith dormitory is delivered by the mth canteen.

$$\min Z = \sum_{i=1}^{N} \sum_{j=1}^{L_i} \left[\sum_{k=1}^{n_{ij}} d_{r_{ij(k-1)} r_{ijk}} + d_{r_{ij} r_{ijl_j}} \delta(n_{ij}) \right] \quad (23.1)$$

s.t.

$$\sum_{k=1}^{n_{ij}} q_{r_{ijk}} \leq Q_{ij} \quad (23.2)$$

$$\sum_{k=1}^{n_{ij}} t_{r_{ijk}} \leq T_{ij} \quad (23.3)$$

$$\sum_{i=1}^{N}\sum_{j=1}^{L_i} n_{ij} = P \tag{23.4}$$

$$R_{ij} \in \{r_{ijk} | r_{ijk} \in [1, 2, \cdots, P], k = 1, 2, \cdots, n_{ij}\} \tag{23.5}$$

$$R_{ij(1)} \cap R_{ij(2)} = \varnothing, \forall ij(1) \neq ij \tag{23.6}$$

$$\delta(n_{ij}) = \begin{cases} 1, n_{ij} \geq 1 \\ 0, otherwise \end{cases} \tag{23.7}$$

In this model, (1) equation is the objective function: the shortest total distribution route; (2) equation shows that the total weights of boxed meals on each road shall not exceed the load capacity of the delivery car; (3) equation indicates that the volume of box lunch on each road shall not exceed the transportation volume of the delivery car; (4) equation indicates that the needs of each dormitory student can be met; (5) equation indicates the composition of students need to be delivered meals to on each path; (6) equation restricts each customer to be delivered by only one distribution vehicle; (7) equation indicates that when the amounts of students served by the jth cars of the ith distribution center is larger than or equal to 1, i.e., $n_{ij} \geq 1$,which indicates that this vehicle is involved in distribution, then δ $(n_{ij}) = 1$;when the jth vehicle of the ith distribution center serves a number of customers larger than or equal to 1, i.e., $n_{ij} < 1$, which means that this vehicle is not involved in the service, then δ $(n_{ij}) = 0$.

Parameters

See Table 23.1.

23.3 Ant Colony Algorithm

Assume that there are m delivery vehicles and each delivery vehicle needs to traverse each dormitory, let that $\tau_{ij}(t)$ denotes the pheromone value between dormitory i and dormitory j at the tth moment. The distribution that vehicles select the next moving dormitory based on the state transfer probability, i.e., Pk denotes the state transfer probability of the kth (k = 1, 2, …, m) distribution vehicle moving from dormitory i to dormitory j, which is obtained by combining several parameters with the following equation:

Table 23.1 Variables

Variables	Definition
N	Denotes the number of delivery canteens
P	Denotes the number of student dormitories
L_i	The number of distribution cars of the *i*th delivery canteen
Q_{ij}	The load of the *j*th car of the *i*th delivery canteen
T_{ij}	Volume of the *j*th car in the *i*th delivery canteen
q_k	Cargo weight of the *k*th student dormitory
t_k	Volume of boxed meals at the *k*th student dormitory
d_{ij}	Distance from student dormitory *i* to student dormitory *j*
n_{ij}	The number of student dormitory delivered by the *j*th car of the *i*th delivery canteen
R_{ij}	The path of the *j*th car of the *i*th delivery canteen
r_{ijk}	The *k*th student dormitory for the *j*th car of the *i*th delivery canteen
r_{ij0}	Denotes the delivery canteen

$$p_i^k(t) = \frac{[\tau_{ij}(t)]^\alpha [\eta_{ij}(t)]^\beta}{\sum_{s \in J_k(i)} [\tau_{is}(t)]^\alpha [\eta_{is}(t)]^\beta}, j \in J_k(i) \tag{23.8}$$

$$\begin{cases} dx(t) = [Ax(t-\delta) + Bu(t) + Cw(t)]dt + DdW(t), t \in (0, T] \\ x(s) = \xi(s), s \in [-\delta, 0] \end{cases} \tag{23.9}$$

$$J_k(i) = 1, 2, \cdots, n - tabu_k \tag{23.10}$$

$\eta_{ij}(t)$ is a heuristic function defined according to its own adaption, and α, β denote the relative importance of the pheromone and the heuristic factor respectively.

$$\eta_{ij} = \frac{1}{d_{ij}} \tag{23.11}$$

The distance between dormitory i and j is denoted by d_{ij}, and d_{ij} and $\eta_{ij}(t)$ are inversely related. The smaller d_{ij} is, the larger $\eta_{ij}(t)$ is, then the state transfer probability p is also the larger so that $\eta_{ij}(t)$ represents the expected value of the transfer of a distribution vehicle from dormitory i to the next dormitory j. After all distribution vehicles complete the task of visiting all dormitories in one iteration, the pheromones need to be updated in a timely manner in order to avoid accumulating too many pheromones resulting in covering the path-inspired information that has not yet been explored.

$$\tau_{ij}(t+n) = (1-\rho)\tau_{ij}(t) + \Delta\tau_{ij} \tag{23.12}$$

$$\Delta\tau_{ij} = \sum_{k=1}^{m} \Delta\tau_{ij}^{k},$$

$$\Delta\tau_{ij}^{k} = \begin{cases} \frac{Q}{L_k}, \text{ if the ant passed the requirement points i and j in this iteration} \\ 0, \mathit{otherwise} \end{cases} \tag{23.13}$$

ρ represents the pheromone volatility factor, which directly affects the global search ability and convergence speed of the whole ACO, and (1 − ρ) is the pheromone residual factor, which reflects the strength of the mutual influence between individuals of the ant mosquito. The range of values of ρ is $\{\rho \mid 0 < \rho < 1\}$ and initial $\tau_{ij}(0)=0$.$\Delta\tau ij$ is the increment of pheromone in this iteration, which can be expressed as:

$$\Delta\tau_{\sigma} = \frac{Q}{L_k} \tag{23.14}$$

Q is a constant that represents the fixed value of the pheromone carried by the mosquito itself; L_k represents the path traveled by the delivery vehicle k in this iteration in the formula.

23.4 Improved Ant Colony Algorithm

(1) The next dormitory to be chosen is selected by the roulette wheel method of the genetic algorithm;
(2) The optimal solution is updated using the 2-opt method;
(3) Comparison with dynamic programming method and ant colony algorithm without 2-opt.

Roulette Algorithm

The roulette selection method selects individuals based on the selection percentage of each individual in proportion to its fitness value, and the larger the fitness, the larger the selection percentage. Assuming the fitness value of an individual x_i as f(x_i), the percentage of the individual being selected as p(x_i), and the cumulative percentage as q(x_i), the formula are as follows:

$$p(x_i) = \frac{f(x_i)}{\sum_{j=1}^{N} f_j}, q(x_i) = \sum_{j=1}^{i} p(x_j) \tag{23.15}$$

The cumulative probability in the above equation represents the sum of the selection probabilities of all individuals before each individual, and the specific procedure is as follows:

(1) Calculate the selection percentage $p(x_i)$ for each delivery canteen
(2) Calculate the cumulative percentage $q(x_i)$ of each delivery canteen
Generate a random array m with elements ranging from 0 to 1, and sort them from minimum to maximum. If the cumulative percentage $q(x_i)$ is larger than the element $m(i)$ in the array, the delivery canteen $x(i)$ is selected;If it is smaller than $m(i)$, the next delivery canteen $x(i+1)$ is compared until an delivery canteen is selected.
(3) If N delivery canteens need to be rotated to, then repeat step (3) N times.

2-opt Algorithm

In this study, the 2-opt algorithm is added to the ant colony algorithm to adjust the optimal solution obtained by the ant colony algorithm each time in a multi-strategy way to help the ant colony algorithm jump out of the local optimum.2-opt heuristic algorithm starts from an arbitrary feasible solution and repeatedly replaces two edges in the solution with other two edges outside the solution. In each improvement step, the 2-opt algorithm selects two edges u_1, u_2 and v_1, v_2 from the tour so that u_1, u_2, v_1, v_2 are different and appear in this order in the tour, and it replaces these edges with edges u_1, u_2, v_1, v_2 as long as this change reduces the length of the tour.

Algorithm Comparison

The dynamic programming algorithm is an algorithmic strategy to solve optimization problems by decomposing problem instances into smaller, similar sub-questions and storing the solutions of the sub-questions while avoiding the computation of duplicate sub-questions. The specific procedure is as follows:

(1) Assume that we start from city 0, pass through all cities, and return to city 0. The information recorded is: the current city location and the current set of cities not traversed s.
(2) Construct a T-table to record the value of T through the state transfer equation, calculate the distance from the initial point through all cities back to the initial point, and compare the traversed city order, the definition of the state transfer equation is shown below:

$$T(s, init) = \min\{T(s\text{-}i, i) + dist[i][init]\} \tag{23.16}$$

T(s, *init*) stands for the distance from the initial point *init* back to *init* through all the points in the city set s.

(3) Finally, the minimum distance of the shortest path is obtained by comparison.

23.5 Application

In this chapter, the algorithm is written in Python language, and two main experiments are conducted. The first experiment applies the improved ant colony algorithm of this paper to the Northeastern University campus meal distribution problem and adjusts the number of iterations to optimize the shortest path; the second experiment uses the dynamic programming algorithm and the ant colony algorithm without 2-opt to analyze the same meal distribution problem. Finally, the conclusions are obtained by analyzing and comparing with the improved ant colony algorithm. The experimental environment is: Windows 11 system, 12th Gen Intel(R) Core(TM) i5-12500H CPU @ 2.50 GHz.

Parameter Setting

First, the relevant parameters are set: the ant colony size Pop is 60, the number of iterations MAXGEN is 50, the importance factor Alpha is 1, the visibility factor Beta is 1, volatility coefficient Rho is 0.15, information update parameter Q is 15, vehicle load W is 9, vehicle volume T W is 9, the volume of carriage T is 10, the heuristic factor Eta is set as the reciprocal of distance, and the initial pheromone matrix is the unit matrix.

Algorithm Design

Algorithm 1 Improved ant colony algorithm

Input: the number of cycles NC $\rightarrow$ 0;the maximum number of cycles NCmax;the initializa-tion information $\tau_{ij}(t) \rightarrow C$;the initial moment $\Delta\tau_{ij}(t) \rightarrow 0$

Output: Optimal path L

```
while not termination condition do
    for k=1;k<m;k++ do
        Place m ants randomly on n cities
        for s=0;s<n;s++ do
            Calculate its state transfer probability p_ij^k according to (8) rate
            Select the city with the maximum state transfer probability
        Calculate the perimeter length L_k of all m ants walking through
    Calculate Δτ^k according to (11) and (12)
    t=t+n; NC=NC+1
return
```

Comparative Algorithm Design

Algorithm 2 Dynamic programming algorithm

Input: the initial point init point $\rightarrow$ 0;the aggregate of traversing cities $s \rightarrow 0$;the numberof cities $N \rightarrow 0$

Output: the minimum distance of optimal path *distance*

```
if dp[s][init]!= -1 then
    distance←dp[s][init]
    if s == (1 « (N)) then
        distance←dist[0][init]
        for i←1 to N do
            if s and (1 « i) then
                m←TSP(sand( (1 << i)), i, num+1)+dist[i][init]
                if m < sumpath then
                    sumpath← m
                    path[s][init]← i
                    dp[s][init]← sumpath
                    distance←dp[s][init]
                    return distance
```

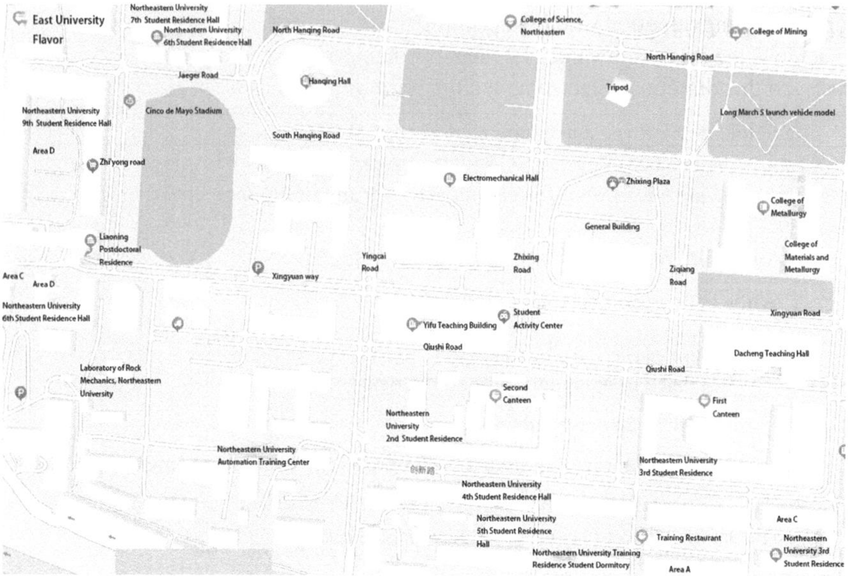

Fig. 23.1 Campus map

Experiment Procedure

Data Collection

First, a map of the Northeasten University campus (scale 1:20 m) was obtained on Baidu Maps as Fig. 23.1; Using the Python CV2 library to display through the canvas, a frame on top of the map was drawn. Then the coordinates of the distribution center cafeteria and customer dormitories were automatically identified on the campus map to read and recorded on the map, the result is shown in the Fig. 23.2.

The horizontal and vertical coordinates of the obtained customer points as well as the dispatch center represent the location information and label them. Considering that the size of the horizontal and vertical coordinates is reduced by 10 times at the same time, the results are as follows (the first 6 rows in the table are the coordinates of the dispatch center and the rest are the coordinates of the customers) (Table 23.2).

Calculation and Results

The path optimization was performed using Python for 50 times to obtain the shortest path lengths starting from 6 distribution centers and the mean value of lengths for each number were calculated as follows (Table 23.3).

The optimal path starting from 6 dispatch centers was obtained as follows Figs. 23.3, 23.4, 23.5, 23.6, 23.7 and 23.8.

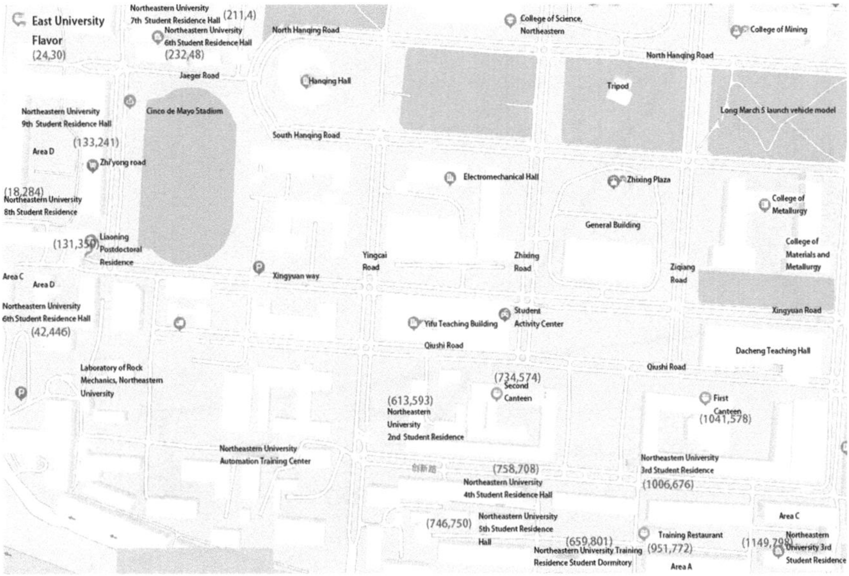

Fig. 23.2 Campus map coordinates

Table 23.2 Coordinates of the points

Number	0	1	2	3	4	5	6	7
x label	95.1	2.4	1.8	73.4	104.1	74.6	61.3	21.1
y label	77.2	3	28.4	57.4	57.8	75	59.3	0.4
Number	*8*	*9*	*10*	*11*	*12*	*13*	*14*	*15*
x label	13.3	100.6	13.1	75.8	4.2	85.9	114.9	23.2
y label	24.1	67.6	35	70.8	44.6	80.1	79.8	4.8

Table 23.3 The length of shortest distance

No	0	1	2	3	4	5
1	334.17	389.79	401.27	333.98	242.62	275.63
2	327.83	239.42	402.27		322.59	338.19
3		265.91			253.56	275.63
4					315.78	
Average	331.00	298.37	401.77	333.98	283.63	296.48

From the number of occurrences of each number in the above table, we can see that the most frequent occurrence of the optional route beginning from point 4 and the least frequent occurrence of the optional route beginning point 3. Among them, comparing the average value of the shortest path length of 6 coordinates as the

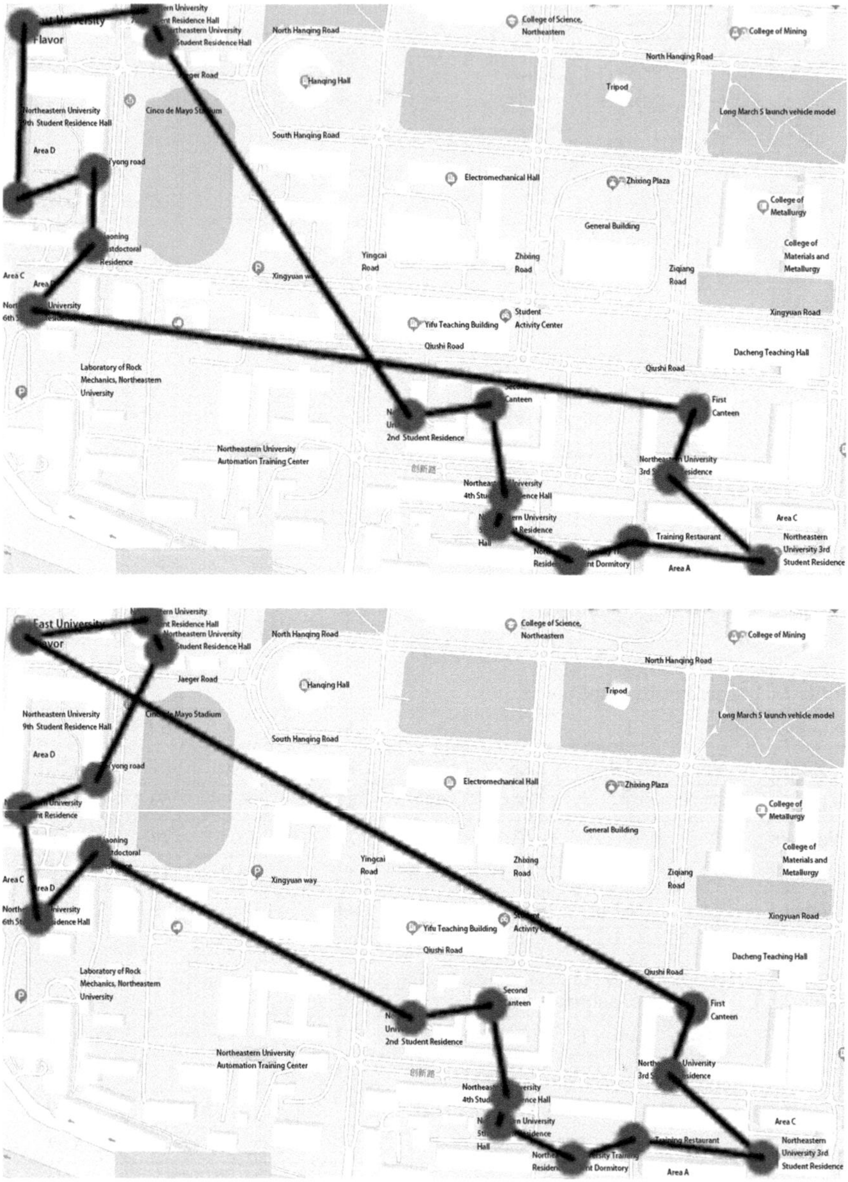

Fig. 23.3 No. 0 path

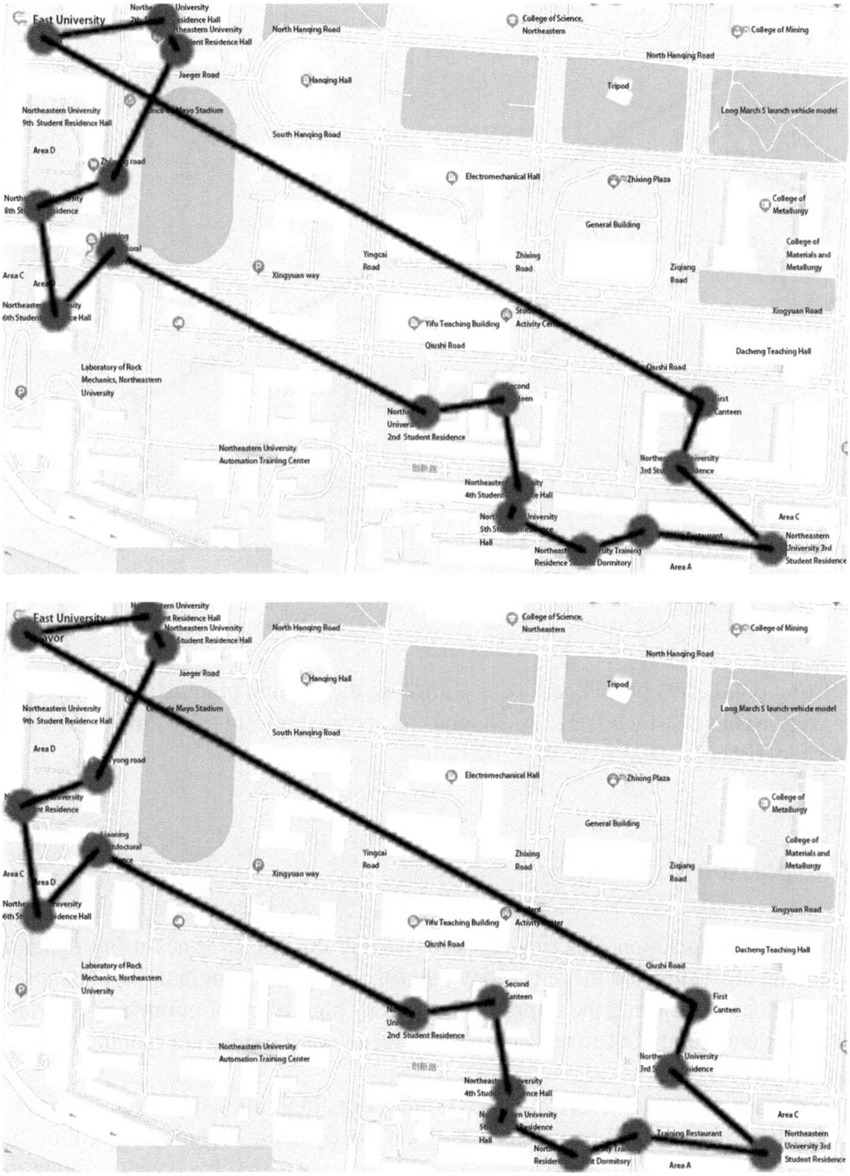

Fig. 23.4 No. 1 path

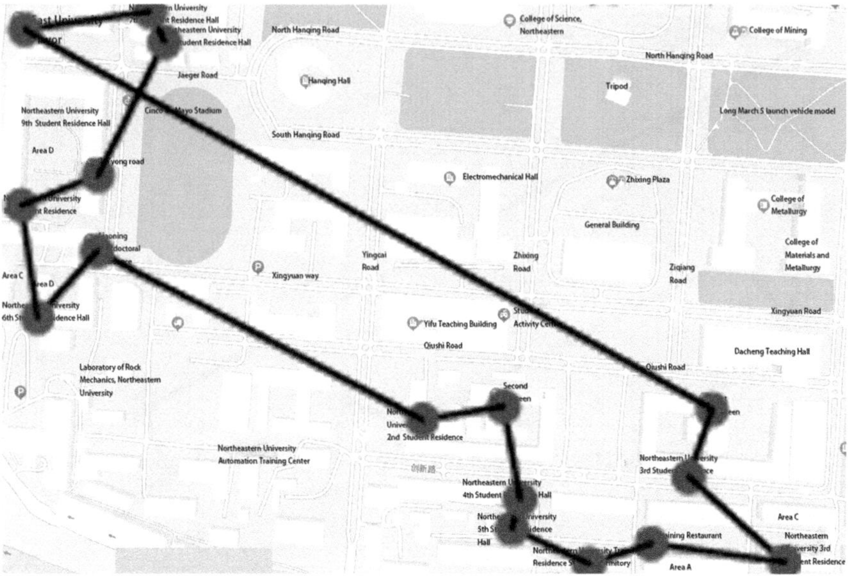

Fig. 23.4 (continued)

starting point, the optimal path with point 4 as the starting point is the shortest, the average value of which is 283.6358, and the actual shortest path is 56.727 km with the coordinate of (104.1, 57.8).

Comparison Experiment

The path optimization experiments were carried out for 20 times using dynamic planning algorithm and the ant colony algorithm without 2-opt in the application of for Python, while using the improved ant colony algorithm for optimizing the route for the same number of times, comparison results of the program running time are as follows:

From above Tables 23.4, 23.5 to 23.6 and Fig. 23.9, it's seen that the improved ant colony algorithm runs much faster than the dynamic planning algorithm, but slightly slower than the ant colony algorithm without 2-opt. It's concluded that the improved ant colony algorithm performs better than the dynamic planning algorithm and slightly worse than the ant colony algorithm without 2-opt. Compared with the dynamic planning algorithm and the ant colony algorithm without 2-opt, in general, the improved ant colony algorithm not only runs faster than the other two algorithms, but also costs the shorter optimization path and time.

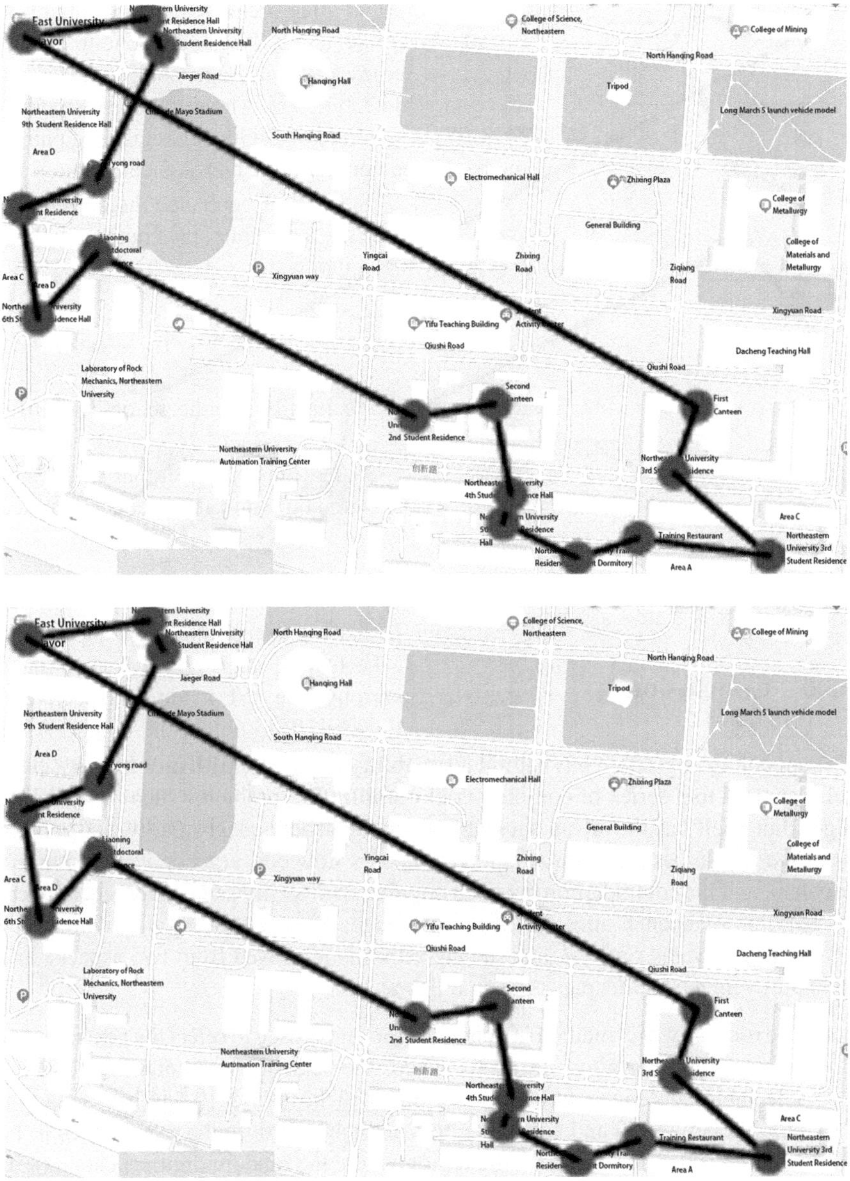

Fig. 23.5 No. 2 path

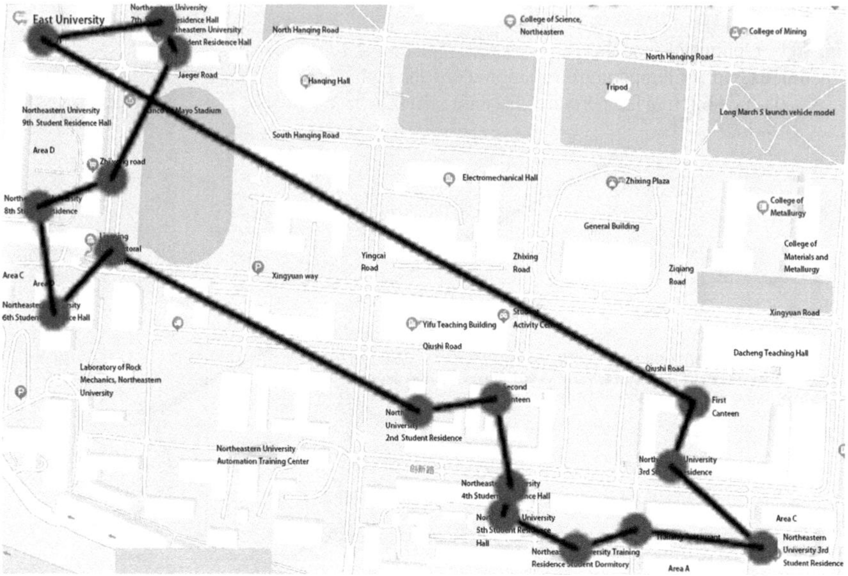

Fig. 23.6 No. 3 path

23.6 Conclusion

The ant colony algorithm has natural adaptability to address path problems successfully applied to a series of combinatorial optimization problems such as TSP. The algorithm itself has the advantages of positive feedback, adapt- ability, and robustness, but it also has many disadvantages such as slow convergence, local optimum problem, and the contradiction between population diversity and convergence speed.

This improvement is aimed at the disadvantages of the ant colony algorithm which is easy to be locally optimal, and the algorithm is improved from two aspects, and the improvements in this paper are as follows:

(1) The roulette wheel method of genetic algorithm is used to select the next delivery cafeteria and dormitory to go to, where the selection percentage p_k of each selectable cafeteria or dormitory is arranged in the interval [0,1]. Then, a random number between 0 and 1 is generated, and the next dormitory or canteen to be selected is decided based on the region where the random digit appears.
(2) Add the 2-opt algorithm to optimize the optimal solution obtained from each iteration, recalculating the path length to obtain a shorter path length than the original one to replace the original path, and cyclically iterating to finally obtain the optimal solution, but affecting the running speed of the program.

In the experimental part, the program is implemented using Python programming all the code, used the model to solve the real-life food delivery problem, and then finally proved that the improved ant colony algorithm has the purpose of achieving

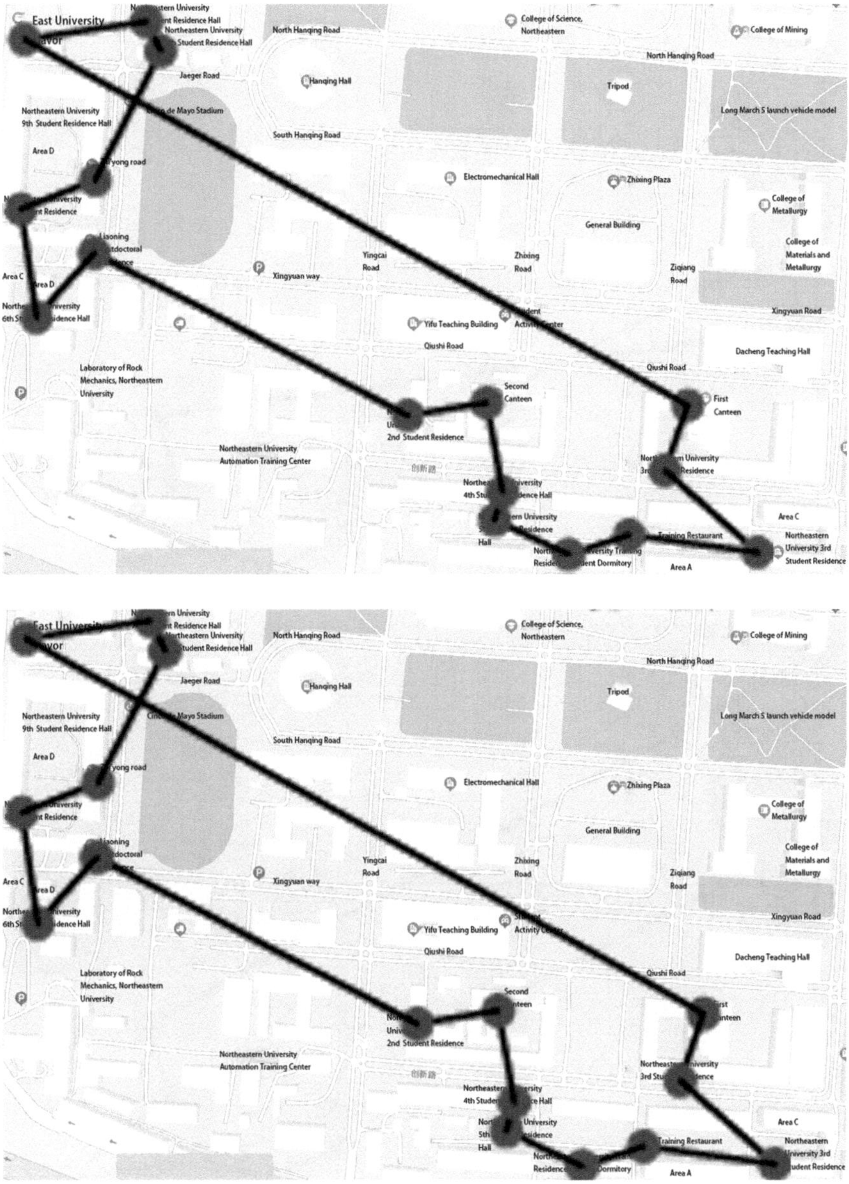

Fig. 23.7 No. 4 path

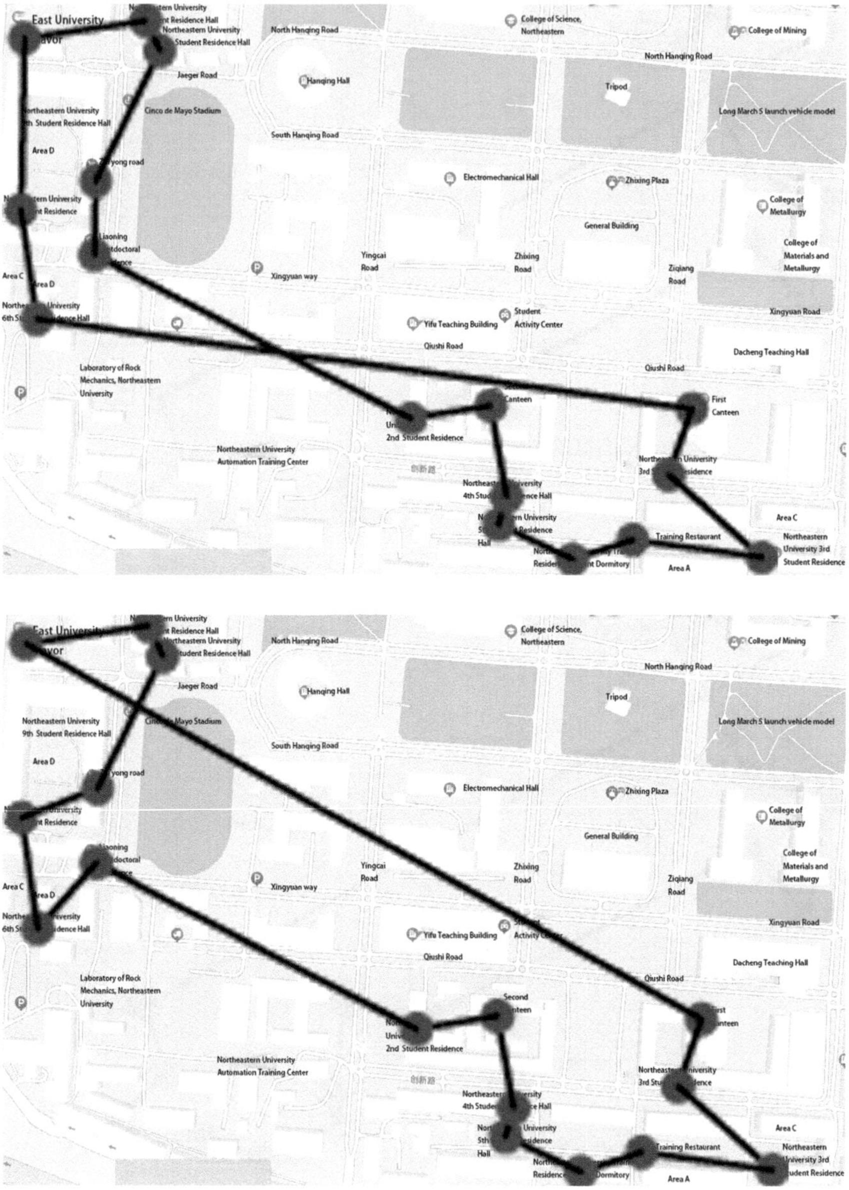

Fig. 23.7 (continued)

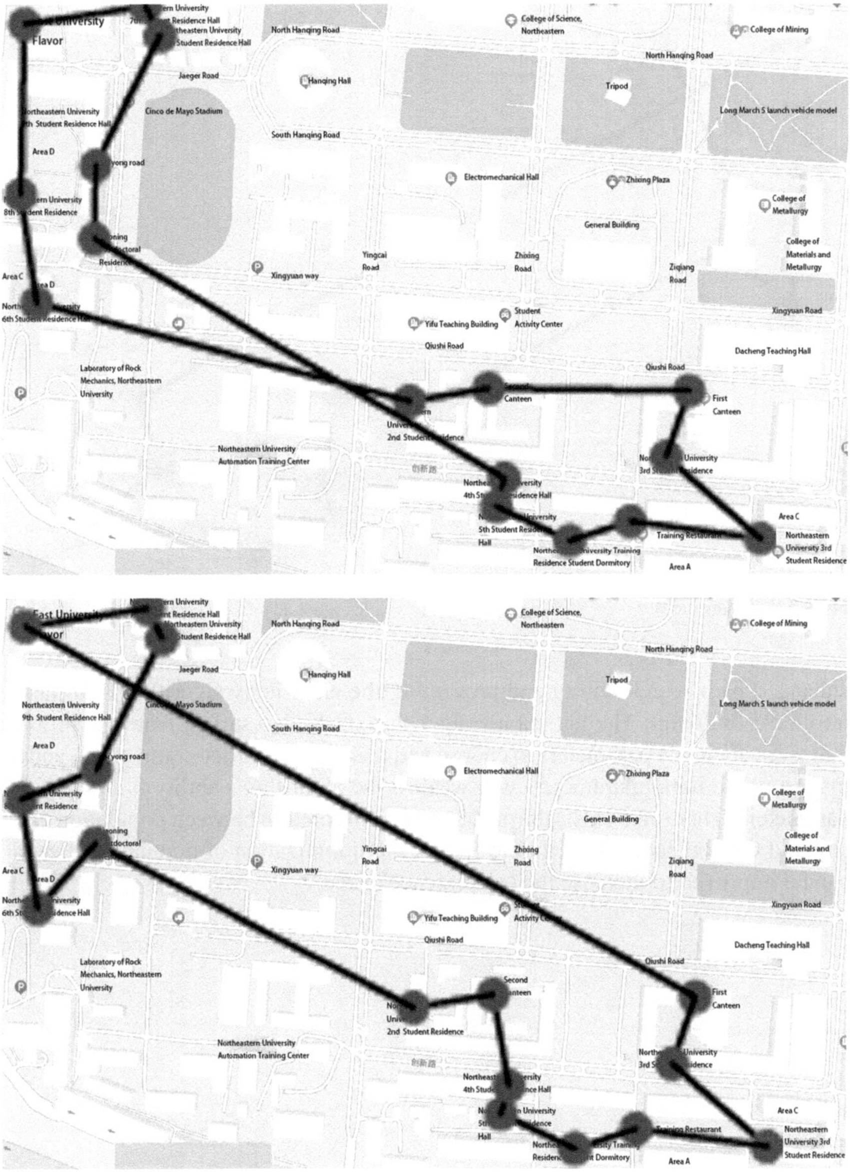

Fig. 23.8 No. 5 path

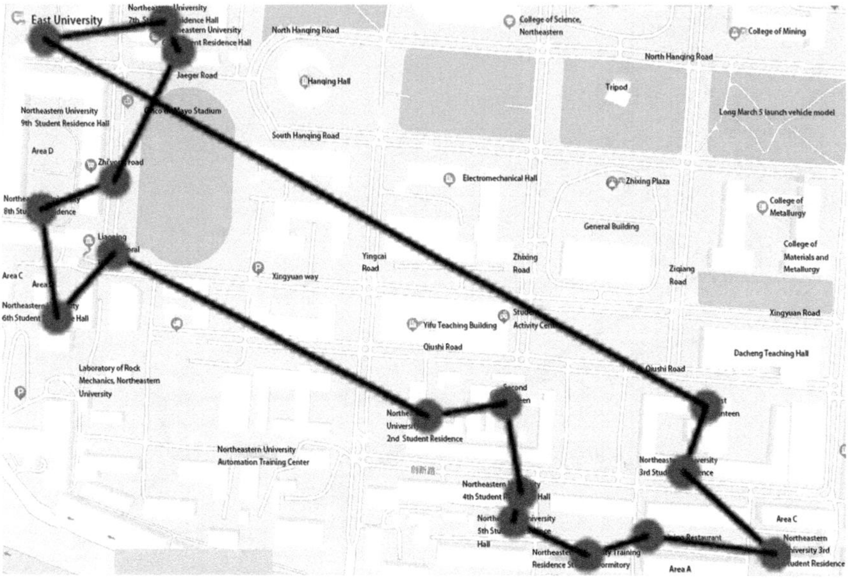

Fig. 23.8 (continued)

enhanced path search ability and preventing the ant colony algorithm from falling into local optimum. The ant colony algorithm itself also suffers from the problem that there are many parameters to choose and it is difficult to determine them, and the inappropriate initial parameters will weaken the algorithm's ability to find the best parameters. There also exists the problem of contradiction between population diversity and convergence speed. In the future, the optimization of population diversity can be improved to increase the convergence speed.

Table 23.4 Comparison of improved ant colony algorithm

Improved ant colony algorithm									
1	2	3	4	5	6	7	8	9	10
9.00E-07	5.00E-07	4.00E-07	6.00E-07	8.00E-07	3.00E-07	4.00E-07	5.00E-07	7.00E-07	4.00E-07
11	12	13	14	15	16	17	18	19	20
4.00E-07	3.00E-07	6.00E-07	7.00E-07	4.00E-07	4.00E-07	5.00E-07	1.30E-06	5.00E-07	5.00E-07
Average 5.63E-07									

Table 23.5 Comparison of dynamic plannin algorithm

Dynamic planning algorithm									
1	2	3	4	5	6	7	8	9	10
1.4059407	1.3590288	1.3674288	1.4164565	1.3815934	1.3339275	1.3575223	1.4675743	1.4050517	1.3637905
11	12	13	14	15	16	17	18	19	20
1.3706238	1.4528359	1.4448846	1.3707678	1.4487196	1.3900669	1.3486401	1.4905277	1.3539059	1.4359551
Average 4.65004E-07									

Table 23.6 Comparison of ant colony algorithm without 2-opt

Ant colony algorithm without 2-opt

1	2	3	4	5	6	7	8	9	10
5.00E-07	3.00E-07	6.00E-07	4.00E-07	2.00E-07	3.00E-07	6.00E-07	5.00E-07	6.00E-07	9.00E-07
11	12	13	14	15	16	17	18	19	20
5.00E-07	3.00E-07	2.00E-07	3.00E-07	6.00E-07	5.00E-07	6.00E-07	4.00E-07	4.00E-07	6.00E-07

Average
4.65004E-07

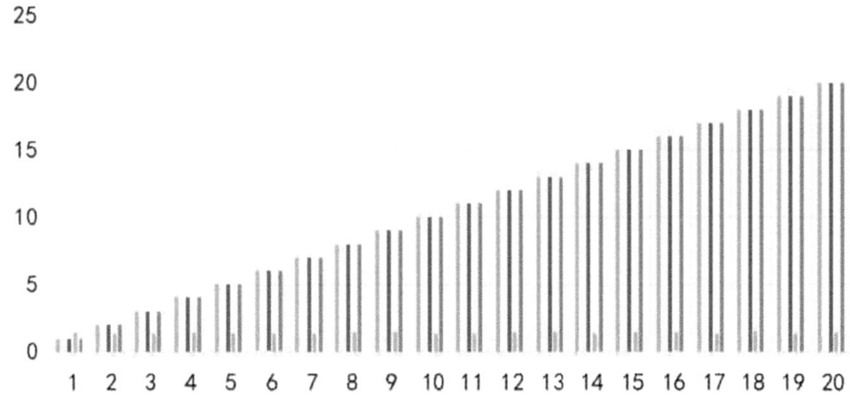

Fig. 23.9 Comparison between ant colony algorithm and improved ant colony algorithm, dynamic planning algorithm and ant colony algorithm without 2-opt

References

1. Zhou, J.: Impact of the new coronary pneumonia outbreak on the transportation industry and policy recommendations. Transp. Res. **1**, 13–18 (2020)
2. Jia Y.T.: How to judge the challenges of the new crown pneumonia epidemic for universities. High. Sci. Educ. (4), 2 (2020)
3. Zhang, H.F., Ge, H.W., Yang, J.L., Tong, Y.B.: Review of vehicle rout- ing problems: models, classification and solving algorithms. Arch. Comput. Methods Eng. 1–27 (2021)
4. Sun, Y., Li, Y.S., Yang, Y.Y., Yue, H.D.: Differential evolution algorithm with population knowledge fusion strategy for image registration. Complex Intell. Syst. **8**(2), 835–850 (2022)
5. Han, Y.J., Peng, Y.F., Wei, H., Shi, B.L.: Superheuristic genetic algorithm for solving vehicle path problems with soft time windows. Comput. Integr. Manuf. Syst. **25**(10), 2571–2579 (2019)
6. Wang, Y., Wang, L., Chen, G.C., Cai, Z.Q., Zhou, Y.Q., Xing, L.N.: An improved ant colony optimization algorithm to the periodic vehicle routing problem with time window and service choice. Swarm Evol. Comput. **55**, 100675 (2020)
7. Pang, Y., Luo, H.H., Xia, Y.K.: Path optimization of waste furniture recycling vehicles based on taboo search algorithm. Comput. Integr. Manuf. Syst. **26**(5), 1425 (2020)
8. Chen, J.M, Zhang, S.Y., Dan, B.: An improved particle swarm optimization algorithm for solving multi-compartment vehicle path problems. Comput. Integr. Manuf. Syst. **25**(11), 2952 (2019)
9. Dorigo, M., Gambardella, L.M.: Ant colony system: a cooperative learning approach to the traveling salesman problem. IEEE Trans. Evolut. Comput. **1**(1), 53–66 (1997). https://doi.org/10.1109/4235.585892
10. Le, T.D.C., Nguyen, D.D., Oláh, J., Pakurár, M.: Optimal vehicle route schedules in picking up and delivering cargo containers considering time windows in logistics distribution networks: a case study. Prod. Eng. Arch. **26**(4), 174–184 (2020)
11. Zaghdoud, R., Amara, M., Ghedira, K.: Improved particle swarm ap- proach for dynamic automated guided vehicles dispatching. Int. J. Adv. Comput. Sci. Appl. **13**(6) (2022)
12. Wang, X.F.: Research on logistics vehicle routing optimization based on improved ant colony algorithm. In Proceedings of 2019 3rd International Conference on Artificial Intelligence, Systems, and Computing Technology (AISCT 2019), pp. 482–486. Francis Academic Press, UK (2019)
13. Wang, Y.J., Lu, X., Zuo, Z.Q.: Autonomous vehicles path planning with en- hanced ant colony optimization. In 2019 Chinese Control Conference (CCC), pp. 6633–6638. IEEE (2019)

14. Pala, O., Aksarayli, M., et al.: An ant colony optimization algorithm approach for solving multi-objective capacitated vehicle routing problem. Alphanumeric J. **6**(1), 37–48 (2018)
15. Li, Y.Q., Guo, C.S., Zhang, L.Y., Sun, Y.S.: Application of ant colony algorithm in emergency logistics vehicle routing problem. In Proceedings of the 2012 International Conference on Convergence Computer Technology, pp. 137–140 (2012)

Chapter 24
Design and Study of a Management Information System for Engineering Measurement Laboratories

Hualong Cai, Fenghuan Huang, Yifeng Huang, and Caiying Huang

Abstract With the high-quality development of Internet of Things technology and the wide application of computer technology, the construction scale of information management is expanding, and the management of engineering measurement laboratories is also carrying out information construction. Engineering measurement laboratory management in manual operation mode, there are a series of problems such as equipment status detection, experimental progress, experimental archives data can not be real-time query and archiving and other lack of information management. This has led to low efficiency, poor timeliness of information, inadequate management, inaccurate instrument accuracy measurements and high safety risks in the management of engineering measurement laboratories. Therefore, this paper uses computer application technology to design and study the management information system of engineering measurement laboratory. Therefore, this paper uses computer application technology to design and study the management information system of engineering measurement laboratory. It not only guarantees the timeliness and accuracy of information data in engineering measurement laboratories, but also manages relevant information more scientifically and effectively and solves as many problems as possible in the whole process of engineering measurement, improves the management quality of engineering measurement laboratories and reduces various safety risks.

24.1 Introduction

At present, some developed countries have established high-level, large-scale, comprehensive laboratories and have produced world-leading experimental results. In recent years, China has also become aware of the importance, criticality, overall

H. Cai · F. Huang · Y. Huang · C. Huang (✉)
School of Civil Engineering, Jiaying University, Meizhou 514015, Guangdong, China
e-mail: 15118645071@163.com

R. Kountchev (Deceased) et al. (eds.), *AI Methods and Applications in 3D Technologies*, Smart Innovation, Systems and Technologies 388,
https://doi.org/10.1007/978-981-97-2144-3_24

situation and strategic position of laboratories and has formally approved the establishment of national laboratories. However, due to the large scale of engineering measurement laboratories, the crossover complexity of multiple disciplines and the fact that they are not yet mature, there are still problems such as the lack of information technology in the management mechanism of measurement laboratories, low transparency of work, low management efficiency and unreasonable and non-standardized management systems [1]. Based on this status quo, this paper relies on the fast-developing Internet of Things technology and computer application technology to design and study the management information system of engineering measurement laboratory. This facilitates timely, accurate and complete enquiry, collection and effective use of all types of relevant information and data generated during the actual operation of engineering measurement experiments, efficient and procedural management of engineering measurement laboratory data files, and automatic detection of measurement laboratory equipment faults. This is conducive to improving the management of engineering measurement laboratories, improving the accuracy of measurement data, preventing various risks that may arise during the practice of engineering measurement and improving the safety of engineering measurement laboratories.

24.2 Components and Functions

The Engineering Measurement Laboratory Management Information System consists of four main modules, namely the Information Data Upload Module, the Data File Management Module, the Experiment Progress Control Module and the Equipment Safety Warning Module. The functional modules are closely linked, and the information management of the engineering measurement laboratory is realized through the automated transmission and effective use of the information data integration and sharing of the engineering measurement laboratory. The system components are shown in Fig. 24.1.

The Information Data Upload Module

This module has functions to collect, share, optimize and archive data and upload it to the information database. Automated transfer of information data through information data sharing on the Engineering Measurement Laboratory information platform. The system's card reader can quickly transfer key measurement data and related information (e.g. time, specific location, operator, experimental object, instrument used) from the engineering measurement work session to the system database automatically. The module automatically uploads information data including data records of instrument acquisition and testing, progress reports of survey works, results of

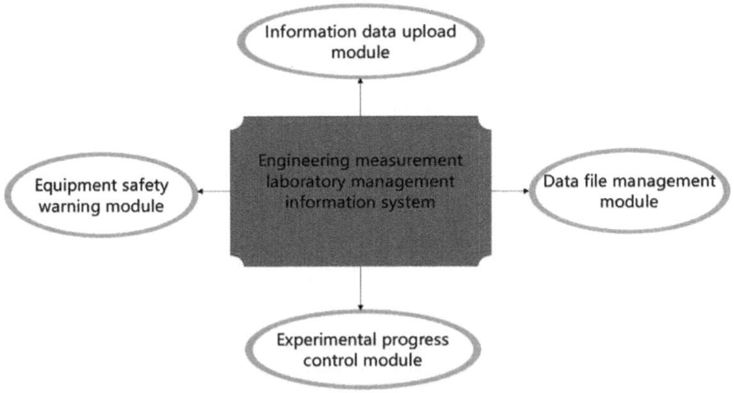

Fig. 24.1 System composition diagram

survey experiments, notes and other information records. The various types of information data uploaded to the system database are optimally integrated and archived in the form of grading, classification and coding, so that the relevant engineering measurement information data can be managed in a procedural, systematic and efficient manner. In addition, the module also has the function of real-time updating of relevant engineering measurement information data and tracking of engineering measurement information data in real time [2].

The Data File Management Module

The data document management module features automatic generation, document enquiry, classification and management. Users can search for relevant information documents by entering keywords, and can also search for the information they need by selecting specific dates, months and years. This module automatically summaries the relevant information documents of the Engineering Surveying Laboratory Information System, mainly relevant personnel information, instrument acquisition and testing information, internship reports, experimental progress reports and results reports, etc. The management of the Engineering Surveying Laboratory can record and modify various relevant data files in real time within the scope of the authorization of the information system, including the time of storage, the storage period, the frequency of use and query, etc. [3]. At the same time, it can also track and archive relevant data in real time and accurately for a series of tedious links (e.g. submission of applications, allocation of funds, approval and acceptance and archiving, etc.), and can make timely adjustments when relevant information changes, providing timely, reliable and complete information for management decisions and work in engineering measurement laboratories and increasing the transparency of work [4]. In addition, the relevant managers can manage the grading, classification and coding integration of engineering measurement laboratory data files, and carry out paperless informative operating procedures for laboratory management.

The Experimental Progress Control Module

The experimental schedule control is mainly to develop the experimental schedule of the engineering measurement project and then to track and control the relevant data and progress in real time for each specific part of the experimental project [5]. When the actual operation of measurement experiments in the process of experimental progress control deviation, the relevant technical personnel can use the module to real-time tracking and control of experimental progress, and the original experimental progress plan to compare timely, understand the actual situation, analyze the possible interference factors in the process of measurement experiments, find the actual situation and the deviation of the plan preparation and take scientific adjustment, optimize the original experimental progress plan. Better control the experimental progress and realize the dynamic management of the whole process of engineering measurement laboratory. In addition, the module can automatically perform experimental progress checks to reduce the risk of managing the progress of engineering measurement experiments [6].

The Equipment Safety Warning Module

Most of the instruments and equipment used for engineering measurement experiments are expensive optical instruments and electronic component equipment. Ageing, improper storage and use of the instrument can easily lead to inaccurate measurement of the data, causing large errors and affecting the results. Unsafe operator behaviour can also lead to safety hazards such as injury to people and damage to equipment [7]. The module is mainly used to warn the operator of unsafe conditions of the equipment and use the voice playback function to remind the operator of relevant precautions when using the equipment, effectively reducing the safety risks to people and equipment. The module automatically transmits to the general operating system of the measurement laboratory to automatically trip the power supply of the faulty equipment when there is a potential fault in the measurement equipment. At the same time, the module also uses the STC15W4K32S4 single chip as the core control device, sensing the presence of people through pyroelectric infrared sensors, and then activating ultrasonic distance measurement in the presence of people, and the corresponding port responds with a voice announcement when the laboratory operator enters the set range [8]. A warning signal is sent out for faulty equipment in the form of a voice announcement to inform the user, and then a warning signal is sent out in the form of a text message which is automatically transmitted to the manager's mobile phone to notify the relevant maintenance personnel to repair the

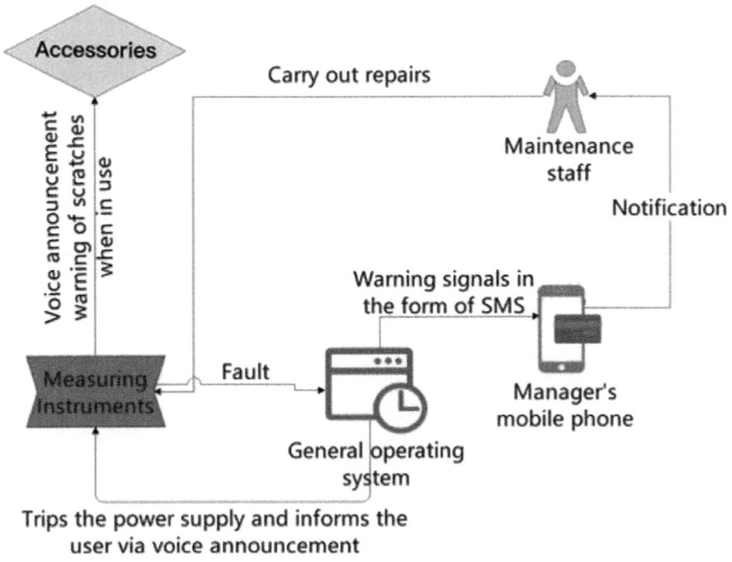

Fig. 24.2 Flow chart for warning of faulty measuring instruments and equipment

faulty instrument and equipment. In addition, there are many sharp accessories in engineering surveying (e.g. steel rulers, pointed tweezers and pocket knives). When using sharp accessories for measurement experiments, this device will send out a voice announcement: careful when using accessories, to remind users to pay attention to safety, to better meet the needs of engineering measurement laboratory safety prevention, to achieve "prevention" and "after the fact to deal with This is to achieve the organic unity of "prevention beforehand" and "treatment afterwards" and to prevent problems before they occur. The early warning process for faulty measuring instruments and equipment, as shown in Fig. 24.2.

24.3 System Architecture

Software Architecture System

The Engineering Surveying Laboratory Management Information System is developed in PHP, the system is designed in B/S mode, and the MySQL database is accessed through the PHP framework. The system is deployed on a Windows Server 2003 R2 server, with a web browser on the user side for system operation [9]. The information system allows you to upload information and data, manage data and documents, control the progress of experiments and provide safety warnings. Data sharing and management through MySQL database ensures that engineering

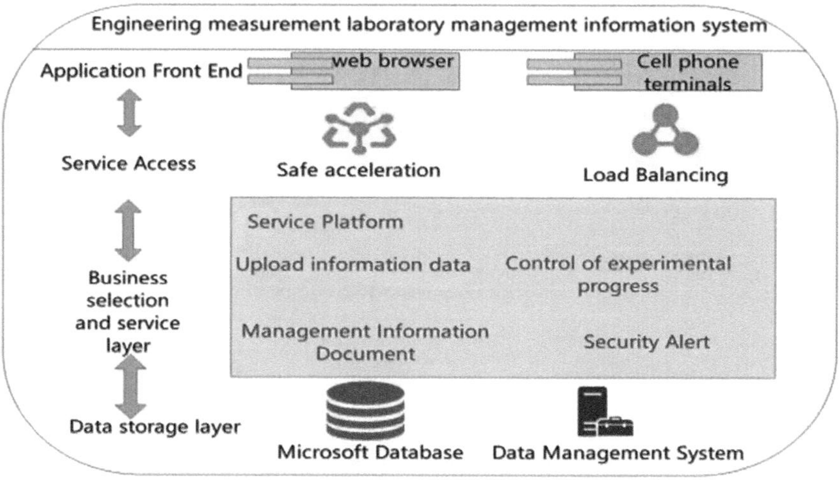

Fig. 24.3 Software architecture system

measurement information data is shared and transmitted in a confidential, efficient and secure environment. The software architecture is shown in Fig. 24.3.

Hardware Architecture

Real-time information data is automatically transmitted to the database with relevant engineering measurement data, and the database server interacts with the web server to provide information data on the real-time nature of engineering measurement. The relevant measurement laboratory managers interact with the information data of engineering measurement through the web server, and can extract the relevant engineering measurement data through the information database for optimization and integration, algorithm research and classification and archiving, etc. [10]. The operator of the engineering measurement laboratory sends an access request to the information database through the browser to the Web server, which interacts with the database and returns it to the Web server after the relevant manager has approved it, and then the Web server returns it to the browser page in the form of a web page, ensuring the security and reliability of the measurement laboratory information data. The hardware architecture is shown in Fig. 24.4.

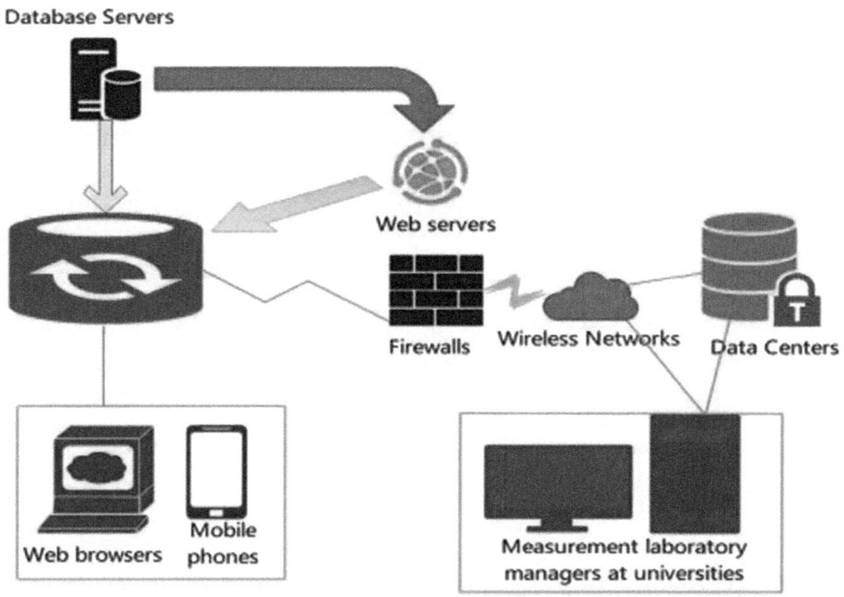

Fig. 24.4 Hardware architecture system

24.4 System Implementation

System Platform Interface

Users are required to undergo face swipe authentication or SMS authentication through their personal identity before they can register their account and set a password. After registering an account and being audited by the backstage administrator and passed, they can log in through the corresponding account and password to access the Engineering Measurement Laboratory Management Information System. The system interface shows four main functional modules, namely the information data upload module, the data file management module, the experiment progress control module and the equipment safety warning module. The user can view, understand and extract the required information data by clicking on the corresponding function module. The search function on the left side of the system allows you to search for relevant documents by date (year, month, day), relevant surveyor and keywords, allowing you to quickly and accurately search for the relevant information and data you need. There are also three main sections to check and understand the status of instrument testing, the progress of instrument acquisition and the progress of measurement experiments. In addition, the system provides functions such as account cancellation, password change, help and settings, which can better enable users to manage the use of their accounts on demand. The interface of the system platform after login is shown in Fig. 24.5.

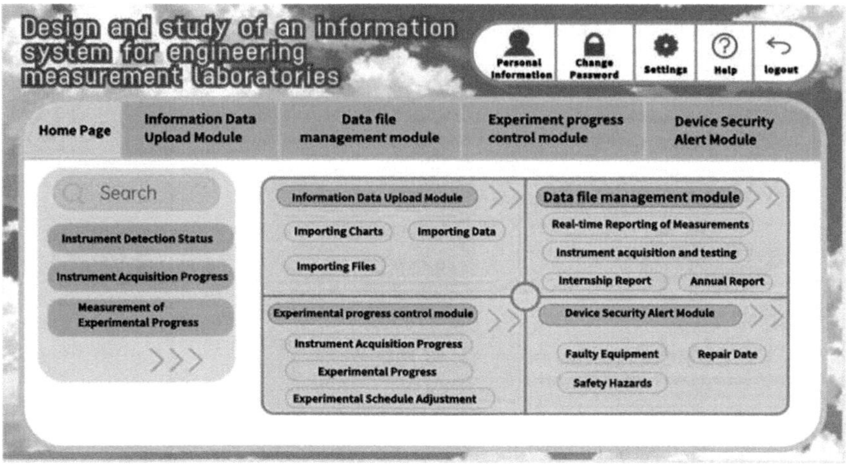

Fig. 24.5 System platform interface diagram

System Related Scree

The data file management module interface shows the engineering measurement experiment project reports and daily work. Engineering surveying laboratory project reporting documents include placement reports, results reports, laboratory project progress and annual reports. The daily work includes measurement summary data, instrument acquisition and testing information, meeting content and relevant personnel information. Users can access the page to view the details by clicking on the appropriate document. In addition, the user can search for relevant data files by date, keyword, file type and relevant experimenter. The system standard and programmable integrates and categorizes and archives a large amount of data in the engineering measurement laboratory. With the continuous development of the Internet of Things and computer application technology, the information management of engineering measurement laboratories is also being improved and perfected. The information of engineering measurement laboratory management has become an important engine for improving the quality of engineering measurement practice results. The data file management module interface, as shown in Fig. 24.6.

System Part Code

The Engineering Surveying Laboratory Management Information System adopts a B/S architecture with HTML/CSS design for the front-end, PHP development for the back-end and MySQL for data management. Part of the system's function code

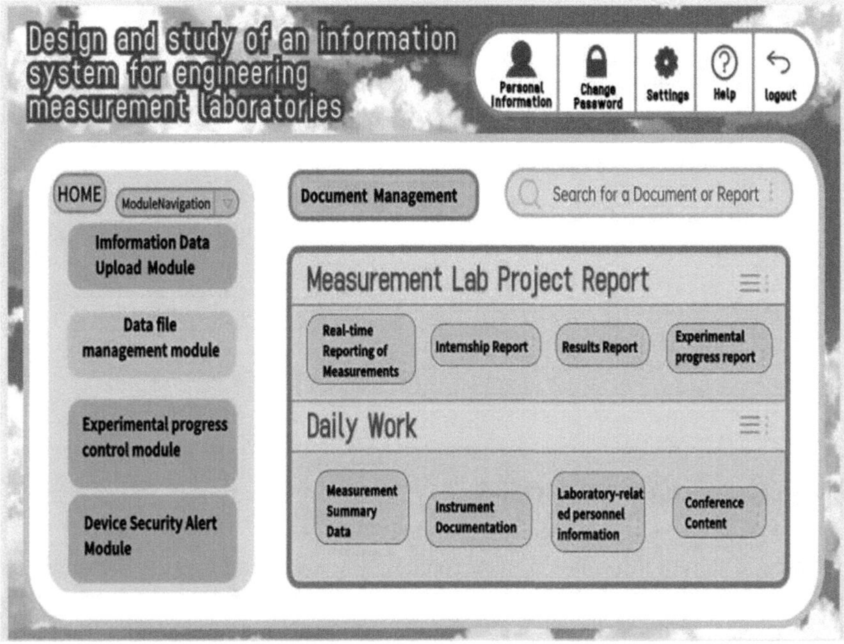

Fig. 24.6 Data file management module

can query the data in the data table. Some of the functional codes of this system are as follows:

```
require "conn.php";
//Step 1: Query the database data using SQL statements
$sql = "select * from stu";
// Step 2: Instantiate the database connection method
using PDO exec()
$result = $pdo->query($sql);
echo "<pre>";
print_r($result);
//Step 3: Use the fetchAll() method of the
(PDO::FETCH_ASSOC)
//Attribute query and get all data, the fetched data is
returned as an array
echo "<pre>";
print_r($row);
//Step 4: Determine if the query data is successfully
//and echos the data in an unordered list
if ($row > 0) {
echo "enquiry successful";
foreach ($row as $stu) {
```

```
echo "<ul>";
echo "<li>{$stu['stu_no']}</li>";
echo "<li>{$stu['stu_name']}</li>";
echo "<li>{$stu['gender']}</li>";
echo "<li>{$stu['telephone']}</li>";
echo "<li>{$stu['age']}</li>";
echo "<li>{$stu['college']}</li>";
echo "</ul>";
}
} else {
echo "enquiry failed.$sql";
}
```

24.5 Research Comparison

In China, different information management systems of engineering measurement have been developed in the field of engineering measurement based on specific measurement work contents, such as engineering deformation monitoring with structural safety monitoring as the core [11], which provides dynamic observation, data processing and early warning prediction for engineering measurement objects. However, nowadays, most of the developed engineering measurement information management systems are not robust enough due to the top-level design and integrated planning, which is more difficult to adapt to the uncertainty of engineering measurement operations by construction or operation and other factors and the vagaries of the environment, leading to the problem of separate construction, and even, each system module is difficult to achieve data integration and information sharing, and the phenomenon of information silos appears. The engineering measurement laboratory management information system researched and designed in this paper can realize the functions of uploading information and data, integrating and classifying relevant data and files, controlling the experiment progress and providing safety warning to the equipment, which can realize the information interaction and interconnection of each module and solve the problem of information silo. In addition, the system also tracks and queries engineering measurement data and experimental progress in real time to better realize the dynamic management of the whole process of engineering measurement work.

24.6 Conclusion and Outlook

Using the Engineering Measurement Laboratory Management Information System, it is possible to manage all types of measurement data, instruments and equipment and documentation reports for engineering measurement in an informative manner.

It facilitates relevant engineering measurement laboratory managers to edit, update and manage relevant information materials in real time in a comprehensive manner, realizes centralized management, real-time transmission and sharing of information data, and improves the management level and quality of engineering measurement laboratories. In addition, the system guarantees the progress and safety performance of engineering measurement experiments and more specifically meets the demand for information technology in the management of engineering measurement laboratories, constantly promoting the development of laboratories in the direction of digital, paperless, procedural and intelligent.

In the era of Internet+, engineering measurement laboratories will also continue to promote and enhance information technology to meet the needs of measurement laboratory management. In the future, we will also continue to improve and perfect the engineering measurement laboratory management information system to bring more convenience to measurement laboratory management and effectively improve the quality, timeliness and safety of measurement laboratory work. In addition, we should pay more attention to the training of engineering measurement laboratory personnel, improve their professionalism, better for the development of engineering measurement laboratory management to lay a solid foundation to promote the sustainable and healthy development of engineering measurement laboratory management.

Acknowledgements The research was supported by Jiaying university Talent Initiation Project. (311A1116). The research was supported by Jiaying College Research Fund. (131/322D43).

References

1. Chang, X.H., Zhong, D.T.: Analysis of the management system of national laboratories and their major science and technology infrastructure. China Soft Sci. **06**, 13–22 (2021)
2. Yuan, X.Q., Liu, C.B., Chen, J.H., Yan, B.S.: Research on the method of uploading operational monitoring data of high-speed rotating machinery. Mechatron. Eng. **40**(03), 377–383 (2023)
3. Yang, Z.L., Zhao, L., Cai, S.H., Lu, H., Xiang, X.: Exploration and practice of 5S+PDCA engineering measurement laboratory management model. Lab. Sci. **25**(02), 182–187 (2022)
4. Wang, Z., Guo, X.C., Wu, L.G.: The development and application of instrument and equipment procurement progress enquiry system in colleges and universities. Lab. Res. Explor. **33**(10), 283–287 (2014)
5. Zhang, J.W., Zhang, W.B., Xie, W.L., Tao, W., Liu, Y.X.: Exploration of quality control and progress control strategies in construction project management. Bulk Cem. **06**, 23–24 (2022)
6. Gong, X.D.: Trial discussion on construction progress management in school teaching laboratory building project. Sichuan Cem. **07**, 197 (2018)
7. Yang, Zh.L., Cao, Y., Wan, L., Zhao, L., Lu, H., Cai, S.H.: Analysis of safety hazards and countermeasures in engineering measurement laboratories. Exp. Sci. Technol. **19**(06), 152–156 (2021)
8. Wang, Z.X., Li, J.G., Cai, Z.Y.: Design of early warning system for equipment safety distance based on fuzzy control. J. Heilongjiang Inst. Technol. (Comprehensive Edition) **18**(04), 65–68 (2018). https://doi.org/10.16792/j.cnki.1672-6758.2018.04.014
9. Zhang, W.J., He, H.X., Xu, H.L.: Design and development of laboratory management information system. Comput. Digital Eng. **38**(05), 147–150 (2010)

10. Wang, J.K.: Research on laboratory management information system of university in B/S mode. Electron. Technol. Softw. Eng. **01**, 159–160 (2021)
11. Wang, D.: A framework for engineering surveying services in the era of intelligence. Eng. Surv. **51**(01), 1–6 (2023)

Chapter 25
Course Control of Unmanned Surface Vehicle Based on Fuzzy Control of Transition Rules

Guangyuan Li and Zhuquan Xiang

Abstract The course control of unmanned boat is the foundation of unmanned boats movement, which often faces complex and changeable environmental interference problems. To further improve the anti-jamming ability of unmanned boat course control, a conversion rule method is proposed based on the fuzzy control algorithm with a strong anti-jamming ability to overcome the problems of fuzzy rules difficult to make and self-adaptation. The mapping relationship between the PID controller and the fuzzy controller is analyzed theoretically, and a set of conversion rule methods is established to convert PID parameters into fuzzy control rules. By virtue of the PID controller's advantages of high steady-state accuracy, and simple self-adaptation, the defects of the fuzzy controller are made up. The simulation experiment of ship motion control based on MATLAB is verified. The results show that there is a correspondence between the PID controller coefficient and the linear fuzzy rules. The ship motion simulation verifies the effectiveness of the conversion rule method and the introduction of deviation integral. The conversion rule method based on PID parameters generating linear fuzzy rules is feasible and effective. The new integral fuzzy controller can effectively suppress the fluctuation of course deviation under wave interference, which is of great significance for the course control of unmanned craft with complex and changeable working environments.

25.1 Introduction

Unmanned surface vehicles (USVs) can replace human beings and play an important role in military reconnaissance, fishery farming, environmental monitoring, and other fields [1–3]. The basis for it to complete preset tasks is the accurate movement of USV in the water, so the research on the course control algorithm of USV is very important [4, 5]. There are many problems in USV course control, such as complex and variable

G. Li · Z. Xiang (✉)
School of Ship, Ocean, Energy and Power Engineering, Wuhan University of Technology, Wuhan 430063, China
e-mail: zuquan_xiang@163.com

R. Kountchev (Deceased) et al. (eds.), *AI Methods and Applications in 3D Technologies*, Smart Innovation, Systems and Technologies 388,
https://doi.org/10.1007/978-981-97-2144-3_25

interference and difficult establishment of an accurate model [6–8], which requires that the control algorithm is independent of the model and has high robustness. As a modeless controller with strong robustness, the fuzzy control algorithm is very suitable for USV course control, but it has some problems such as poor steady-state accuracy and the difficulty in obtaining precise fuzzy rules. Integral links can effectively improve the steady-state performance of the system. The method of adding an integral link to the fuzzy controller to improve the fuzzy control performance has attracted more scholars' attention and has been successfully applied in some fields [9, 10]. However, the number of fuzzy rules increases exponentially with the increase of dimension, which further increases the difficulty of fuzzy rulemaking and self-adaptation. A fuzzy control algorithm urgently needs a systematic, simple, and easy fuzzy rule-making method.

This paper starts with the similarity between the PID controller and fuzzy controller, compares and analyzes the common points of the input–output relationship between the two control algorithms, and analyzes the theoretical feasibility of PID control parameters and fuzzy rule conversion. By analyzing the structure of the counter saturated PID controller and 3D fuzzy controller, the relation between linear fuzzy rules and PID parameters is revealed, and the transformation rule method of systematically obtaining linear fuzzy rules is pointed out. Based on the transformation rule method, an adaptive integral fuzzy controller is designed.

25.2 USV Motion Model

Based on the USV motion model proposed by Fossen, the USV mathematical model used for simulation verification is constructed, and the onboard coordinate system *O-XYZ* and geodetic coordinate system *O-XeYeZe* are established. For the USV motion control, only its motion on the horizontal plane is considered, the ship mass is assumed to be uniform and symmetrical along the central axis, and the hydrodynamic derivative above the second order is ignored. Only the influence of parameters on the main diagonal of the inertial matrix and the damping matrix is considered. The simplified three-degree-of-freedom ship maneuvering motion equation is as follows [11, 12]:

$$\boldsymbol{M}\dot{\boldsymbol{V}}+\boldsymbol{C}\boldsymbol{V}+\boldsymbol{D}\boldsymbol{V}=\tau_{\mathrm{pro}}+\tau_{\boldsymbol{env}} \tag{25.1}$$

τ_{pro} is the controlling force matrix, τ_{env} is the environmental force matrix, $\boldsymbol{V}$ is the velocity matrix in shipboard coordinate system, $\dot{\mathrm{V}}$ is the derivative matrix of $\boldsymbol{V}$ with respect to time, $\boldsymbol{M}$ is the inertia matrix, $\boldsymbol{C}$ is the Coriolis force-centripetal force matrix, and $\boldsymbol{D}$ is the damping matrix. The matrix structure is as follows:

$$M = \begin{bmatrix} m - X_{\dot{u}} & 0 & 0 \\ 0 & m - Y_{\dot{v}} & mX_G - Y_{\dot{r}} \\ 0 & mX_G - Y_{\dot{r}} & I_z - N_{\dot{r}} \end{bmatrix} \tag{25.2}$$

$$C = \begin{bmatrix} 0 & 0 & -mv + Y_{\dot{v}}v \\ 0 & 0 & mu - X_{\dot{u}}u \\ mv - Y_{\dot{v}}v & mu - X_{\dot{u}}u & 0 \end{bmatrix} \tag{25.3}$$

$$D = \begin{bmatrix} -X_u & 0 & 0 \\ 0 & -Y_v & -Y_r \\ 0 & -N_v & -N_r \end{bmatrix} \tag{25.4}$$

$$V = \begin{bmatrix} u & v & r \end{bmatrix}^T \tag{25.5}$$

$$\tau_{\text{pro}} = [\tau_u \tau_v \tau_r]^T \tag{25.6}$$

$$\tau_{\text{env}} = [\tau_{envu} \tau_{envv} \tau_{envr}]^T \tag{25.7}$$

m represents the ship's mass, both of which are hydrodynamic derivatives, Y_v, Y_r, N_r $X_{\dot{u}}$, $Y_{\dot{v}}$, $Y_{\dot{r}}$, $N_{\dot{r}}$, X_u, X_G is the longitudinal distance of the ship's center of gravity; u and v are the longitudinal and transverse velocities of the shipboard coordinate system; r is the yaw velocity.

Considering the difference in the coordinate system established by the ship's motion model, the conversion formula of the velocity between the ship's coordinate system and the fixed coordinate system is as follows:

$$\dot{\eta} = j(\varphi)V \tag{25.8}$$

η is the pose matrix of the ship, representing the position x, y and yaw Angle φ in the geodetic coordinate system of the ship, $j(\varphi)$ is the coordinate transformation matrix, the specific matrix structure is as follows:

$$j(\varphi) = \begin{bmatrix} \cos\varphi & -\sin\varphi & 0 \\ \sin\varphi & \cos\varphi & 0 \\ 0 & 0 & 1 \end{bmatrix} \tag{25.9}$$

$$\eta = \begin{bmatrix} x & y & \varphi \end{bmatrix}^T \tag{25.10}$$

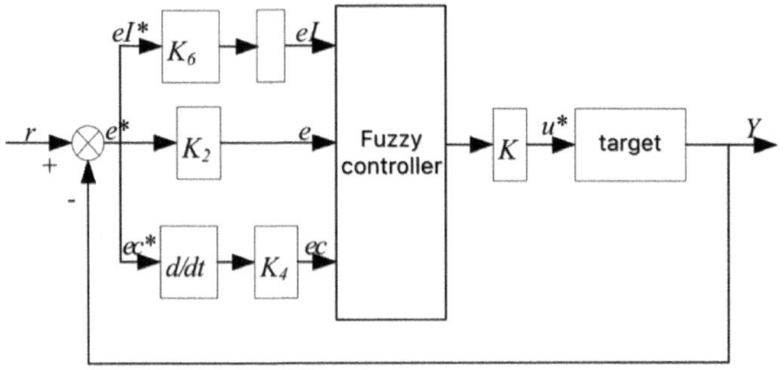

Fig. 25.1 Structure diagram of integral fuzzy controller

25.3 Conversion Rule Method

Basic Structure of 3D Fuzzy Controller

The structure of integral fuzzy controller is like that of PID controller. There are three inputs, *e*, *eI* and *ec*, and one output, *u*. Error of expection and reality is *e* and *ec* is differential of *e* and *eI* is the integral of *e*. The quantization factors k_2, k_4, k_6 are used to process the e^*, ec^* and eI^* of the actual input respectively to the input of controller, *e*, *eI* and *ec*. They are often expressed by positive numbers. The following k_2, k_4, k_6 are the same as defined here. These quantization factors are usually taken as the reciprocal of the corresponding input domain, and the input domain after processing is unified as [0,1]. Then the quantization factor K is used to adjust the output u^* of the controller and transmit it to the controlled object. The structure diagram is shown in Fig. 25.1.

The limit of the upper of *eI* discussion domain coincides with the idea of anti-saturation integral PID. When the accumulated error integral *eI* exceeds the upper limit of the set discussion domain, only the upper limit of the discussion domain is used for calculation until it starts to reverse decrease.

PID Parameter and Linear Fuzzy Rule Conversion

As it is difficult to perceive the deviation integral in the human operation, it is difficult to make conventional fuzzy rules based on operation experience. At present, there is no method to make three-dimensional fuzzy rules systematically. The acquisition of three-dimensional fuzzy rules, especially nonlinear three-dimensional fuzzy rules, greatly hinders the engineering application of three-dimensional fuzzy controllers. PID controller has a wide range of applications in engineering, PID controller only needs to obtain K_p, K_i, K_d three parameters, compared with the construction of fuzzy rules is much simpler. Based on the internal relation between the PID controller and fuzzy controller, the fuzzy rules of the 3D fuzzy controller are constructed by using PID controller parameters, which is undoubtedly a way to solve the difficult problem of making 3D fuzzy rules.

For the ideal response function, the input–output relation of the fuzzy controller is interpolation function:

$$Y(e, ec) = \sum_{i=-n}^{n} \sum_{j=-m}^{m} A_i(e) B_i(ec) u_{ij} \tag{25.11}$$

$\boldsymbol{A} = \{A_i | i = -n,\ldots,0,1\ldots,n\}$, $\boldsymbol{B} = \{B_j | j = -m,\ldots,0,1\ldots,m\}$ is the linear primitive corresponding to e and ec, usually in the form of triangular wave function. $A_i(e)$ and $B_j(ec)$ are the membership functions corresponding to A_i and B_j in the primitive group. The point ei whose membership function value is 1 is called the peak point corresponding to the primitive e_i, which is called the predecessor peak point. $\boldsymbol{C} = \{C_{ij} | i = -n,\ldots,0,1\ldots,n; j = -m,\ldots,0,1\ldots,m\}$ is the primitive of the output control, $u_{ij}(i = -n,\ldots,0,1\ldots,n;\ j = -m,\ldots,0,1\ldots,m)$ is the peak point of primitive C_{ij}, also known as the peak point of the post-component. The fuzzy rules stored in the computer are the matrix formed by the peaks of the front and rear parts.

Suppose the reasoning rule is expressed in the form: "If A_i and B_j, then C_{ij}", which essentially uses a set of data pairs $((e_i, ec_j), u_{ij})$ composed of the peak points of the preceding and the peak points of the following to do the interpolation function to approximate the real response function [13], "If A_i and B_j then C_{ij}" can be regarded as the combination of "if A_i then C_i" and "if B_j then C_j", namely:

$$\begin{cases} u_{ij} = f(u_i, u_j) \\ u_i = q_1(e_i) \\ u_j = q_2(ec_j) \end{cases} \tag{25.12}$$

When function $f(u_i, u_j)$ is the linear combination of peak point u_i and u_j corresponding to C_{ij}, and interpolation functions $q_1(e_i)$, $q_2(ec_j)$ are linear mappings, fuzzy rules are linear rules, then:

$$\begin{cases} f(u_i, u_j) = k_2 \cdot u_i + k_4 \cdot u_j \\ u_i = k_1 \cdot e_i \\ u_j = k_3 \cdot ec_j \end{cases} \tag{25.13}$$

That is:

$$f(u_i, u_j) = k_1 \cdot k_2 \cdot e_i + k_3 \cdot k_4 \cdot ec_j \tag{25.14}$$

In the formula, k_1, k_2, k_3, k_4 are constants, and the concepts of k_2, k_4 are the same as those in Fig. 25.1. In order to make the quantization factor of the upper limit of the input domain structured as 1, they are taken as the reciprocal of the upper limit of the corresponding domain.

K_p, K_d are taken as PID controller coefficients, and the input–output relation of PID controller is as follows:

$$Y'(e, ec) = K_p \cdot e + K_d \cdot ec \tag{25.15}$$

When function $f(u_i, u_j)$ is the linear combination of data pair (e_i, u_i) and (ec_j, u_j), interpolation function $q_1(e_i)$, $q_2(ec_j)$ are nonlinear mapping, fuzzy rules are nonlinear rules, and:

$$f(u_i, u_j) = k_1 \cdot q_1(e_i) + k_3 \cdot q_2(ec_j) \tag{25.16}$$

Equation (25.14) can be rewritten as:

$$f(u_i, u_j) = q_1'(e_i) \cdot e_i + q_2'(ec_j) \cdot ec_j \tag{25.17}$$

The input–output relation of PD controller with variable parameters, such as PD controller with variable proportions, is as follows:

$$Y'(e, ec) = Kp(e) \cdot e + Kd \cdot ec \tag{25.18}$$

When function $f(u_i, u_j)$ is the nonlinear combination of data pair (e_i, u_i) and (ec_j, u_j), interpolation function $q_1(e_i)$, $q_2(ec_j)$ are nonlinear mappings, fuzzy rules are nonlinear rules, and the input–output relation is Eq. (25.12). By comparing the input–output relation between the PD controller and the fuzzy controller. It can be found that the fuzzy controller is almost equivalent to the ordinary PD controller when interpolation is infinitely dense. However, in real life, it is difficult to obtain infinitely many accurate interpolation points. Set the sampling value of the deviation as e* and the sampling value of the deviation change rate as ec*. The triangular membership function is adopted, with input $(2m + 1)$ fuzzy partitions and output $(2n + 1)$ fuzzy partitions (m and n are positive integers not less than 1). Set the fuzzy rule base as linear rule base, e^* and ec^* treated by quantization factors k_2 and k_4 fall between e_i and e_{i+1}, and the input falls between ec_j and ec_{j+1}. The corresponding non-0 membership function values are $\mu_i(e^*)$, $\mu_j(ec^*)$, $\mu_{i+1}(e^*)$, $\mu_{j+1}(ec^*)$, respectively. Membership function of corresponding output using Zadeh fuzzy logic:

$$\mu_{ij}(e*, ec*) = \min(\mu_i(e*), \mu_j(ec*)) \tag{25.19}$$

The output control u^* is obtained by using the nonlinear weighted average decision method.

$$u* = k \cdot \frac{\sum\limits_{\mu_{ij}(e*,ec*)\neq 0} \mu_{ij}(e*, ec*) \cdot u_{ij}(e*, ec*)}{\sum\limits_{\mu_{ij}(e*,ec*)\neq 0} \mu_{ij}(e*, ec*)} \tag{25.20}$$

The integral fuzzy controller has more input *eI* than the conventional fuzzy controller, and the membership function and input–output relation are as follows:

$$\mu_{ijk}(e*, ec * eI*) = \min\big(\mu_i(e*), \mu_j(ec*), \mu_k(eI*)\big) \tag{25.21}$$

$$u* = k \cdot \frac{\sum\limits_{\mu_{ijk}(e*,ec*,eI*)\neq 0} \mu_{ijk}(e*, ec*, eI*) \cdot u_{ijk}(e*, ec*, eI*)}{\sum\limits_{\mu_{ij}(e*,ec*)\neq 0} \mu_{ijk}(e*, ec*, eI*)} \tag{25.22}$$

To visually display the calculation of the membership function, a three-dimensional input variable cube is constructed with the center of the interval as the origin, as shown in Fig. 25.2.

The first quadrant represents that the membership functions of the input of Eq. (25.21) are μ_{i+1} (e^*), μ_{j+1} (ec^*) and μ_{k+1} (eI^*), and EHCG region represents that the quantization input values fall on the side biased to the (i + 1), (j + 1) and (k + 1) interpolation. And μ_i $(e^*) \geq \mu_k$ $(eI^*) \geq \mu_j$ $(ec^*) \geq \mu_{j+1}$ $(ec^*) \geq \mu_{k+1}$ (eI^*) $\geq \mu_{i+1}$ (e^*). So the calculated results of Eq. 21 in the EHCG area with first quadrant should be μ_{j+1} (ec^*); Similarly, in the EHCG region of the second quadrant, the membership functions representing the input are μ_i (e^*), μ_{j+1} (ec^*) and μ_{k+1} (eI^*), and the quantized input values fall on the side biased to the $(i + 1)$, $(j + 1)$ and $(k + 1)$ interpolation. And μ_i $(e^*) \geq \mu_k$ $(eI^*) \geq \mu_j$ $(ec^*) \geq \mu_{j+1}$ $(ec^*) \geq \mu_{k+1}$ (eI^*) $\geq \mu_{i+1}$ (e^*), the results should be $\mu_i(e^*)$. 3D area of the cube is divided into 48 regions, corresponding to 8 quadrants, a total of 48*8 combinations. Take the 6*8

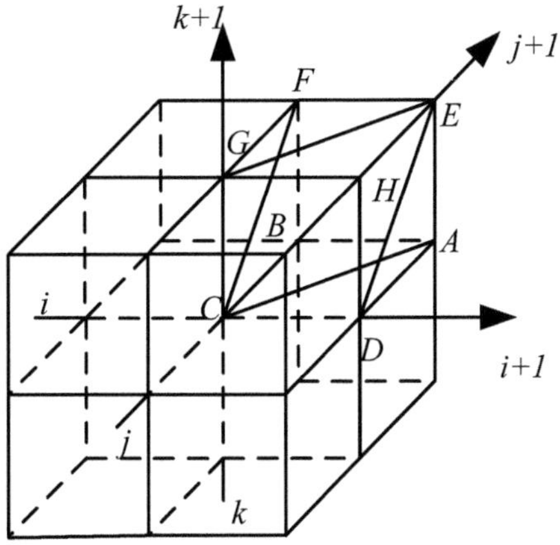

Fig. 25.2. 3D input variable cube

combinations corresponding to the 6 regions of $i+1, j+1$ and $k+1$ as an example, and the results of the membership synthesis of the output are shown in Table 25.1.

The output results of each region can be obtained by taking the fuzzy weighted average decision formula as the representative. The output can be obtained by taking the EHCG region as the representative

$$\begin{cases} u* = u_1 + u_2 \\ u_1 = \left[k_1\left(\dfrac{e_i + e_{i+1}}{2}\right) + k_3\left(\dfrac{ec_j + ec_{j+1}}{2}\right) + k_5\left(\dfrac{ei_k + ei_{k+1}}{2}\right)\right]K \\ u_2 = \dfrac{k_3\left[\frac{ec*}{E_2} - \left(\frac{ec_j + ec_{j+1}}{2}\right)\right] + 2k_5\left[\frac{ei*}{E_3} - \left(\frac{ei_k + ei_{k+1}}{2}\right)\right]}{4 + 2\left[\left(\frac{e_i + e_{i+1}}{2} - \frac{e*}{E_1}\right) + 2\left(\frac{ei_k + ei_{k+1}}{2} - \frac{ei*}{E_3}\right)\right]} E_3 \end{cases} \quad (25.23)$$

In the function form of analogy formula 13, the linear three-dimensional fuzzy rules constituted by the front and rear peaks are expressed as follows:

$$u_{ijk} = k_1 \cdot e_i + k_3 \cdot ec_j + k_5 \cdot eI_k \quad (25.24)$$

It can be seen from the output formula that when the fuzzy rules take linear rules, the output of the fuzzy controller consists of two parts. One part is the nonlinear part related to the input distance falling into the center of the gap. The other part is a PID-like linear structure with an interpolation gap center as input. The output of the nonlinear part is determined by the position of the quantized input relative to the interpolation gap center, and the output of the linear part is the corresponding output value of the core of the region into which the quantized input falls. It can be considered that the output of the fuzzy controller is determined by the linear part and then modified by the nonlinear part.

When the interpolation is not infinitely dense, the input–output relation between the fuzzy controller and the PID controller is somewhat different, but the main part of the output function of the fuzzy controller is still consistent with the input–output

Table 25.1 Calculation results of IFLC output membership function

μ	I	II	III	IV	V	VI	VII	VIII
EHCG	$\mu_{j+1}(ec)$	$\mu_i(e)$	$\mu_i(e)$	$\mu_j(ec)$	$\mu_k(ei)$	$\mu_k(ei)$	$\mu_k(ei)$	$\mu_k(ei)$
EHCD	$\mu_{j+1}(ec)$	$\mu_i(e)$	$\mu_i(e)$	$\mu_j(ec)$	$\mu_k(ei)$	$\mu_i(e)$	$\mu_i(e)$	$\mu_k(ei)$
EACD	$\mu_{k+1}(ei)$	$\mu_i(e)$	$\mu_i(e)$	$\mu_j(ec)$	$\mu_k(ei)$	$\mu_i(e)$	$\mu_i(e)$	$\mu_j(ec)$
CBAE	$\mu_{k+1}(ei)$	$\mu_i(e)$	$\mu_j(ec)$	$\mu_j(ec)$	$\mu_k(ei)$	$\mu_i(e)$	$\mu_j(ec)$	$\mu_j(ec)$
CGEF	$\mu_{i+1}(e)$	$\mu_i(e)$	$\mu_j(ec)$	$\mu_j(ec)$	$\mu_k(ei)$	$\mu_k(ei)$	$\mu_k(ei)$	$\mu_k(ei)$
CBEF	$\mu_{i+1}(e)$	$\mu_i(e)$	$\mu_j(ec)$	$\mu_j(ec)$	$\mu_k(ei)$	$\mu_k(ei)$	$\mu_j(ec)$	$\mu_j(ec)$

relation of the PID controller. It is feasible to convert the fuzzy rules by using PID parameters.

When the PID parameters are known, the PID input–output relation can be used to generate discrete peak point pairs, to form fuzzy rules. In essence, the nonlinear I/O relationship in practice is simplified into a linear I/O relationship, and then it is discretized. Some errors are bound to occur in this process, which makes the steady-state accuracy of the fuzzy control algorithm after transformation reduced compared with the PID control algorithm before the transformation.

Adaptive Integral Fuzzy Controller

Self-couping Proportional Integral Differe, Fig. 25.3 shows the structural diagram of the Self Couping Adaptative Integral Fuzzy Logic Controller (SCAIFLC) designed with conversion rules and variable-domain fuzzy control strategies. Firstly, dimensionality factor Zc is generated based on the input deviation, and then the parameters of the SCPID controller are generated by Zc. Meanwhile, quantitative factors $K_2(e)$, $K_4(ec)$, $K_6(ei)$ are used to scale the theoretical domain based on the input deviation, deviation differential, and deviation integral. Finally, the fuzzy rules are adjusted based on the expanded upper limit of the domain and the generated SCPID parameters, and the output is calculated based on the new rules.

The control law, gain setting rule, adaptive law and large range asymptotic stability characteristics of SCPID when $Zc > 0$ have been defined and proved in literature [14–16]. The control law and gain setting rule are as follows:

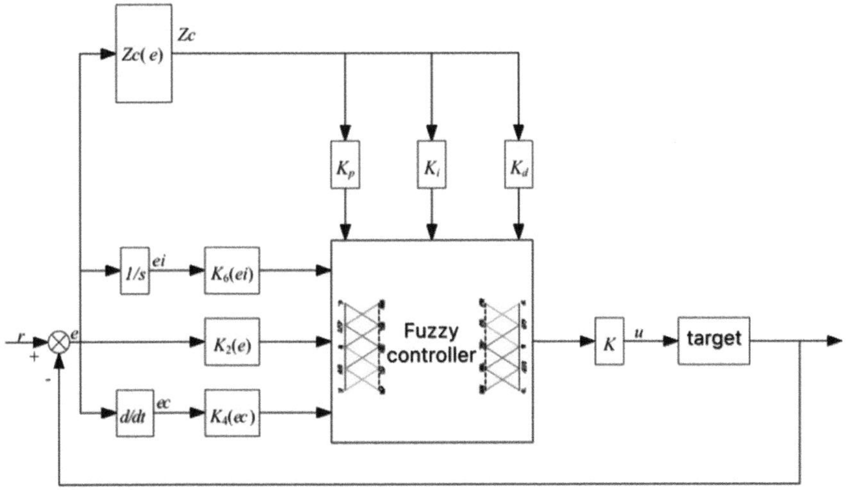

Fig. 25.3 Structure diagram of SCAIFLC3 fuzzy controller

$$u = \frac{Zc^3 \cdot eI + 3Zc^2 \cdot e + 3Zc \cdot ec}{b_0} \tag{25.25}$$

$$\begin{cases} K_i = Zc^3 \\ K_p = 3Zc^2 \\ K_d = 3Zc \end{cases} \tag{25.26}$$

The adaptive law of dimension factor Zc of SCPID controller is as follows:

$$\begin{cases} Zc = \frac{10a}{T_s} \exp\left(\frac{|e|}{e_{\max}}\right), |e| < e_{\max} \\ Zc = \frac{10a}{T_s} \exp(1), |e| > e_{\max} \end{cases} \tag{25.27}$$

where b_0 is the gain coefficient of the controlled object system, $e_{\max}$ is the upper limit of the theoretical domain of deviation e, α is a constant and its value range is $0 < \alpha < 100$, $\alpha > 10$ for systems with high response speed requirements, $\alpha < 10$ for systems with low response speed requirements but high system stability requirements, and T_s is the expected time for the system to transition from dynamic to steady state.

The design of the scaling factor in the IFLC3 domain should conform to the concept of SCPID controller, take dimensional influence into consideration, and choose quantization factor function as scale type after comprehensive consideration, specifically:

$$K(x) = \left(\frac{|x|}{E}\right)^{\tau} + \varepsilon \tag{25.28}$$

In the formula, τ and ε are constants greater than 0 and ε is close to zero. The value of τ control the shrinking and stretching speed of $K(x)$, usually in (0,10].

25.4 Ship Path Following Simulation

Environment and Hull Parameter Settings

To test the performance of IFLC3 in the complex and changeable ship motion system, the SCPID controller before conversion and SCAIFLC after conversion are respectively used as the speed controller based on MATLAB(R2018a) to compare the course control effect and path following effect. The expected path is a square-wave trajectory with a wavelength of 100 m, a wave height of 50 m, and a target speed of 1 m/s. Navigation is carried out based on switching LOS algorithm [17]. Ship parameters [18] are shown in Table 25.2.

Table 25.2 Hull parameters of unmanned boat

Parameter name	Parameter value	Parameter name	Parameter value
m	30.000 kg	$X_{\dot{u}}$	−6.856
L	7.000 m	$Y_{\dot{v}}$	−17.500
Iz	0.300 kg·m^2	$N_{\dot{r}}$	−1.200
X_u	−3.021	$X_{u\|u\|}$	−12.906
Y_v	−20.500	$Y_{v\|v\|}$	−24.200
N_r	−1.252	$N_{r\|r\|}$	−0.330

Static Water Experiment Simulation

To verify the steady-state accuracy changes of the SCPID controller and SCAIFLC controller before and after conversion, the SCPID controller and SCAIFLC controller were used to carry out static water path following experiments respectively under the same other conditions. The track curve of unmanned craft is shown in Fig. 25.4, and the course angular deviation curve of unmanned craft is shown in Fig. 25.5.

Under the action of SCAIFLC and SCPID controllers, the UAV successfully completed the path following, and the track almost overlaps after it entered the steady state. The track deviation overshoot of SCAIFLC was small (0.100 m), and the track deviation overshoot of SCPID control was 0.191 m. This shows that the conversion rule method is feasible and effective.

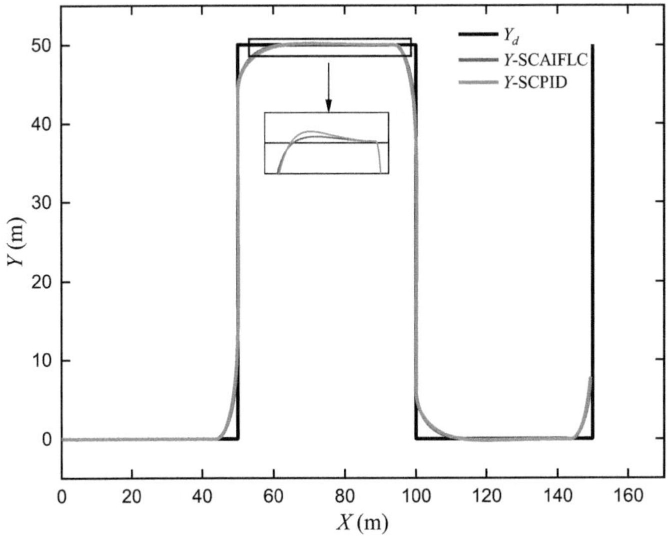

Fig. 25.4 Track curve of unmanned boat simulation in still water experiment

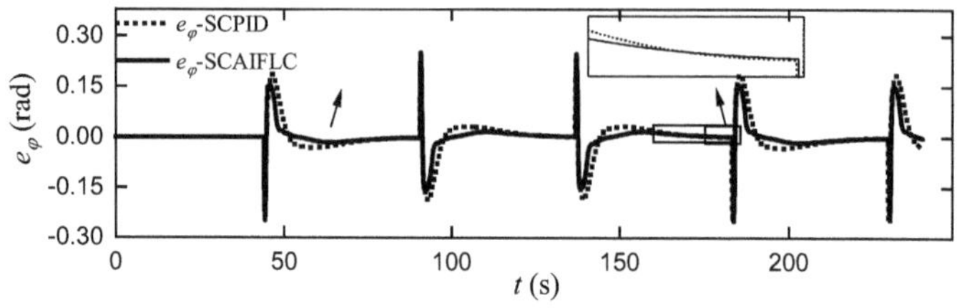

Fig. 25.5 Variation curve of course Angle deviation of unmanned craft

After entering steady-state, the steady-state deviation, of course, Angle was 0.001 rad (about 0.057°) for the SCPID controller and 0.002 rad (about 0.115°) for SCAIFLC. The SCAIFLC based on the conversion rule method can successfully complete the course control task. The conversion rule method is feasible and effective, but the steady-state accuracy is slightly reduced, which is consistent with the derived results.

Simulation of Wave Disturbance

According to the p-m spectrum, the superposition of harmonics is used to generate irregular simulation waves, the initial phase of harmonics is generated by a random function, and each harmonic wave energy is distributed by arithmetic sequence. The simulation wave interferes with about second-order sea state, and the wave Angle is 45°, pointing to the northeast. The wave height variation curve is shown in Fig. 25.6.

The variation curve of course Angle deviation of unmanned craft is shown in Fig. 25.7. It can be seen from the figure that the variation fluctuation of course Angle deviation under the control of SCPID is larger, and the peak value of the specific deviation fluctuation is about 0.031 rad, that is, 1.776°. The SCAIFLC controller has better robust stability, and its peak value is 0.010 rad, that is 0.573°. The robustness of the SCAIFLC controller is better than that of the SCPID controller.

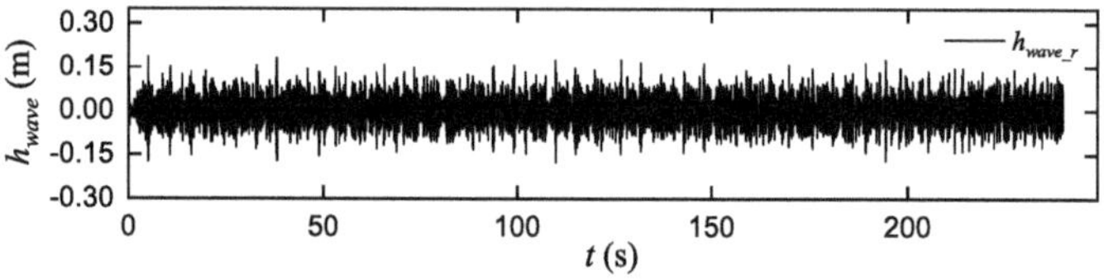

Fig. 25.6 Variation curve of simulated wave height

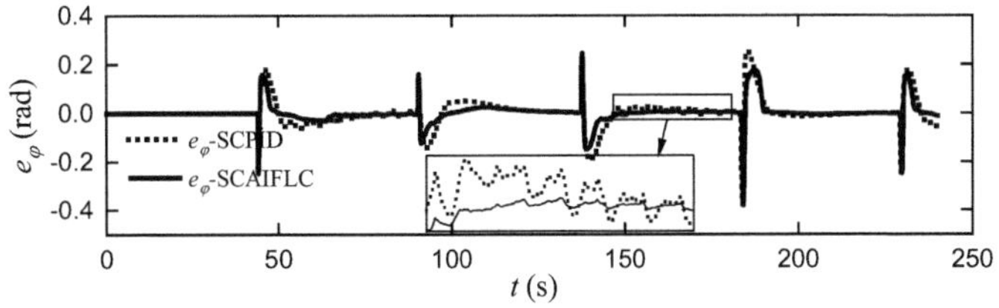

Fig. 25.7 Course angle deviation curve of unmanned craft

25.5 Conclusion

The motion control system of an unmanned boat needs a control method that can resist disturbance effectively, which is what fuzzy control is good at. However, the formulation of fuzzy rules and adaptive problems restricts the application of fuzzy controllers. The formulation of fuzzy rules based on operational experience is bound to be subject to human sense, but human perception is difficult to find to a certain degree and human perception of signals other than e and ec is weak, which leads to the dilemma that fuzzy control is difficult to formulate high-dimensional fuzzy rules.

However, the parameters of PID controllers commonly used in engineering are also established based on experience, and there must be an internal relationship between PID parameters and fuzzy rules. For example, PID parameters contain integral parameters that conventional fuzzy control does not have, and the integral link is an important link related to the steady-state accuracy of PID control. Therefore, introducing an integral link into a fuzzy controller can make up for the poor steady-state performance of a fuzzy controller.

By comparing the input–output relations of PID controller and integral fuzzy controller, this paper proposes a feasible conversion rule method to generate fuzzy rules. The transition rule method uses fewer the PID controller parameters to generate a fuzzy rule peak matrix with more parameters, which greatly reduces the difficulty of making high-dimensional fuzzy rules and the difficulty of self-adaptation. The simulation results verify the feasibility of the conversion rule method, and the SCAIFLC controller generated by the conversion rule in both the static water and wave experiments successfully completes the course control task of the unmanned boat. In the static water simulation, the steady-state accuracy of SCAIFLC decreased by 0.001 rad, about 0.057°, compared with that of the SCPID controller. In the wave simulation, the peak value of the course Angle deviation of SCAIFLC was only one-third of that of the SCPID controller, about 0.573°, which was consistent with the conclusion obtained in the theoretical derivation process. Moreover, the steady-state accuracy of SCAIFLC generated by the conversion rule method is not greatly reduced, which has practical application value.

This paper completes the rule conversion method of linear fuzzy rules based on PID parameters, reveals the connection between PID controller and fuzzy controller, and constructs SCAIFLC. The steady-state accuracy of SCAIFLC based on the conversion rule method is similar to that of the corresponding PID controller, but it has strong anti-interference ability. SCAIFLC, which does not need accurate mathematical model, has strong anti-interference ability, fuzzy rule acquisition and self-adaptive simple, fits well with the ship course control which is difficult to accurately model due to the variable environmental interference. The proposal of the conversion rule method is of great significance to the development and application of fuzzy control in the control technology of unmanned craft.

The current transformation rule method can only generate linear fuzzy rules, because the transformation base, PID controller, is a linear combination of deviation, deviation differential and deviation integral. In practice, the relationship between the three is much more complex, and the nonlinear transformation rule method needs to be further studied.

References

1. Andrija, L., Vedran, S., Juraj, B.: Unmanned surface vehicle–tritor. Brodogradnja **73**(3), 135–150 (2022)
2. Wu, J.S., Li, R.H., Li, J.W., et al.: Cooperative unmanned surface vehicles and unmanned aerial vehicles platform as a tool for coastal monitoring activities. Ocean Coast. Manag. **232**, 106421 (2023)
3. Li, Y., Zhang, J., Wang, H., et al.: A control strategy for unmanned surface vehicles flocking. Int. J. Pattern Recognit. Artif. Intell. **36**(09), 2259020 (2022)
4. Peng, Z.H., Liu, E.N., Pan, C., et al.: Model-based deep reinforcement learning for data-driven motion control of an under-actuated unmanned surface vehicle: path following and trajectory tracking. J. Franklin Inst. **15**, 1371–1383 (2022)
5. Putri, N., Farhanah, M., et al.: Development of navigation system for unmanned surface vehicle by improving path tracking performance. J. Eng. Sci. Technol. **15**(2), 1371–1383 (2020)
6. McCue, L.: Handbook of marine craft hydrodynamics and motion control. IEEE Control. Syst. **36**(1), 78–79 (2016)
7. Wang, G.X., Xu, G.H., Liu, G., et al.: Fuzzy iterative sliding mode control applied for path following of an autonomous underwater vehicle with large inertia. Math. Probl. Eng. **2019**(1), 1–14 (2019)
8. Sivaraman, S., Suresh, R., Lokukaluge, P.P.: Data driven control based on deep Q-network algorithm for heading control and path following of a ship in calm water and waves. Ocean Eng. **259**, 111802 (2022)
9. Ye, L., Hu, L., Zhang, M.Y., et al.: Optimization of fuzzy control strategy for heat pump air conditioning system of pure electric vehicle. Control Eng. China **28**(8), 1526–1533 (2021)
10. Zhao, Z.Z., Yang, Z., Zhao, S.H.: Electric spring decoupling control strategy based on error integral fuzzy control. Renew. Energy Resour. **39**(9), 8 (2021)
11. Fossen, T.I.: Handbook of Marine Craft Hydrodynamics and Motion Control. Wiley, America (2011)
12. Fossen, T.I., Breivik, M., Skjetne, R.: Line-of-sight path following of underactuated marine craft. IFAC Proceed. Vol. **36**(21), 211–216 (2003)
13. Guo, H.G., Li, H.X., Hu, K.: A novel variable universe adaptive fuzzy controller. Fuzzy Syst. Math. **25**(6), 9 (2011)
14. Zeng, Z.Z., Liu, W.Y.: Self-coupling PID controllers. Acta Autom. Sinica **47**(2), 404–422 (2021)
15. Zeng, Z.Z., Chen, Z.Y.: On control theory of PID and auto-coupling PID. Control Theory Appl. **37**(12), 2654–2662 (2020)

16. Su, J., Zeng, Z.Z.: Auto-coupling PID control method for nonlinear time varying systems. Control Theory Appl. **39**(2), 299–306 (2022)
17. Zeng, J.F, Wan, L., Li, Y.M, et al.: Switching-line-of-sight-guidance-based robust adaptive path-following control for underactuated unmanned surface vehicles. Acta Armamentarii **39**(12), 11 (2018)
18. Oh, S.R., Su, J.: Path following of underactuated marine surface vessels using line-of-sight based model predictive control. Ocean Eng. **2010**(37), 289–295 (2010)

Chapter 26
NVGAN: A GAN-Based Night Vision Image Enhancement Model

Ming Gao

Abstract Night vision image enhancement is the process of improving the perception and clarity of an image by processing it at night or under low light conditions. To achieve this effect, the mainstream approach is to adopt traditional image processing methods or to adopt different learning strategies based on deep learning. In this paper, we propose a GAN-based night vision image enhancement model, NVGAN, which consists of two networks: a generative network and a discriminative network. The generative network is used for illumination enhancement and deep and shallow feature extraction and enhancement for night vision images, and the discriminative network is selected as DnCNN for suppressing the effect of noise on night vision images. Meanwhile, the loss function is further improved to obtain a better performance for the GAN network in order to obtain a better enhancement effect.

26.1 Introduction

Night vision images are images acquired under low light (dark illumination) conditions and usually need to be realized with the help of night vision devices [1], such as infrared thermal imagers. Analyzed from the perspective of human vision, in order to see objects under low-light conditions, the human pupil physiologically and automatically dilates, thus allowing more light to enter the eye. However, in some cases, even if the pupil of the human eye is dilated to its limit, it still cannot get enough light to see the object. Based on the above, it is often necessary to use night vision technology to assist in enhancing to achieve night vision.

Night vision technology is highly dependent on night vision devices, which enable people to see around them in complete darkness or low-light conditions by converting weak signals in low light into visible images or videos [2]. However, night vision images are usually black and white or green in color for the reason that they can better highlight the contours and details of objects at night. This effect is achieved by the

M. Gao (✉)
College of Optics and Electronic Technology, China University of Metrology, Zhejiang, China
e-mail: P21040854019@cjlu.edu.cn

R. Kountchev (Deceased) et al. (eds.), *AI Methods and Applications in 3D Technologies*, Smart Innovation, Systems and Technologies 388,
https://doi.org/10.1007/978-981-97-2144-3_26

working principle of night vision devices, i.e., the use of optoelectronic conversion devices to convert light signals into electrical signals, which are then processed to convert the electrical signals into images or video signals for output [3]. Figure 26.1 shows also the night vision scene diagram.

Night vision technology uses different enhancement methods to enhance light to obtain night vision images. These methods include infrared night vision images, thermal imaging night vision images, low-light night vision images, and visible night vision images. Table 26.1 shows a comparison of the four night vision techniques.

Fig. 26.1 Night vision scene effect

Table 26.1 Comparison of the three night vision technologies

Night vision technology	Characteristics
Infrared night vision images	Use infrared radiation technology to capture infrared radiation from objects and display them by converting them into visible light signals, suitable for observation in dark or low light environments
Thermal imaging night vision images	Use the thermal radiation of an object to generate an image by detecting and displaying the difference between different temperatures of the object to form a color or grayscale image
Low-light night vision images	By enhancing light transmission or increasing the sensitivity of the image sensor to capture images in low-light environments, the displayed image will have some noise, but the surrounding environment can still be observed
Visible night vision images	Uses light in the visible spectrum to capture images for brighter nighttime viewing

These night vision techniques in Table 26.1 can be applied in various fields, including military, security, and nocturnal wildlife observation.

At present, infrared night vision devices, thermal cameras and micro-optical night vision devices have mature theories and technologies, but the night vision images they obtain are monochromatic grayscale images, and it is easy to have other objects to cause interference with the target, which causes difficulties for the human eye for the recognition of the target. The human eye recognizes only a few dozens of gray levels, but the recognition of colors can reach thousands, and color information can greatly enhance the recognition probability of the human eye to the target. Experiments show that color images can improve recognition rate compared to grayscale images. In applications such as surveillance, reconnaissance and security applications, color images have two main advantages over monochrome images. First, color can improve contrast, which leads to better scene segmentation and object detection. This contrast improvement can be applied to both true-color and pseudo-color images, where pseudo-color images can be formed by fusing images from cameras with different spectral sensitivities (e.g., visible light). A 2nd benefit of color is that it provides more information. Access to color knowledge stored in the brain or computer databases can be used for better object recognition and scene understanding, an advantage that applies only to true color images. Therefore, color night vision techniques are receiving increasing attention for better application of color information in night vision images.

26.2 Literature Review

GAN Networks

GAN (Generative Adversarial Networks), a generative adversarial network, was first proposed by Ian Goodfellow et al. in 2014 [4], with a core design of two neural networks against each other. Specifically, one generator network is mainly responsible for generating fake data, while the other discriminator network tries to distinguish the real from the fake data. Through continuous training, the generator can gradually generate realistic data, while the discriminator gradually becomes more accurate, thus facilitating the development of generative models.

In 2015, Radford et al. [5] proposed the Deep Convolutional Generative Adversarial Network (DCGAN), which enabled GAN to better process image data for the first time. Next, Mirza et al. proposed Conditional Generative Adversarial Network (CGAN) [6], which introduces conditional vectors to generate specific types of images for better feature representation. Zhu et al. [7] proposed Cyclic Consistent Generative Adversarial Network (CycleGAN), which allows image transformation without the need for paired training data. With the development of GAN, it has a wide range of applications in image generation, image restoration, speech synthesis, and video generation [8].

Image Enhancement

Image enhancement refers to a method of manipulating an image to improve its quality, enhance specific information or improve its visual effect. Common image enhancement techniques include grayscale transformation, histogram equalization, filters, noise cancellation and color enhancement.

Grayscale transformation is to achieve enhancement by adjusting the gray level of an image to further increase the contrast or brightness of the image. Histogram equalization is the redistribution of the image's histogram to extend the range of gray levels to enhance the image's contrast. Filtering is the use of different types of filters to enhance specific features in an image. Noise suppression is the use of filters or other techniques to reduce noise in an image in order to improve the clarity of the image. Color enhancement, on the other hand, specifically refers to the adjustment of color images to change the color balance or increase saturation in order to enhance the visual effect of the image.

Supervised Learning

Supervised Learning (SL) is a common approach to machine learning, which aims to learn a model from labeled data in order to predict or classify unlabeled data [9]. In supervised learning, the model learns a function from the input data that makes the most accurate mapping relationship between the input and the output. This function can be linear or nonlinear, or probability-based.

Supervised learning is used in a wide range of applications, such as image classification, speech recognition, natural language processing, and recommendation systems. Common supervised learning algorithms include linear regression, logistic regression, decision trees, etc. [10, 11].

26.3 Deep Learning Scheme for Night Vision Image Enhancement Based on GAN

NVGAN for Generative Adversarial Networks

The training process of GAN is a dynamic balancing process. During training, the generator tries to generate more and more realistic samples to fool the discriminator. The discriminator, on the other hand, strives to improve its ability to distinguish generated samples from real samples. Through this adversarial process, the generator and discriminator continuously improve their performance and eventually reach an equilibrium, allowing the generator to generate samples very close to the real data distribution.

Therefore, in order to better restore low-light image details, we propose the NVGAN method based on Attention, RDN and GAN structures. In this paper, NVGAN consists of a generator network and a discriminator network.

To be more specific, the generator network contains a global illumination estimation module, a convolutional residual module and a channel attention-based dense residual module, which are used to guide the subsequent modules for illumination enhancement, extracting shallow features and deep features respectively. From another perspective, we adopt DnCNN as the discriminator network, which improves the noise suppression effect.

Technical Solution

Night vision image enhancement is a technique that can make the image clearer and brighter by enhancing the image brightness, contrast and sharpness in low light conditions. Night vision image enhancement techniques usually use some special sensors and algorithms, such as infrared sensors and computer vision algorithms, to improve the image quality. Night vision image enhancement technology is widely used in military, public security, traffic monitoring and other fields, which can help people to better observe and identify targets in low-light environment and improve safety and efficiency.

Traditional night vision image enhancement solutions usually use image enhancement algorithms, infrared light source illumination, thermal imaging techniques and deep learning techniques. Among them, image enhancement algorithms include histogram equalization, grayscale stretching, median filtering and other algorithms that can improve the contrast and clarity of images. Infrared light source illumination Using infrared light source illumination can improve the brightness and clarity of nighttime images, but it is necessary to ensure that the intensity and wavelength of the infrared light source are suitable for the current environment. Thermal imaging technology generates images by measuring thermal radiation from the target surface, independent of light and ambient light, and can achieve true night vision. Deep learning techniques, on the other hand, can improve image quality and clarity by training models that can be optimized for different application scenarios.

In this paper, we select a night vision image enhancement scheme using deep learning techniques and propose an innovative GAN-based deep learning scheme for night vision image enhancement.

Among them, GAN (Generative Adversarial Network) is a deep learning model that is commonly used for image generation tasks. GAN-based night vision image enhancement can also be regarded as an image generation task, i.e., generating a low-quality night vision image into a high-quality image. The specific implementation steps are as follows.

1. Dataset preparation: collect and label the night vision image dataset, which can include low-quality night vision images and high-quality daytime images.

2. Build GAN network: build the generator and discriminator. The generator inputs low-quality night vision images and outputs high-quality daytime images; the discriminator is responsible for judging whether the images generated by the generator are real or not, i.e., distinguishing between the images generated by the generator and the real daytime images.
3. Training the GAN network: Using the dataset for training, through continuous iterations, the generator gradually learns the skill of converting night vision images into high quality daytime images, while the discriminator gradually becomes more accurate.
4. Test and evaluation: the generator is tested using a test set to evaluate the quality of the daytime images it generates.

The first step is to train the discriminator so that it can distinguish between the real data and the night vision images generated by the generator. In each training iteration, the discriminator receives a batch of real data and a batch of fake night vision images generated by the generator. The discriminator needs to distinguish these two data sets and adjust their weights and biases accordingly.

The generator is then trained, and once the discriminator has been trained well enough, it is time to start training the generator. The generator will try to generate night vision images that are as close to the real data as possible. In each training iteration, the generator generates a batch of fake night vision images, which are then evaluated for quality by the discriminator. If the fake night vision images are considered by the discriminator to be very similar to the real night vision images, then the weights and biases of the generator are adjusted accordingly. After completing the above steps, the above training steps are repeated until the GAN network generates data of the desired quality.

Loss Function Design

In order to obtain better enhancement effect by providing GAN network, we improve the loss function, which consists of two loss functions of generative network and discriminative network:

$$L_{total} = L_{Advl} + L_{Gen} \tag{26.1}$$

Among them, the generative network loss function consists of MSE, MAE, and MAE plays a role in suppressing the night vision image noise. Meanwhile, the discriminant network loss function adopts the least squares loss function, which is used to guide the Nash equilibrium between the generative network and the discriminant network.

To be more specific, Loss total represents the overall loss function of the NVGAN model, Loss Advl represents the loss function of the discriminator for GAN in this paper, and Loss Gen represents the loss of the generator.

Network Model Architecture Design

The GAN network in this paper is redesigned based on NVGAN, and this model consists of two networks, the generative network and the discriminative network. The generative network is used for illumination enhancement and deep and shallow feature extraction and enhancement for night vision images, and the discriminative network is selected as DnCNN to suppress the effect of noise on night vision images. As shown in Fig. 26.2.

In Fig. 26.2, we designed two core modules, Core Region Module (CRM) and Core Adversarial Region Design Module (CARDM), for night vision image enhancement. The CRM module contains a series of cascaded CRB extractors for better feature propagation. Meanwhile, the input X0 of CRM is also used as the input of CARDM and enters a series of consecutive CARDB extractors. The two modules are ultimately combined for night vision enhancement.

As the most important network architecture in the GAN, the generator network in this paper incorporates attention mechanisms mainly for extracting key information, forming the innovative ConvAttBlock that is used for better restoration of night vision image details. The architecture diagram of the generator network is shown in Fig. 26.3.

In Fig. 26.3, we can see that the hierarchical structure of the generator network is “Input - > Attention map - > Max Pooling & MixAttBlock - > Max Pooling & MixAttBlock - > Max Pooling & MixAttBlock - > ConvAttBlock & ConvTranspose - > ConvAttBlock & ConvTranspose - > ConvAttBlock & ConvTranspose - > Output”. Adopting the above cascading mode helps the NVGAN model better extract and enhance night vision image features.

The network architecture of ConvAttBlock is shown in Fig. 26.4, and the architecture of MixAttBlock is shown in Fig. 26.5.

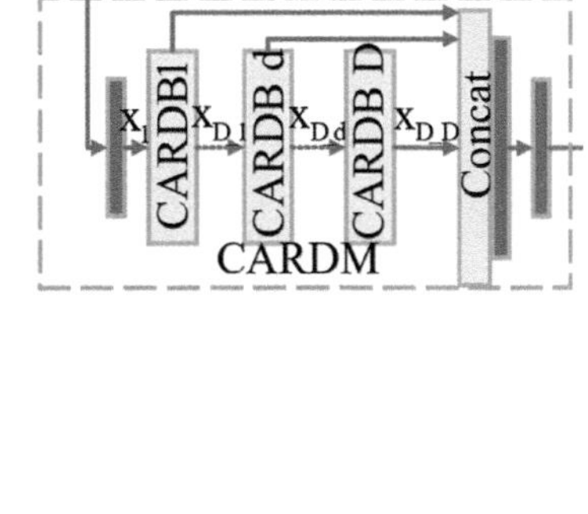

Fig. 26.2 A GAN-based night vision image enhancement model NVGAN proposed in this paper

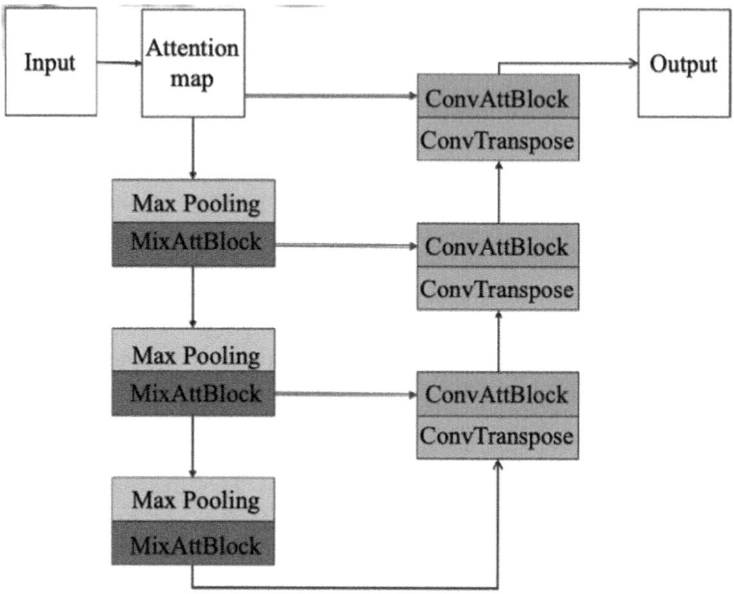

Fig. 26.3 The architecture diagram of the generator network in NVGAN

Experiments

In the experiments, we used Adam optimizer with an initial learning rate of 1e-3 and applied learning rate decay by 1% per epoch. The epoch was set to 100, and the batch size was set to 16.

In this section, we first visualized the loss function curves of the generator and discriminator during training. It can be observed that NVGAN tended to converge at 100 epochs and achieved optimal performance, as shown in Fig. 26.6.

Next, we demonstrated the night vision enhancement performance of NVGAN. As shown in Fig. 26.7, it can be seen that NVGAN has strong enhancement capabilities.

As can be seen from Fig. 26.7, both the proposed ConvAttBlock and MixAttBlock modules achieved good night vision image enhancement results, which can be better applied in real-life scenarios.

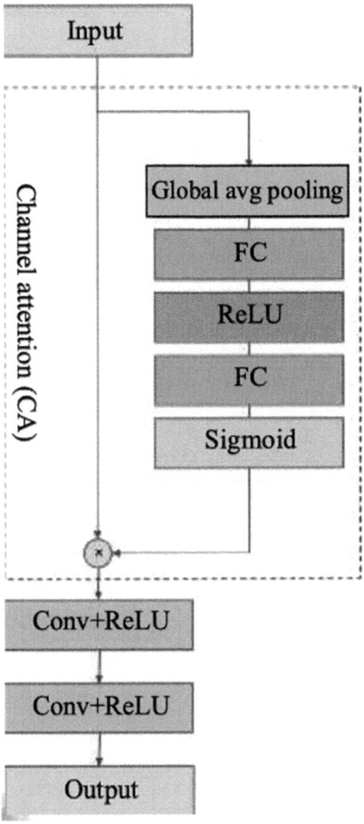

Fig. 26.4 The ConvAttBlock of NVGAN

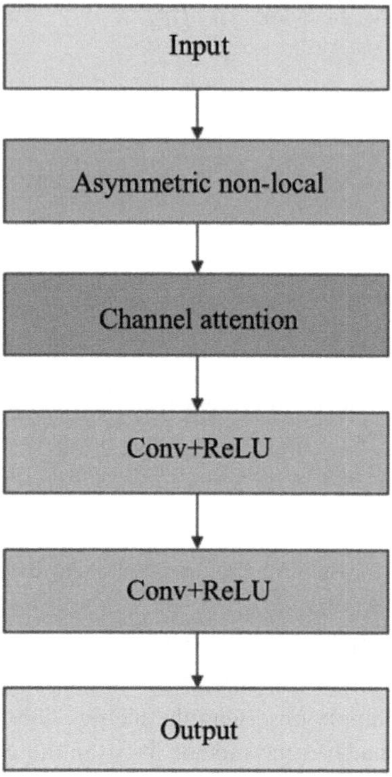

Fig. 26.5 The ConvAttBlock of MixAttBlock

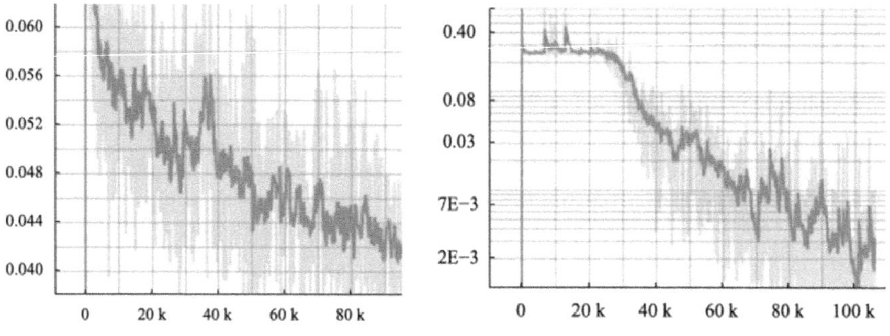

Fig. 26.6 The loss function in the training phase

Fig. 26.7 Night vision enhancement comparison results

26.4 Conclusion

In order to achieve the effect of night vision image enhancement more efficiently, a GAN-based night vision image enhancement model NVGAN is proposed in this paper. The specific technical solution is that NVGAN consists of two networks: generative network and discriminative network, in which the generative network is mainly used for illumination enhancement and deep and shallow feature extraction and enhancement for night vision images, while the discriminative network is selected as DnCNN for suppressing the effect of noise on night vision images. The discriminant network is selected to suppress the effect of noise on the night vision images. We further improve the loss function to obtain better performance. This model is important for night vision image enhancement.

References

1. Haque, M.J., Muntjir, M.: Night vision technology: an overview. Int. J. Comp. Appl. **167**(13), 8887–8937 (2017)
2. Toet, A., Hogervorst, M.A.: Progress in color night vision. Opt. Eng. **51**(1), 010901–010901 (2012)
3. Grigoryan, A.M., Agaian, S. S. Thermal and night vision image visibility and enhancement. Mobile Multimedia/Image Processing, Security, and Applications 2020. SPIE, vol. 11399, pp. 169–178 (2020)
4. Goodfellow, I., et al.: Generative adversarial nets. Neural Information Processing Systems MIT Press (2014)
5. Radford, A., Metz, L., Chintala, S.: Unsupervised representation learning with deep convolutional generative adversarial networks(2015). arXiv preprint arXiv:1511.06434 (2015)
6. Mirza, M., Osindero, S.: Conditional generative adversarial nets. arXiv preprint arXiv:1411.1784 (2014)
7. Zhu, J.Y., Park, T., Isola, P., et al.: Unpaired image-to-image translation using cycle-consistent adversarial networks. In: Proceedings of the IEEE International Conference on Computer Vision, pp. 2223–2232 (2017)
8. Din, N.U., Javed, K., Bae, S., et al.: A novel GAN-based network for unmasking of masked face. IEEE Access **8**, 44276–44287 (2020)
9. Cunningham, P., Cord, M., Delany, S.J.: Supervised learning. In: Machine Learning Techniques for Multimedia: Case Studies on Organization and Retrieval, pp. 21–49 (2008)
10. Hastie, T., Tibshirani, R., Friedman, J., et al.: Overview of supervised learning. In: The Elements of Statistical Learning: Data Mining, Inference, and Prediction, pp. 9–41 (2009)
11. Learned-Miller, E.G.: Introduction to Supervised Learning, p. 3. I: Department of Computer Science, University of Massachusetts (2014)

Chapter 27
Oilfield Staff Workwear Detection and Monitoring System

Yuting Guo, Zhijia Zhang, YunPeng Liu, and Shixian Wang

Abstract In this paper, deep learning algorithms are applied to oilfield work scenarios, and YOLOv5s is selected as the backbone network for staff workwear detection and analysis of existing problems: To improve the small-scale objective detection capability, the second layer of the backbone network is added for feature output and the SimAM attention mechanism module is introduced. To improve the obscuration objective detection capability, the G-Transformer encoder module is introduced, and the Weighted boxes fusion (WBF) candidate frame screening method is selected for post-processing. After that, the above-improved method is designed in general, and the G-YOLOv5s detection network is proposed. Experiments show that G-YOLOv5s improves detection accuracy by 5.31% compared to the initial network, while the speed is only reduced by 20 frames/second. Finally, the PyQt tool with "OpenCV library + frame loop" as the core control method for monitoring system development can realize real-time monitoring of workers' dressing in oilfield scenes, which has high practical value.

Y. Guo (✉) · Z. Zhang · S. Wang
School of Artificial Intelligence, Shenyang University of Technology, Liaoning, Shenyang 110870, China
e-mail: yyuting_guo@163.com

Y. Guo · Z. Zhang
Shenyang Key Laboratory of Information Perception and Edge Computing, Shenyang, China

Y. Liu
Shenyang Institute of Automation, Chinese Academy of Sciences, Beijing, China

Y. Liu · S. Wang
Liaoning Provincial Key Laboratory of Optoelectronic Information and Simulation Technology, Liaoning Province, China

R. Kountchev (Deceased) et al. (eds.), *AI Methods and Applications in 3D Technologies*, Smart Innovation, Systems and Technologies 388,
https://doi.org/10.1007/978-981-97-2144-3_27

27.1 Introduction

With the development of deep learning technology, intelligent monitoring based on deep learning has gradually replaced manual labor. Staff workwear monitoring in oilfield work scenes has become a vital application research direction. Although the detection method has achieved greater accuracy than the method of machine vision, the detection effect is still not ideal in such a complex scenario as an oilfield. Moreover, because there is no system installation, the practical value is low. This study combines the oilfield work scenario, network detection defect analysis, and targeted optimization. Finally, PyQt is used to visually develop the design of a monitoring system based on the G-YOLOv5s algorithm in an oilfield working scenario.

27.2 System Architecture Design

In this paper, we use PyQt [1] tool to develop and design the monitoring system. After system initialization, the PyQt interface selects the input path of the file to be detected, such as local video, local image, or call camera path. The system process is controlled by the core method "OpenCV [2] + Frame Loop", as shown in Fig. 27.1. When the user clicks the "Start Detection" button, and the first frame loop starts. The system intercepts the input video stream into a single frame by calling the OpenCV library and sends it to the workwear detection network for detection frame by frame. When the detection is completed, the result is drawn to the display interface and goes to the next frame loop, and the process is the same as above. Until the "End" button is clicked, the system jumps out of the loop, the process ends, and the system stops detecting.

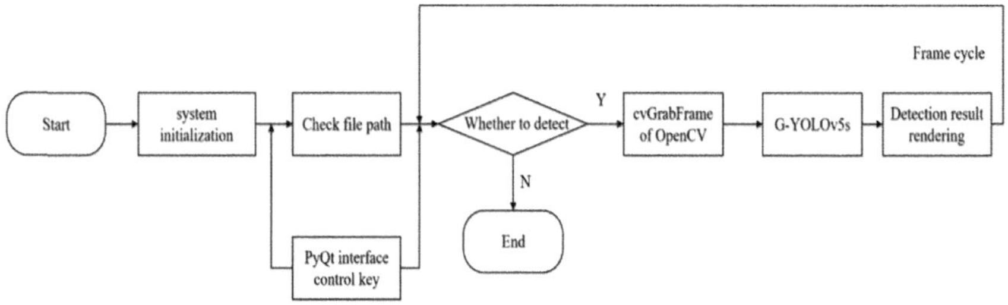

Fig. 27.1 Flow chart of the monitoring system

27.3 Datasets

This design uses a homemade dataset of workers' attire in oilfield work scenes, and 6000 images are obtained through network collection and field photography. LabelImg software was used to calibrate the candidate frame, including one human body label and two labels each to determine whether to wear the safety hat, work clothes and rubber shoes. In addition, the dress label is positioned on the human body label. When the two optional boxes Iou > 0.8, it is defined as the dress is valid, and the result is output. Instead, the label is invalid. Finally, the training and verification set were divided according to the ratio of 8:2, namely, 4800 training sets and 1200 verification sets, and subsequent experimental analysis was carried out.

27.4 Workwear Detection Network Design

In this paper, YOLOv5s [3] is selected as the backbone network for training, and the reasons for the low accuracy of this network are analyzed by the validation set into two categories: (1) Due to the complex oil field environment, the camera head is often far away from the workers, and most of the workers in the sample have smaller targets. YOLOv5s has missed or mistakenly detected targets with more minor scales. (2) In oil field work scenarios, workers are densely populated, and there are many obstacles, making it easy for personnel to be obstructed. YOLOv5s may also miss or mistakenly detect such targets. Therefore, this article should focus on improving the above situation to optimize the network detection accuracy.

Optimization of Workwear Detection Model Under Small-Scale Objectives

In order to expand the target receptive field, the design adds the second layer of YOLOv5s backbone network for feature output [4] so that the network can obtain more detailed information on samples, And introduce the SimAM attention mechanism module [5] calculate 3D attention weights through the Energy function, taking into account the similarity of different features in channel and space without introducing additional parameters, enabling the network to extract small target features more comprehensively, thereby improving detection performance, the principle is shown in Fig. 27.2.

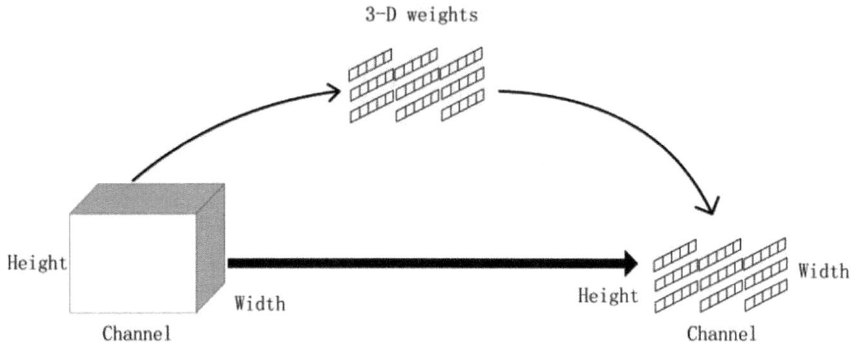

Fig. 27.2 SimAM attentional mechanism

Optimization of Workwear Detection Model Under Obscuration Objectives

In order to improve the representation mining ability of salient features of occluded target samples, the Transformer encoder [6] module is commonly used to replace some convolutional modules and CSP (Cross Stage Partial Network) bottleneck blocks in YOLOv5s. Transformer encoder module is a part of the Transformer model, which is used to convert input sequences into a vector representation, with good parallel computing capabilities and long sequence processing capabilities. As shown in Fig. 27.3, the Transformer encoder module consists of multiple stacked attention mechanism layers and fully connected layers, where each attention mechanism layer consists of a self-attention mechanism and a feedforward neural network, with residual links between the two layers to reduce network complexity. However, adding too many parameters can affect the network detection speed and cause problems such as gradient vanishing and network dispersion.

To improve this problem, we propose the G-Transformer encoder module, which introduces an improved SENet subnet in the residual link to reduce noise in sampling by automatically learning soft thresholds, as shown in Fig. 27.4. Soft threshold [7] by setting a threshold range, the features whose absolute value is greater than the threshold are indented to zero, and the features whose absolute value is less than or equal to the threshold are zeroed out, so that the network eliminates a lot of useless information and retains the target feature information that needs to be learned. The computational complexity of full connection layer and multi-attention mechanism is reduced, the error caused by feature degradation is reduced, and the ability of mining significant feature information and obtaining context information is improved. And because there are identical paths across layers in the module, the network can better train parameters, improving the potential for feature representation.

In the post-processing section, this article selects the WBF [8] (Weighted boxes fusion) candidate box screening method, as shown in Fig. 27.5. It avoids the situation

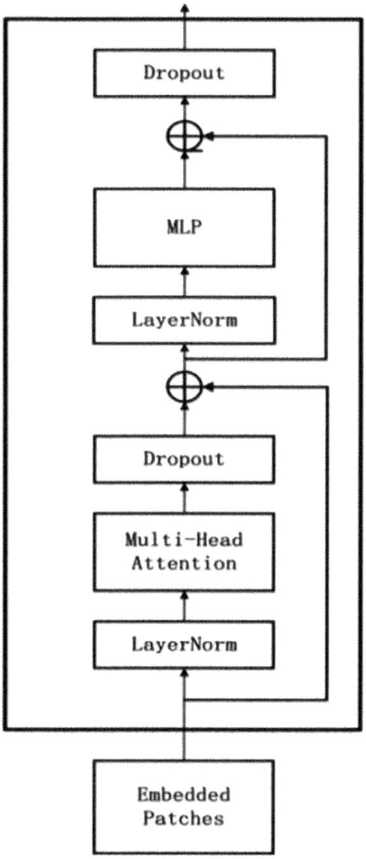

Fig. 27.3 Transformer encoder module. (MLP–Multilayer Perceptron)

of missed detection caused by traditional candidate boxes that have high overlap due to personnel density in oilfield work scenarios.

Overall Network Design

Based on the above-improved modules for overall design analysis, the network structure of G-YOLOv5s is shown in Fig. 27.6.

The network structure of G-YOLOv5s is divided into four parts: input, Backbone, Neck and Prediction. At the input end, Mosaic data enhancement [9] and adaptive picture scaling are used to send images to the backbone network. The Focus modulec in the backbone network is responsible for slicing operations; the C3 [10] module contains three convolution layers and several Bottleneck layers; Conv module contains convolution layer, the BN layer and Silu activation function; the SPP module [11] is responsible for the feature fusion of images of different scales,

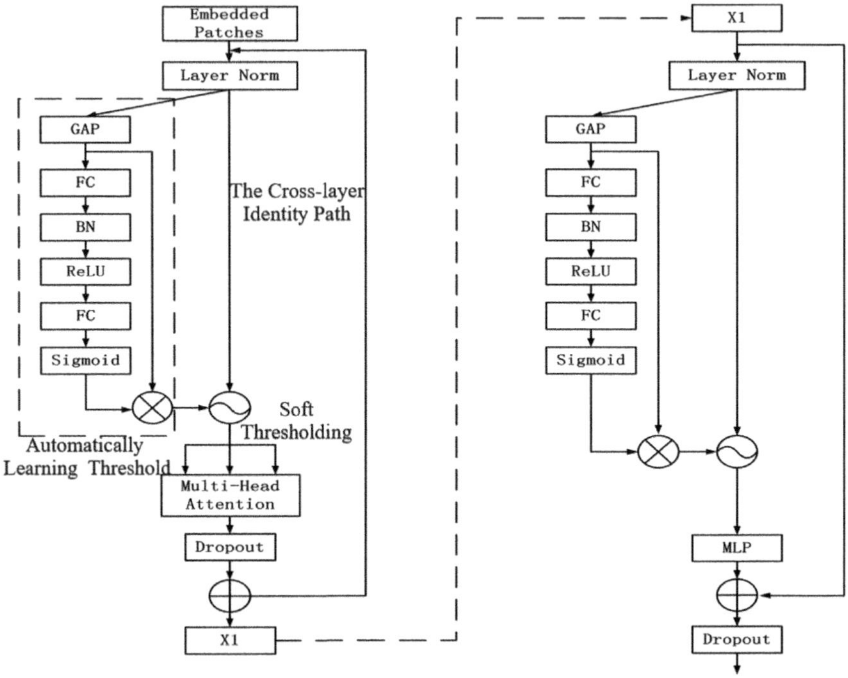

Fig. 27.4 G-Transformer encoder module

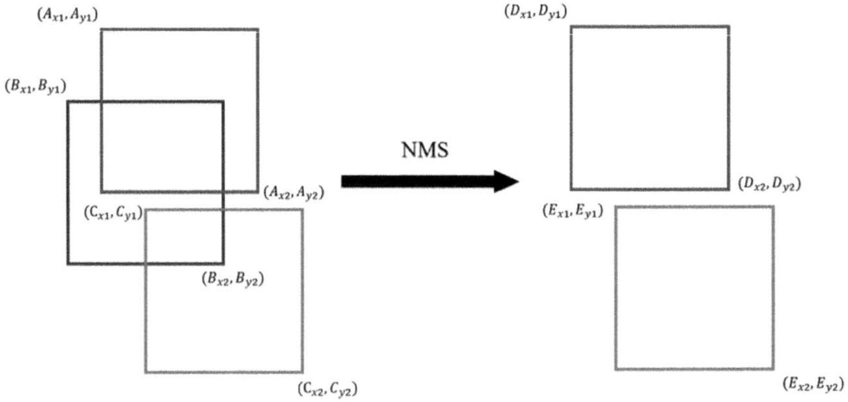

Fig. 27.5 NMS (Non-maximum suppression) filtering diagram

which is conducive to optimizing the situation where the object size difference is large in the image to be detected; the G-Transformer encoder module is responsible for improving feature characterization potential. Part of Neck network adopts FPN + PAN [12] structure, the SimAM module is responsible for improving the

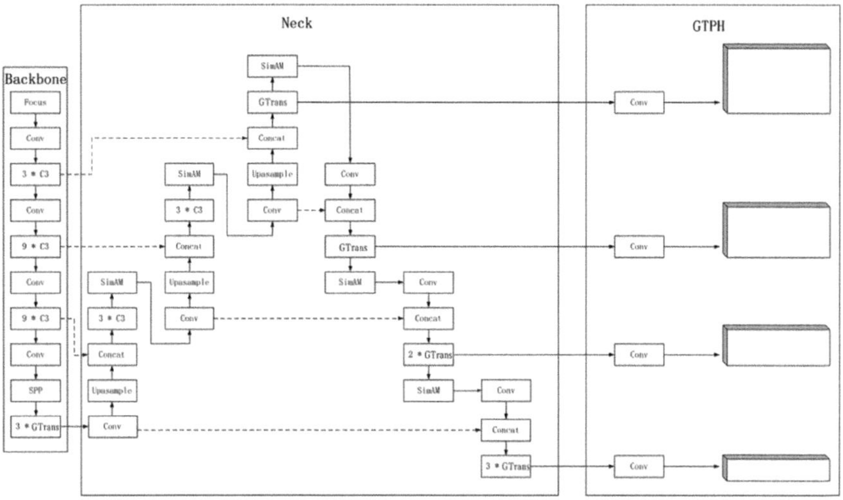

Fig. 27.6 G-YOLOv5s network structure diagram

network's focus on the target; the Upsample module is responsible for upsampling; the Concat module is responsible for concatenating feature maps of different scales on the channel. Finally, the feature map is output by WBF post-processing.

27.5 Results Analysis and System Testing

The experimental compilation environment is Python 1.10.1 + Python 3.6, with 3000 epochs of training. The experiment used mAP, mFPR, mFNR, the number of frames detected per second and, the number of parameters as evaluation criteria. The IoU detection threshold was set to 0.55. To ensure the accuracy of the experiment, the conclusion data is the average of three experiments.

Ablation Experiments

Ablation experiments are performed on an oil field dataset to verify the effectiveness of the improvements. The test results are shown in Table 27.1. From the data in the table, Increase Second Detection Layer to Improve Small Target Detection Accuracy, and the optimization method of introducing the SimAM attention mechanism also improves the detection ability of oil field workwear targets (including multi-scale and occlusion targets) in real application scenarios; The G-Transformer encoder module introduced to improve the detection accuracy of occlusion targets replaces part of fast convolution and CSP bottleneck blocks, and the optimization

Table 27.1 Results of ablation experiments

Model	Adding detection layers	SimAM	G-Trans	WBF	mAP	mFPR	mFNR	Speed	Parm
YOLOv5s	–	–	–	–	86.31	6.48	7.13	81	16.37
R1	√	–	–	–	87.06	6.25	6.77	79	16.79
R2	–	√	–	–	88.41	6.07	6.46	75	17.94
R3	√	√	–	–	89.17	5.79	5.92	72	18.31
R4	–	–	√	–	88.35	4.93	5.28	69	19.26
R5	√	√	√	–	90.55	4.26	4.54	63	20.47
R6	–	–	–	√	87.82	6.17	6.18	75	17.18
GYOLOv5s	√	√	√	√	91.62	3.85	4.27	61	21.18

method of selecting WBF candidate box filter mode also improves the detection accuracy of oil field workwear targets in the actual scene. G-YOLOv5s is a combined model of the above-improved methods, which improves the detection accuracy of the network further than a single module or a part of the combined model. Its overall optimization optimizes the fusion process of multi-scale features of the network, reduces computational redundancy, and improves the mining of target representations by the attention mechanism. In practice, compared with the original YOLOv5s, G-YOLOv5s improves the situation of small-scale and obscuration objectives' false detection and miss detection significantly, the detection accuracy is improved by 5.31%, but the speed is reduced by only 20 frames/second. Figure 27.7a, b shows the comparison examples before and after optimization.

Comparative Experiments

In order to fully reflect the effect of network improvement, G-YOLOv5s is compared with existing networks ZHENGL-YOLOv5, WX-YOLOv5, and YOLOv7s which have better target detection performance in this field in the original dataset. The same configuration environment and data sets were used for all the experiments, and the comparison results are shown in Table 27.2.

According to the data results in the table, G-YOLOv5s has higher accuracy, better effect on error detection and miss detection than other networks, smaller amount of parameters, faster calculation speed, and can meet the requirements of workwear detection in oil field scenarios.

(a) Optimization Comparison Chart 1

(b) Optimization Comparison Chart 2

Fig. 27.7 Overall optimization before and after comparison chart

Table 27.2 Comparison of experimental results

Model	mAP	mFPR	mFNR	Speed	Parm
ZHENGL-YOLOv5s	88.62	5.21	6.18	59	21.74
WX-YOLOv5s	87.94	5.84	6.56	71	18.42
YOLOv7s	89.21	4.92	5.83	57	22.31
G-YOLOv5s	91.62	3.85	4.27	61	21.18

Fig. 27.8 System interface display diagram

System Testing

Finally, the G-YOLOv5s network is deployed as shown in Fig. 27.8. The system can accomplish the expected function and has a good detection effect. Because it only needs to load the trained parameters during initialization and does not require a large amount of calculation during detection, it has fewer hardware limitations. So this design can be transplanted on most devices that are equipped with the Pycharm environment to monitor the attire of oil field workers.

27.6 Summary

This paper presents workwear detection algorithm for oilfield workers based on G-YOLOv5s to improve the detection capability of existing networks for small and obscure targets in oilfield scenarios. Compared with the original YOLOv5s network, G-YOLOv5s provides a 5.31% improvement in accuracy with only a 20 frames/second reduction in detection speed. The monitoring system designed based on this network can also meet the needs of workwear monitoring in oil field work scenes and has high practical value. Next, we plan to deploy the above system on a mobile detection vehicle in the oil field, which can achieve an automatic perception of the detection vehicle for monitoring the standardized clothing of oil field workers in the field of view.

Acknowledgements This work was supported by the Liaoning Province Applied Basic Research Program in 2023, “Key Technology Research on Vehicle Vision Perception and Trajectory Planning for ADAS"(2023JH2/101300237).

References

1. Lu, A.Q., Qin, C.Y., Hu, S.B., et al.: Experimental platform design of digital image processing based on PyQt5. Electr. Comp. Inform. Technol. **5**(02), 158–161 (2021)
2. Culjak, I., Abram, D., Pribanic, T., et al.: A brief introduction to OpenCV. In: Proceedings of the 35th International Convention(MIPRO), Beijing, pp. 1725–1730 (2012)
3. Chu, W.J.: Research on Key technologies of Tank and Armored vehicle target detection based on YOLOv5. Beijing Jiaotong University, Beijing (2021)
4. Leng, J.X.: Research on small target detection methods in complex scenes. University of Chinese Academy of Sciences (School of Computer Science and Technology, Chinese Academy of Sciences), Beijing (2020)
5. Yang, L.X., Zhang, R.Y., Li, L.D., et al.: SimAM: a simple, parameter-free attention module for convolutional neu. ral networks. In: International Conference on Machine Learning(PMLR), Seattle, pp. 11863–11874 (2021)
6. Zhu, X.K., Lyu, S.C., Wang, X., Zhao, Q.: TPH-YOLOv5: improved YOLOv5 based on transformer prediction head for object detection on drone-captured scenarios[J]. IEEE COMPUTER SOC 2778–2788 (2021)
7. Li, X.M., Chen, L., Yan, G.P.: Image denoising based on adaptive soft thresholding and edge enhancement. In: Electronic Measurement Technology, pp. 4–25 (2008)
8. Solovyev, R., Wang, W., Gabruseva, T.: Weighted boxes fusion: ensembling boxes from different object detection models. ELSEVIERRADARWEG, p. 107 (2019)
9. Wang, Y.L.: Construction site helmet wearing monitoring algorithm based on YOLOv5. Inform. Technol. Inform. **7**, 4 (2022)
10. Redmon, J., Farhadi, A.: YOLOv3: An incremental improvement. In: Conference on Computer Vision and Pattern Recognition (CVPR), The city of Saline Lake, pp. 89–95 (2018)
11. Woo, S., Hwang, S., Kweon, I.S.: StairNet: Top-Down Semantic Aggregation for Accurate One Shot Detection, pp. 1093–1102. IEEE-INST Electrical Electronics Engineers INC (2018)
12. He, M., Liao, M., Yang, Z., et al.: Most: a multi-oriented scene text detector with localization refinement. In: Proceedings of the IEEE/CVF Conference on Computer Vision and Pattern Recognition, pp. 8813–8822 (2021)

Chapter 28
HEANet: Hierarchical-Feature Enhanced Attention Network for Remote Sensing Change Detection

Feng Mu, Yongzhuo Pan, Jianan Li, Haolin Qin, Ning Shen, Xin Xu, Zhenxiang Chen, and Tingfa Xu

Abstract Change detection enables the detection of changes in objects from multi-temporal images. Recently, deep learning plays an important role in the field of change detection. Current methods perform multi-stage feature extraction from the input images to obtain high-level and low-level features, but ignoring the relationship between high-level features and low-level features. To deal with the above problem, this paper proposes a hierarchical-feature enhanced attention Network (HEANet), which integrates a hierarchical-feature enhanced attention (HEA) module for strengthening the association of hierarchical-feature and an adaptive scale

F. Mu · J. Li (✉) · H. Qin · N. Shen · X. Xu · Z. Chen · T. Xu (✉)
School of Optics and Photonics, Beijing Institute of Technology, Beijing, China
e-mail: lijianan@bit.edu.cn

T. Xu
e-mail: ciom_xtf1@bit.edu.cn

F. Mu
e-mail: bitmufeng@bit.edu.cn

H. Qin
e-mail: 3303143157@163.com

N. Shen
e-mail: shennbit@163.com

X. Xu
e-mail: sue_xuxin@qq.com

Z. Chen
e-mail: bitchenzx@163.com

T. Xu
Chongqing Innovation Center, Beijing Institute of Technology, Chongqing, China

Y. Pan
Chongqing Institute of Geology and Mineral Resources, Chongqing, China
e-mail: rayphennypan@sina.cn

R. Kountchev (Deceased) et al. (eds.), *AI Methods and Applications in 3D Technologies*, Smart Innovation, Systems and Technologies 388,
https://doi.org/10.1007/978-981-97-2144-3_28

enhancement (ASE) module for better feature representation. Extensive experiments show that our method achieves state-of-the-art performance compared to other methods on SYSU dataset.

28.1 Introduction

In the field of remote sensing technology, the demand for high-precision change detection methods is increasing. Remote sensing image change detection achieves targeted recognition of differences in multi-temporal remote sensing images [1], which is widely used in environmental monitoring, urbanization monitoring, land coverage monitoring, etc. In recent decades, Urbanization become a worldwide trend. Therefore, monitoring urban development through remote sensing change detection is of great significance.

Traditional change detection methods mainly apply spectral information in the remote sensing image. However, these methods typically require manual selection of the optimal threshold during the decision-making stage. In recent years, deep learning methods have performed well in computer vision tasks, such as semantic segmentation [2], object detection [3], and change detection [4]. Researchers have proposed various effective change detection algorithms for multi-temporal images, such as FTN [5], DSAMNet [6], ChangeFormer [7], etc. Generally, deep learning-based methods first perform multi-stage feature extraction from the input images to obtain high-level and low-level features, subtract the feature maps of each level to obtain the difference maps, and decode the difference maps to obtain the result of change prediction. But current methods treat each feature of multi-stage independently, ignoring the relationship between high-level features and low-level features, resulting in poor change detection accuracy. Therefore, it is significant to strengthen the correlation between hierarchical features and maintain their specificity.

In light of the above, this paper proposes a hierarchical-feature enhanced attention network (HEANet) to strengthen the relationship of different multi-stage features. A hierarchical feature enhancement attention module was designed in the encoder part to achieve multi-level feature correlation; In addition, attention mechanisms have been added to the hierarchical feature enhancement attention module to highlight the specificity of the current hierarchical features. In the decoder part, the characteristics of the large Receptive field can be obtained by using the expansion convolution. According to the input characteristics at different stages, the adaptive matching of the size of the Receptive field captures multi-scale information, and the gradual summation method is used to obtain the decoding map to better obtain the change information in the features. At the same time, the supervision attention module is added to ensure that the change detection has better performance.

Specifically, given an input feature map, in the encoder part, the model extracts multi-stage features from different levels of bi-temporal images by means of a siamese feature extractor and assigns the hierarchical-feature enhanced attention module (HEA) to strengthen the correlation of hierarchical features. In the decoder part, the model applies an adaptive scale enhancement module (ASE) and a supervised attention module (SA) to realize the decoding of the feature map. Experiments show that the proposed method recognizes more refined edge changes and performs better in some regions of coherent changes. The contributions of this paper are summarized as follows:

(1) We provide a new change detection network, HEANet, which integrates a HEA module for strengthening the correlation of hierarchical features and an ASE module for better feature representation. (2) We verify the effectiveness of the HEA and ASE module and prove the gain effect of their combination. (3) The proposed method achieves SOTA (state-of-the-art) performance on SYSU dataset [8], with Kappa of 78.78%, Rec of 83.27%, IoU of 72.05% and F1 of 83.75%.

28.2 Proposed Method

Overview

For the change detection task, the input is a set of images with a time-series relationship, denoted as t^1 and t^2. The proposed network extracts output feature by encoder and decoder, and calculates the loss function between the prediction and the groundtruth. The architecture of our network is given in Fig. 28.1.

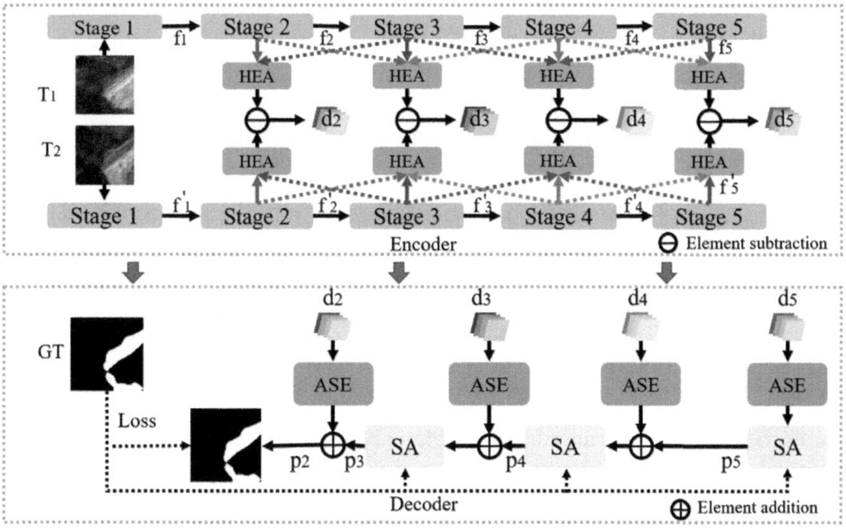

Fig. 28.1 The overall structure of HEANet, which consists of encoder and decoder

The network includes encoder, decoder and Loss function. In the encoding stage, two registered remote sensing mineral images t^1 and t^2 with time-series relationship are independently extracted for feature extraction. Firstly, through a backbone network with five levels, the output features of the last four levels are selected for hierarchical feature enhancement; Subsequently, the features of each adjacent level are used as input for the next stage, and the enhanced level features are obtained through the level feature enhancement module (HEA module). The elements of each level feature are subtracted to obtain the difference feature maps $\boldsymbol{d}_2, \boldsymbol{d}_3, \boldsymbol{d}_4, \boldsymbol{d}_5$ of each level. In the decoding stage, the differential feature maps $\boldsymbol{d}_2, \boldsymbol{d}_3, \boldsymbol{d}_4, \boldsymbol{d}_5$ enter the adaptive scale enhancement module (ASE module) for decoding, followed by the supervised attention module (SA module). Finally, the output is decoded by adding elements step by step. In the Loss function calculation link, the mixed Loss function is used to calculate the difference between the predicted change graph $\boldsymbol{p}_2, \boldsymbol{p}_3, \boldsymbol{p}_4, \boldsymbol{p}_5$ and the real change graph in the code understanding process. The following will provide a detailed introduction to the various parts of the hierarchical feature enhanced attention network.

Encoder

The encoder consists of a backbone network and a hierarchical enhanced attention module. We adopt a lightweight network MobileNetV2 [9] as backbone to extract bi-temporal features. Simultaneously, we removed the average pooling layer and fully connected layer to adapt to the change detection task. The backbone network is divided into five stages with a down-sampling stride of 2. Given a registered pair of images t^1 and t^2, going through the backbone network, five hierarchical features are obtained, denoted as $\boldsymbol{f}_i^{\mathrm{j}}, i \in \{1, 2, 3, 4, 5\}, j \in \{1, 2\}$. To enhance the correlation between the features of different stages, we take the latter four stage output features as input to the HEA module, expressed by the equation as:

$$\boldsymbol{p}_i = \varphi\left(\boldsymbol{f}_{i-1}^{\mathrm{j}}, \boldsymbol{f}_i^{\mathrm{j}}, \boldsymbol{f}_{i+1}^{\mathrm{j}}\right), i \in \{2, 3, 4, 5\}, j \in \{1, 2\}, \tag{28.1}$$

where $\varphi(\cdot, \cdot, \cdot)$ is function of HEA module. Specifically, for the second and fifth stages, the input to HEA module is reduced by $\boldsymbol{f}_{i-1}^{\mathrm{j}}, i = 2$ and $\boldsymbol{f}_{i+1}^{\mathrm{j}}, i = 5$. Details of the HEA module will be described in Sect. 28.2.2. Finally, the output features of each stage are element-wise subtracted to obtain the output of the encoder, denoted as $\boldsymbol{d}_2, \boldsymbol{d}_3, \boldsymbol{d}_4, \boldsymbol{d}_5$.

Decoder

The decoder consists of an adaptive scale enhancement module and a supervised attention module to realize the decoding of the feature map. Those feature maps $\boldsymbol{d}_2, \boldsymbol{d}_3, \boldsymbol{d}_4, \boldsymbol{d}_5$, obtained from the encoder, go through the ASE module, respectively. Then we refer to the idea of the residual network [10] to obtain the output. After encoding and decoding, we will use multi-stage outputs to calculate the loss function.

ASE module, taking a set of inflated convolutions to obtain a larger perceptual field, is designed to obtain the temporal change information. The required receptive field is diverse for different stages, therefore, ASE module discards the idea of sharing a perceptual field at each stage. Details of the ASE module will be described in Sect. 28.2.3. The SA module is introduced in [11].

HEA Module

In the change detection task, we extract multi-stage features of input graphics through the backbone network to obtain high-level features and low-level features. Low-level features focus on fine-grained boundaries and texture information of objects [12], while high-level features contain information about object localization. We aggregate multilevel features to strengthen the correlation between multi-stage features. At the same time, we incorporate a self-attention mechanism to ensure the specificity of the current stage feature. The structure of the HEA module is given in Fig. 28.2a.

Here, we take the third stage of $\boldsymbol{t^1}$ as an example, $\boldsymbol{f}_2^1$, $\boldsymbol{f}_3^1$, $\boldsymbol{f}_4^1$ as input to illustrate the structure of the HEA module. $\boldsymbol{f}_2^1$ goes through max-pooling layer and 3×3 convolution layer to obtain a feature map with the same size as $\boldsymbol{f}_3^1$. Similarly, for $\boldsymbol{f}_4^1$, we adopt 3×3 convolution operation and bilinear up-sample operation to obtain a feature map with the same size as $\boldsymbol{f}_3^1$. For $\boldsymbol{f}_3^1$, we take a 3×3 convolution operation to adjust the channels. Then, we aggregate the three feature maps mentioned above to obtain features $\boldsymbol{c}_2^1$, $\boldsymbol{c}_3^1$, $\boldsymbol{c}_4^1$, expressed by the equation as:

$$\boldsymbol{c}_2^1 = Conv_{3\times3}\{m(\boldsymbol{f}_2^1)\}, \quad (28.2)$$

$$\boldsymbol{c}_3^1 = Conv_{3\times3}(\boldsymbol{f}_3^1), \quad (28.3)$$

$$\boldsymbol{c}_4^1 = Up\{Conv_{3\times3}(\boldsymbol{f}_4^1)\}, \quad (28.4)$$

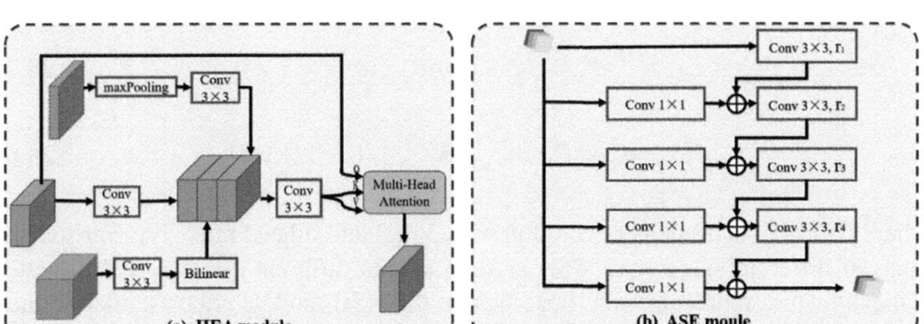

Fig. 28.2 The structure of HEA module and ASE module. **a** HEA module: hierarchical-feature enhanced attention module. **b** ASE module: adaptive scale enhancement module

where $m(\cdot)$ is max-pooling operation, $Conv_{3\times3}$ is 3×3 convolution operation, $Up\{\cdot\}$ is bilinear upsample operation. After completing the size normalization, we aggregate those multi-stage features and adopt 3×3 convolution to obtain $\boldsymbol{h}_3^1$. After completing multi-level feature enhancement, we adopt a self-attention mechanism to maintain the specificity of features, formulated as:

$$\boldsymbol{j}_3^1 = f_{soft}\left(\frac{\boldsymbol{Q}\boldsymbol{K}^{\mathrm{T}}}{\sqrt{d}}\boldsymbol{V}\right), \tag{28.5}$$

where $f_{soft}(\cdot)$ is the softmax activation function, $\boldsymbol{h}_3^1$ is the input of $\mathbf{K}$ and $\mathbf{V}$, $\boldsymbol{f}_3^1$ is the input of $\mathbf{Q}$,$\sqrt{d}$ is the scale factor. Finally, the output of the encoder $\boldsymbol{d}_i$ is obtained by doing an element-wise subtraction operation on $\boldsymbol{j}_i^{\mathbf{j}}$, $i \in \{2, 3, 4, 5\}$, $j \in 1, 2$.

ASE Module

To better capture temporal change information, we adopt a set of inflated convolutions and residual connections. After completing five stages of feature extraction in the backbone network, feature map will decrease by 32 times. Based on the size characteristics of feature maps at different stages, this paper proposes an ASE module to obtain better feature representation. The structure of the ASE module is given in Fig. 28.2b.

Here, we take $\boldsymbol{d}_2$ as input to illustrate the function of the ASE module. Firstly, the ASE module adopt four convolutions with 1×1 for feature transition as:

$$\boldsymbol{d}_2^k = Conv_{1\times1}(\boldsymbol{d}_2), k \in \{1, 2, 3, 4\}, \tag{28.6}$$

where k represents the number of convolutions with 1×1. Secondly, the ASE module handles the results of ordinary and dilated convolution as residual connects:

$$\boldsymbol{d}_2^1 = \boldsymbol{d}_2^1 + Conv_{3\times3}^{r_1}(\boldsymbol{d}_2), \tag{28.7}$$

$$\boldsymbol{d}_2^{k'} = \boldsymbol{d}_2^k + Conv_{3\times3}^{r_k}\left(\boldsymbol{d}_2^{k-1}{}'\right), k \in \{2, 3, 4\}, \tag{28.8}$$

where $Conv_{3\times3}^{r_k}$ is dilated convolution with 3×3 and dilated rate $= r_k$. For feature maps at different stages, the ASE module adopts different dilated rates to better achieve feature representation. Specifically, the ASE module chooses dilated rates as (7,5,3,1) for $\boldsymbol{d}_2$, chooses dilated rates as (5,3,1,1) for $\boldsymbol{d}_3$, chooses dilated rates as (3,1,1,1) for $\boldsymbol{d}_4$, chooses dilated rates as (1,1,1,1) for $\boldsymbol{d}_5$. By doing those operations, we obtain the outputs of the ASE module $\boldsymbol{d}_i'$, $i \in \{2, 3, 4, 5\}$.

28.3 Experiments

Experimental Details

We conduct experiments using the Python language based on the Pytorch 1.9.0 framework. HEANet is implemented in the PyTorch toolbox and trained for 40 k iterations on a single NVIDIA GTX 3090 GPU, with an initial learning rate of 0.01 and a mini-batch size of 32. We use an Adam optimizer with a poly strategy, with an initial momentum setting of 0.9 and a weight attenuation setting of 0.0001. The backbone is initialized with parameters of MobileNetV2 [9], pre-trained on ImageNet [13]. We adopt a hybrid loss, including a binary cross-entropy loss and a dice loss. To evaluate the performance of the proposed method, we utilize four typical metrics: namely, Kappa, recall (Rec), IoU and F1-score.

We use SYSU dataset as benchmark dataset. SYSU consists of 20,000 pairs of bi-temporal remote sensing image patches with 256 $\times$ 256 spatial sizes and 0.5-m spatial resolution, and contains various types of complex change scenes. The ratios of training, validation, and test are officially set to 6:2:2. The dataset is available at https://github.com/liumency/SYSU-CD.

Comparison with the State-Of-The-Art

Quantitative Results

The quantitative results for metrics of all methods are showd in Table 28.1. In particular, the results of A2Net are obtained by re-evaluating with reference to previous work [11]. As Table II shows, FC-diff obtains the lowest quantitative results, which obtain Kappa of 61.42%, Rec of 55.89%, IoU of 52.06%, and F1 of 68.47% among the compared methods. Our method achieves the highest results, which obtain Kappa of 78.78%, Rec of 83.27%, IoU of 72.05%, and F1 of 83.75% among all the compared methods. Compared to the A2Net, our method improves Kappa by 1.06%, Rec by 3.23%, IoU by 1.41%, and F1 by 0.96%, to achieve SOTA performance.

Visual Comparisons

Figure 28.3 provides the results of some methods on SYSU dataset. Generally speaking, our proposed method can recognize more refined edge changes and performs better in some regions of coherent changes. This demonstrates that the HEA module can help to integrate high-level features and low-level features for improving accuracy.

Table 28.1 Quantitative comparison of Kappa, Rec, Prec, IoU and F1-score on SYSU dataset

Model	Kappa (%)	Rec (%)	IoU (%)	F1 (%)
FC-diff [14]	61.42	55.89	52.06	68.47
FC-ef [14]	67.45	74.63	60.11	75.08
FC-cat [14]	72.06	78.00	64.75	78.60
STANet [15]	71.00	74.30	63.32	77.54
L-Unet [16]	73.98	78.08	66.62	79.96
SNUNet [17]	73.91	79.79	66.73	80.44
DSIFN [18]	72.05	75.11	64.42	78.36
BIT [4]	73.33	76.68	65.84	79.40
A2Net [11]	77.72	80.04	70.64	82.79
HEANet (Ours)	**78.78**	**83.27**	**72.05**	**83.75**

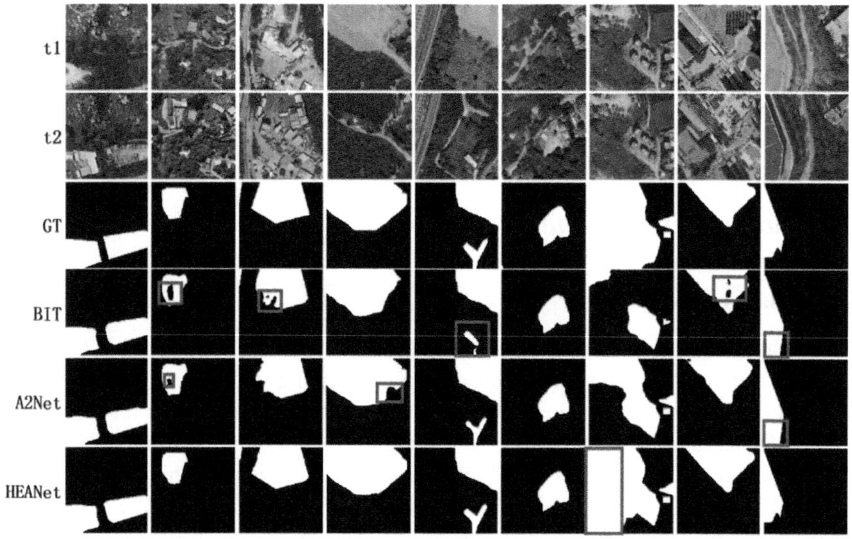

Fig. 28.3 Visual comparisons of the proposed method and some state-of-the-art approaches, GT means groundtruth

Ablation Study

HEANet integrates both the HEA module and the ASE module for accurate change detection. We design ablation experiments on HEANet to verify both the validity of the HEA module and the ASE module. In our experiment, the "Baseline" represents the basic model without the HEA module and the ASE module.

Table 28.2 Ablation study on SYSU dataset

Model	Kappa (%)	Rec (%)	IoU (%)	F1 (%)
Baseline	77.72	80.04	70.64	82.79
Baseline + HEA	78.56	81.51	71.66	83.49
Baseline + ASE	78.08	80.43	69.91	82.83
HEANet	**78.78**	**83.27**	**72.05**	**83.75**

Effect of HEA Module

As can be seen from Table 28.2, when integrating the HEA module to "Baseline" Algorithm, the "Baseline + HEA" increases Kappa by 0.84%, Rec by 1.47%, IoU by 1.02%, F1 by 0.7%, which proves that strengthening the correlation of hierarchical features is effective.

Effect of ASE Module

We report results of integrating the ASE module into "Baseline" Algorithm in Table 28.2, the "Baseline + ASE" increases Kappa by 0.36%, Rec by 0.39%, F1 by 0.04%, which demonstrates that the ASE module can obtain better feature representation when decoding.

The incorporation of both HEA module and ASE module can improve the model performance. The improvement of HEANet proves the effectiveness of the HEA module and ASE module in improving detection performance, and also proves the effective gain of their combination.

28.4 Conclusion

For the task of remote sensing image change detection, a hierarchical-feature enhanced change detection method is proposed, which cleverly combines deep and shallow features based on the characteristics of multi-stage feature extraction in change detection. In the encode part of HEANet, a hierarchical-feature enhanced attention module is proposed, which aggregates multi-stage features and adds a self-attention mechanism to strengthen the correlation of hierarchical features and ensure the specificity of the current stage feature. In the decoder part of HEANet, an adaptive scale enhancement module is designed based on the characteristics of different stages of features, making the decoding more scientific and effective. Experiments show that the method exhibits SOTA performance in the change detection tasks. In the future, we will expand our method in the following directions: 1) we will create our own remote sensing change detection dataset and validate our method in more semantic segmentation tasks. 2) further investigate the effectiveness of hierarchical feature enhancement methods in object detection tasks.

References

1. Singh, A.: Review article digital change detection techniques using remotely-sensed data. Int. J. Remote Sens. **10**(6), 989–1003 (1989)
2. Mu, F., Li, J.N., Shen, N., et al.: Pixel-adaptive field-of-view for remote sensing image segmentation. IEEE Geosci. Remote Sens. Lett. **19**, 1–5 (2022)
3. Peng, P., Mu, F., Yan, P., Song, Li, H., et al.: GCANet: A cross-modal pedestrian detection method based on gaussian cross attention network. In: Intelligent Computing: Proceedings of the 2022 Computing Conference, vol. 2. Springer International Publishing, Cham, pp. 520–530 (2022)
4. Chen, H., Qi, Z.P., Shi, Z.W.: Remote sensing image change detection with transformers. IEEE Trans. Geosci. Remote Sens. **60**, 1–14 (2021)
5. Yan, T.Y., Wan, Z.F., Zhang, P.P.: Fully transformer network for change detection of remote sensing images. In: Proceedings of the Asian Conference on Computer Vision (ACCV), pp. 1691–1708 (2022)
6. Chen, J., Yuan, Z., Peng, J., et al.: Dasnet: dual attentive fully convolutional siamese networks for change detection of high resolution satellite images[J]. IEEE J. Select. Top. Appl. Earth Observ. Remote Sens. (2020)
7. Bandara, W.G.C., Patel, V.M.: A transformer-based siamese network for change detection[M]. (2022)
8. Shi, Q., Liu, M.X., et al.: A deeply supervised attention metric-based network and an open aerial image dataset for remote sensing change detection. IEEE Trans. Geosci. Remote Sens. **60**, 5604816 (2022)
9. Sandler, M., Andrew, H., et al.: MobileNetV2: inverted residuals and linear bottlenecks. In: 2018 IEEE/CVF Conference on Computer Vision and Pattern Recognition, pp. 4510–4520 (2018)
10. He, K.M., Zhang, X.Y., Ren, S.Q., et al.: Deep residual learning for image recognition. In: 2016 IEEE Conference on Computer Vision and Pattern Recognition (CVPR), pp. 770–778 (2016)
11. Li, Z.L., Tang, C., Liu, X.W., et al.: Lightweight remote sensing change detection with progressive feature aggregation and supervised attention. IEEE Trans. Geosci. Remote Sens. **61**, 1–12 (2023)
12. Zhang, P.P., Wang, D., Lu, H.C., et al.: Amulet: aggregating multi-level convolutional features for salient object detection. 2017 IEEE International Conference on Computer Vision (ICCV), pp. 202–211(2017).
13. He, K.M., Zhang, X.Y., et al.: Delving deep into rectifiers: surpassing human-level performance on ImageNet classification. In: IEEE International Conference on Computer Vision, pp. 1026–1034 (2015)
14. Rodrigo, C.D., Bertrand, L.S., et al.: Fully convolutional siamese networks for change detection. In: 2018 25th IEEE International Conference on Image Processing (ICIP), pp. 4063–4067 (2018)
15. Chen, H., Shi, Z.W.: A spatial-temporal attention-based method and a new dataset for remote sensing image change detection. Remote Sens. **12**(10), 1662 (2020)
16. Maria, P., Maria, V., et al.: A deep multitask learning framework coupling semantic segmentation and fully convolutional LSTM networks for urban change detection. IEEE Trans. Geosci. Remote Sens. **59**(9), 7651–7668 (2021)
17. Fang, S., Li, K.Y., Shao, J.Y., et al.: SNUNet-CD: a densely connected siamese network for change detection of VHR images. IEEE Geosci. Remote Sens. Lett. **19**, 1–5 (2021)
18. Zhang, C.X., Yue, P., Tapete, D., et al.: A deeply supervised image fusion network for change detection in high resolution bi-temporal remote sensing images. ISPRS J. Photogramm. Remote. Sens. **166**, 183–200 (2020). https://doi.org/10.1016/j.isprsjprs.2020.06.003

Chapter 29
A Security Management Mode of Electrical Power System Through Zero Web of Trust System

Dongmei Bin, Chunyan Yang, Xin Li, Ying Ling, and Jieke Lu

Abstract With the continuous growth of industrial field, the structure of electrical power system (EPS) is more diverse, and the boundary uncertainty is gradually increasing. In the dispatching work, the operation of EPS is adjusted and controlled by the operators. At the same time, under the basic condition of ensuring their personal safety, the operators need to follow the operation rules of EPS and various dispatching rules and regulations to the greatest extent. This article constructs an EPS security management model of network system based on zero trust, and puts forward a security application solution based on zero trust architecture, thus establishing a truly effective EPS information security architecture. The simulation results show that the error of the proposed algorithm can be stabilized at around 1.3; and the accuracy of the algorithm is as high as 95.76%, which is about 9.87% higher than the compared algorithm. The results show that this method is practical and reliable. The EPS security management mode based on zero trust proposed in this article can effectively reduce data security risks and improve data security and business security of EPS.

29.1 Introduction

With the rapid growth of information technology, emerging technologies such as cloud computing, big data, IoT, mobile internet and artificial intelligence have brought new productivity to the informatization development and modernization of government departments and various enterprises, but at the same time they have also brought new challenges to information security [1]. Due to the continuous growth of industrial field, the structure of information system is more diversified, and the boundary uncertainty is increasing gradually. In order to meet the technical requirements of security protection of new infrastructure, the traditional industrial data security protection methods need to be iteratively updated [2]. Driven by the national "new

D. Bin (✉) · C. Yang · X. Li · Y. Ling · J. Lu
Electric Power Research Institute of Guangxi Power Grid Co. Ltd., Nanning 530023, Guangxi, China
e-mail: bin_dm.sy@gx.csg.cn

R. Kountchev (Deceased) et al. (eds.), *AI Methods and Applications in 3D Technologies*, Smart Innovation, Systems and Technologies 388,
https://doi.org/10.1007/978-981-97-2144-3_29

infrastructure" policy and the wave of digital transformation, the electronic information industry system based on the innovation and integration of technologies such as AI, big data, IoT, cloud computing, 5G, information and Internet is undergoing transformation and reorganization, and people's daily life and work patterns have undergone tremendous changes [3]. For any subject who accesses the enterprise network, including the person who initiated the access, the device who initiated the access and the accessed application, it is untrusted by default at each access, and it is necessary to build the trust of dynamic access through continuous authentication and access authorization [4]. In order to meet the commercial needs of customers' interconnection, the current industrial products pay more attention to cross-domain collaboration, focusing on the needs of horizontal interactive integration and interactive scenarios, while the vertical security service capability needs to be improved [5].

The cycle of industrial data includes the whole life cycle from customer demand to design, R&D and manufacturing. Because the life cycle involves more users and business development links, the data involved in different links is more diverse and complex. Therefore, with the diversification and complexity of industrial data, there are new requirements for data classification, storage, cleaning and analysis [6]. The growth of power IoT and the extensive connection of a large number of heterogeneous terminals bring about the fuzziness of network boundaries and the diversification of access scenarios. The existing border-based security protection system faces security risks, and there are shortcomings in security authentication and trust assessment [7]. With the increasing diversity of Internet application scenarios, the network architecture and business scenarios of telecom operators have also undergone major changes, and the traditional network boundaries have become blurred or even disappeared [8]. How to better adapt to the network environment in today's world, the concept of zero trust security came into being, which holds that both internal users and external users cannot be trusted, and it is necessary to build a dynamic access control process [9]. In view of the fact that the traditional network security model cannot meet the needs of enterprises in the digital transformation stage, the network security model based on zero trust came into being. The experiment selected event work order data, monthly job plan data, job change data, defect work order data, and alarm details data for testing. The experiment simulates various network attacks, such as forging inverters to attempt terminal information collection servers to establish connections and upload data, with the goal of verifying the protection ability of the zero trust architecture based power IoT terminal security access method in power grid business. Facing the security threat of EPS network, this article constructs an EPS security management model of network system based on zero trust, and puts forward a security application solution based on zero trust architecture.

29.2 Methodology

Principle of Zero Trust Security Architecture

With the wide application of new technologies such as cloud computing, containers and big data, more and more network assets need to be included in network connection control, and the number and complexity of security policies increase exponentially, which makes management and maintenance more and more difficult. Zero-trust security is an identity-centric dynamic access control process, and its core technologies are identity-centric, business security access, continuous risk assessment and dynamic access control. Its technical components can be abstracted into trusted agent, dynamic access control engine, identity analysis engine, identity and authority management, so as to build a dynamic and credible secure access platform. The zero-trust secure network model defaults to untrusted every connection initiated by each user and each device in the enterprise network, and every service request applied [10] before the authentication policy is decided. After the trusted agent intercepts the access request, the access subject is authenticated by the dynamic access control engine, and the authority of the access subject is dynamically judged. Only access requests that have passed the authentication and have access rights will be released.

Under the traditional network partition architecture, enterprise networks are generally divided into external Internet and internal production networks. Dynamic access control is based on identity base, authority base and trust base. Among them, the identity library provides the identity attribute of the access subject, the authority library provides the basic authority baseline, and the trust library is continuously maintained by the identity analysis engine through real-time multi-dimensional risk correlation and trust assessment [11]. In the zero-trust security network model, the object of identity authentication is not only the same as the traditional network security model. At the same time, user information, host information of access devices, behavior characteristics of accessing business systems and user rights will be taken as the objects of identity authentication, and multi-dimensional information will be fused to construct a user portrait. At present, the security protection system of EPS network is mainly based on border isolation protection. Figure 29.1 shows the EPS security protection framework.

When the network boundary cannot be clearly defined, the original network security policy established through the firewall can play a less and less role, so the risk of network security attacks on the system is also increasing. The authentication of zero-trust security network model is not a one-stop service, but a careful and credible judgment of every service request. All behavioral entities in enterprise business access are evaluated by using the zero-trust security network model to ensure the trustworthiness of each behavioral entity. With the growth of enterprise scale, the personnel, equipment and authority of the enterprise will become more and more huge, and the construction of modern identity management system will help to better manage identity and authority. The access behavior in the enterprise network is supervised in the whole life cycle, and the trust score of network access behavior in the

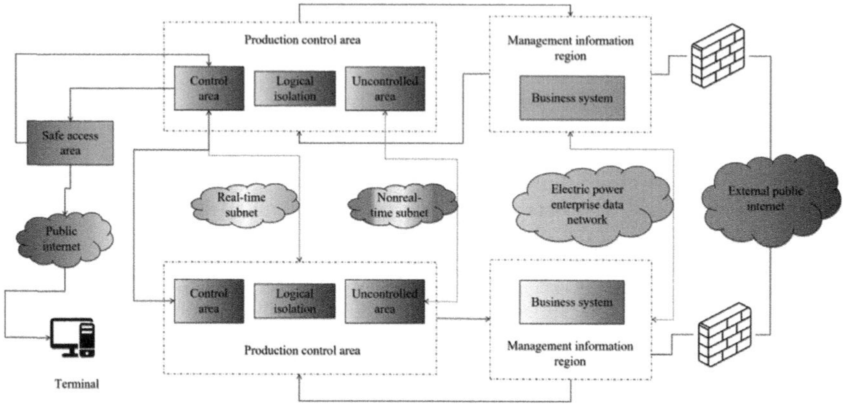

Fig. 29.1 EPS safety protection framework diagram

zero-trust network model is continuously calculated, and different access rights are given to entities according to the real-time score value, thus realizing the variable trust management of access entities.

EPS Security Management Model

According to the general principle of "security zoning, dedicated network, horizontal isolation and vertical authentication", the system divides production control area and information management area, and the production control area is divided into control area and non-control area. Horizontal isolation devices are deployed between network areas, and encryption and authentication devices [12] are deployed at vertical boundaries. At present, the protection system adopts the mechanism of "one-time authentication and continuous trust", and considers the devices in the intranet to be reliable. The access control strategy of the zero-trust complete network model is no longer fixed, and its access control strategy is based on the dual access control of identity authentication and authorization reconstruction. It confirms the risk of current access by judging the multi-dimensional information of network access subjects in real time.

Different from the traditional access strategy based on pre-allocation of users and permissions, zero-trust network adopts a more detailed trust assessment access mechanism. Indirect acquisition of internal state information of the power system through analysis of external observations:

$$\theta_k = f(\theta_{k-1}, v_k) \tag{29.1}$$

$$y_k = h(\theta_k, n_k) \tag{29.2}$$

Among them, θ_k is the N_θ-dimensional state vector at the t_k moment; y_k is the N_y-dimensional observation vector at the t_k moment; v_k and n_k represent the state transition noise vector and the observation noise vector, respectively.

Let the probability distribution of random variable set $X = \{X_1, X_2, \ldots, X_n\}$ be $P(X_1, X_2, \ldots, X_n)$. If all variables are $\{0,\ 1\}$, $2^n - 1$ parameters are needed to determine the joint distribution. The joint distribution can be written as:

$$P(X_1, X_2, \ldots, X_n) = \prod_{i=1}^{n} P(X_i | X_1, X_2, \ldots, X_{i-1}) \tag{29.3}$$

For $\forall X_i \in X$, if $\pi(X_i) \subseteq \{X_1, X_2, \ldots, X_{i-1}\}$ exists, the conditions of X_i assimilation $\{X_1, X_2, \ldots, X_{i-1}\}/\pi(X_i)$ are independent when $\pi(X_i)$ is given, and the above formula can be changed to:

$$P(X_1, X_2, \ldots, X_n) = \prod_{i=1}^{n} P(X_i | \pi(X_i)) \tag{29.4}$$

In order to ensure network security and strengthen the prevention and control of network security risks in the process of digital transformation, electric power enterprises have carried out technical research to prevent the security risks of new business applications [13]. Then the zero-trust security network model is used to replace the traditional network security model, thus improving the network security protection ability of the new generation digital service platform. In a large number of complex policy maintenance work, the policies created in the early days are often not sorted out regularly and recovered in time, resulting in a large number of old policies, affecting the normal operation of security equipment, and even leading to the failure of security policy conflicts, making security products fail to play their due protective role and laying hidden dangers for the overall network security. Trusted access gateway realizes unified management of user identity, dynamic authority control and policy execution through linkage with terminal environment awareness, unified identity management system and unified authority management system.

29.3 Result Analysis and Discussion

In order to ensure network security and strengthen the prevention and control of network security risks in the process of digital transformation, electric power enterprises carry out technical research to prevent the security risks of new business applications. Then the zero-trust security network model is used to replace the traditional network security model, thus improving the network security protection ability of the new generation digital service platform. In a large number of complex policy maintenance work, the policies created in the early days are often not sorted out regularly and recovered in time, resulting in a large number of old policies, affecting

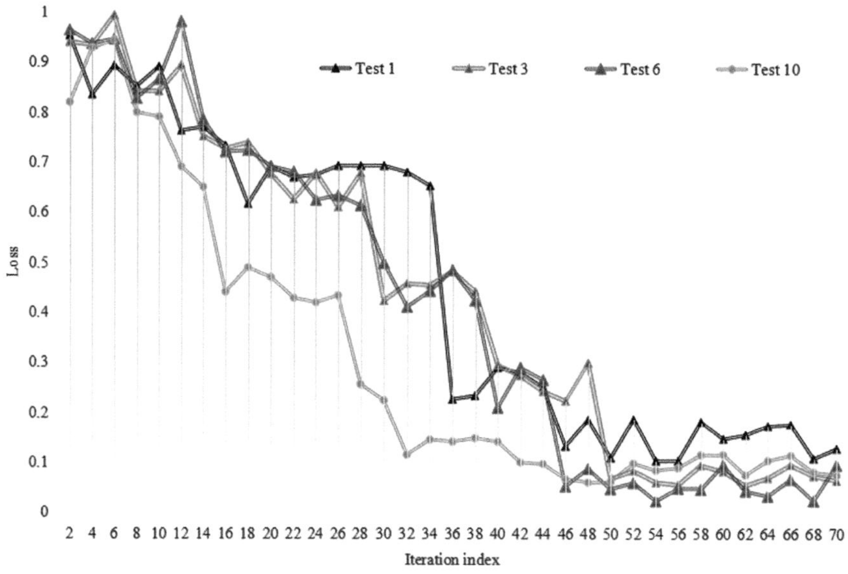

Fig. 29.2 Training situation of the algorithm

the normal operation of security equipment, and even leading to the failure of security policy conflicts, making security products fail to play their due protective role and laying hidden dangers for the overall network security. Trusted access gateway realizes unified management of user identity, dynamic authority control and policy execution through linkage with terminal environment awareness, unified identity management system and unified authority management system. In order to verify the performance of this algorithm, this section carries out various experimental tests. Figure 29.2 shows several training cases of the algorithm.

It is not difficult to find that the proposed algorithm tends to be stable after more than 46 iterations, and can basically achieve the ideal effect. Under the framework of "zero trust", people, equipment, applications, services, posts, institutions, identity types and other natural entities must register in the identity management center and form digital identities, so as to ensure that all entities correspond to a unique digital identity, and associate a series of identity attributes to issue digital certificates for each digital identity. The core logic of zero-trust identity security is to establish a security protection system with "identity" as the logical boundary by establishing a security access control area between the access subject and the access object. Only the trusted subject can access the object. Fine-grained access control is provided for any access request. Access control points are deployed in depth to provide access control for application systems and service interfaces. Only access requests with access rights and satisfactory trust assessment are allowed to access applications and obtain data. Figure 29.3 shows the comparison of F1 values of several algorithms.

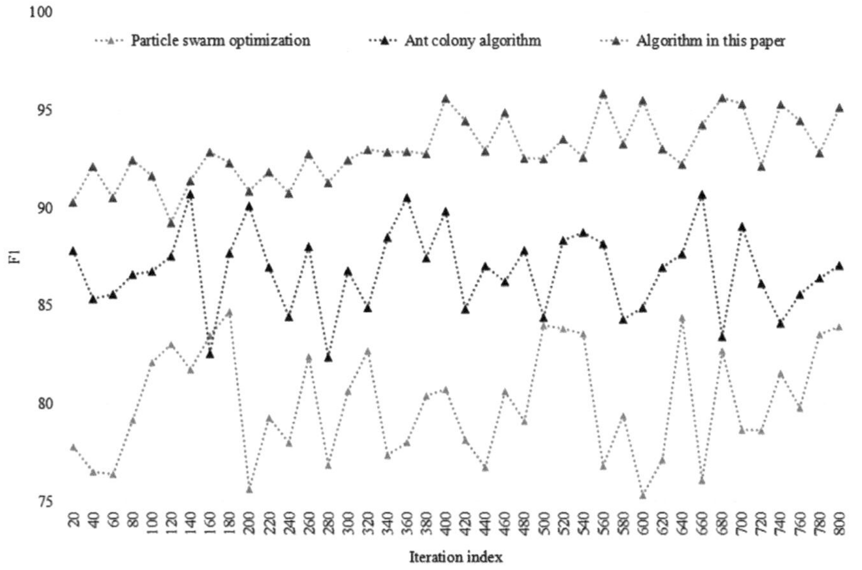

Fig. 29.3 Comparison of F1 values of several algorithms

Compared with the traditional particle swarm optimization algorithm and ant colony algorithm, the F1 value of this method is obviously improved, which proves the superiority of this algorithm in performance. As an optional component in zero-trust identity security solution, intelligent identity analysis system can provide advanced security capabilities in zero-trust architecture. If the intelligent identity analysis system is not deployed in the early stage of enterprise construction, the trusted access console will directly link with the trusted environment awareness system and the trusted API proxy platform, and use the terminal awareness information and API to call behavior log analysis to realize dynamic access rights judgment, and then deploy the intelligent identity analysis system as needed. Figure 29.4 shows the comparison of prediction errors of several algorithms.

The network state prediction accuracy results of different algorithms are shown in Fig. 29.5.

In this section, many experiments are carried out, and the experimental results show that the error of the algorithm proposed in this article can be stabilized at around 1.3; And the accuracy of the algorithm is as high as 95.76%, which is about 9.87% higher than the compared algorithm. The results show that this method is practical and reliable. In order to ensure network security and strengthen the prevention and control of network security risks in the process of digital transformation, electric power enterprises carry out technical research to prevent the security risks of new business applications. Then the zero-trust security network model is used to replace the traditional network security model, thus improving the network security protection ability of the new generation digital service platform.

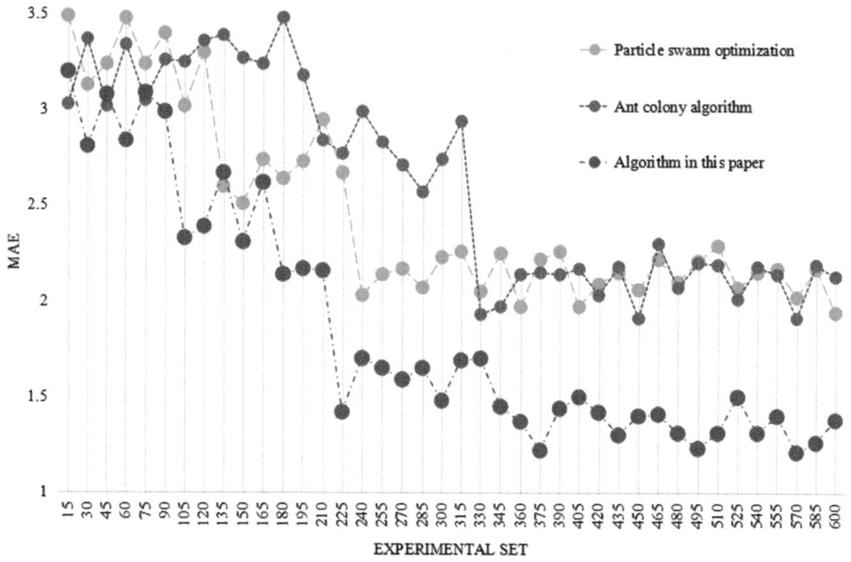

Fig. 29.4 Algorithm error comparison

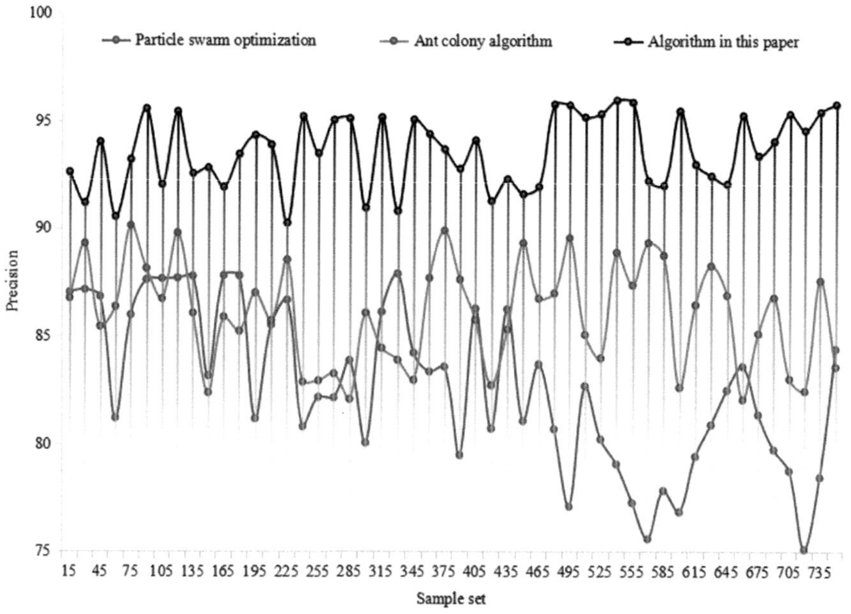

Fig. 29.5 Accuracy results of the algorithm

29.4 Conclusions

The essence of zero-trust security is identity-centered dynamic access control, and full identity is the premise and cornerstone of realizing zero-trust security architecture. With the growth of power IoT and the extensive connection of massive heterogeneous terminals, the network boundaries are blurred, and the access scenarios are diversified. The existing boundary-based security protection system faces security risks, and there are shortcomings in security authentication and trust assessment. This article constructs an EPS security management model of network system based on zero trust, and proposes a security application solution based on zero trust architecture. To verify the performance of the algorithm proposed in this article, many experiments are carried out. The experimental results show that the error of the proposed algorithm can be stabilized at about 1.3; and the accuracy of the algorithm is as high as 95.76%, which is about 9.87% higher than the compared algorithm. The results show that this method is practical and reliable. The proposed zero-trust network security model architecture is based on comprehensive identity, which establishes a unified digital identity identification and governance process for people, equipment, applications, systems and other physical entities in the zero-trust network, and further constructs a dynamic access control system, extending the security boundary to identity entities, and realizing the advancement of the security architecture. Zero-trust network and traditional network will work together for a long time, and combine the risk management methods in the zero-trust network security model, identity authentication, life cycle monitoring and other technologies to jointly solve the threats faced by enterprises in network access.

Acknowledgements Guangxi Power Grid Corporation Science and Technology Project Funding (Research and Application Project on Network Mutual Trust Support System for New Power System Business Entities (GXKJXM20210257)).

References

1. Campbell, M.: Beyond zero trust: trust is a vulnerability. Computer **53**(10), 110–113 (2020)
2. Barry, S.: How a zero-trust approach can help to secure your AWS environment. Netw. Secur. **2018**(3), 5–8 (2018)
3. Greenwood, D.: Applying the principles of zero-trust architecture to protect sensitive and critical data. Netw. Secur. **2021**(6), 7–9 (2021)
4. Gu, Z., Tang, Y., Sun, H., et al.: Study on framework of comprehensive defense architecture for power system security and stability. Proc. Chinese Soc. Electr. Eng. **39**(4), 943–952 (2019)
5. Hasanvand, S., Fallahzadeh-Abarghouei, H., Mahboubi-Moghaddam, E.: Power system security improvement using an OPA model and IPSO algorithm. Simulation **96**(3), 325–335 (2019)
6. Liu, Y., Ye, M., Gu, X., et al.: Probability distribution column based security margin analysis of power system with wind power participating in black start. Trans. China Electrotechn. Soc. **33**(21), 5059–5068 (2018)

7. Xiao, J., Zhang, H., Zhang, B., et al.: Dimension of distribution system security region. Automat. Electr. Power Syst. **42**(1), 16–22 (2018)
8. Hudson, F.D.: Enabling trust and security: TIPPSS for IoT. IT Prof. **20**(2), 15–18 (2018)
9. Majid, M., Reza, G., Mohsen, G.: Power system security assessment with high wind penetration using the farms models based on their correlation. IET Renew. Power Gener. **12**(8), 893–900 (2018)
10. Benasla, M., Allaoui, T., Brahami, M., et al.: Power system security enhancement by HVDC links using a closed-loop emergency control. Electr. Power Syst. Res. **168**(3), 228–238 (2018)
11. Ramya, K., Teekaraman, Y., Kumar, K.: Fuzzy-based energy management system with decision tree algorithm for power security system. Int. J. Comp. Intell. Syst. **12**(2), 1173 (2019)
12. Zhang, J., Yang, Y., Li, X., et al.: Coordinated control model of power system cascading failure considering security and economy. Proc. Chinese Soc. Electr. Eng. **38**(16), 4784–4791 (2018)
13. Chen, H., Chen, Z.: Steady-state security region characterization and operation margin calculation of AC/DC power system. High Volt. Eng. **44**(4), 1328–1335 (2018)

Chapter 30
Research and Implementation of an Intelligent System for Evaluating Financial Management Performance

Ting Cao

Abstract The financial performance evaluation results refer to the overall conclusion issued by the financial performance evaluation department, with the performance evaluation report as the main carrier, which comprehensively reflects the budget performance completion status of the performance evaluation object. Performance evaluation results are an important basis for arranging government budgets, improving policies, and improving management. From this, it can be seen that performance evaluation itself is not an end in itself. The fundamental purpose and value of performance evaluation lies in promoting government departments at all levels to manage and use financial funds more efficiently, continuously improving government public management and service capabilities, and comprehensively improving government performance efficiency. This article proposes a financial management performance evaluation method and designs and implements an intelligent system to achieve performance evaluation. The evaluation objects mainly include three modules: budget execution rate, output indicators, benefit indicators, and satisfaction indicators. The project performance self-evaluation table and self-evaluation report are output. This system provides self-evaluation basis and process for standardizing budget units and further standardizing budget performance management work.

30.1 Introduction

To strengthen the performance evaluation of financial expenditure in administrative institutions through institutionalized and legalized management methods, it is necessary to establish a unified and standardized financial expenditure performance evaluation mechanism from the central government to local governments. In order to optimize the performance workflow and work rules of financial management work, it is first necessary to develop performance evaluation methods, and then clarify the rights

T. Cao (✉)
Financial Information Management Center of Jiangsu Provincial Department of Finance Nanjing, Jiangsu 210024, China
e-mail: caotiiing@163.com

R. Kountchev (Deceased) et al. (eds.), *AI Methods and Applications in 3D Technologies*, Smart Innovation, Systems and Technologies 388,
https://doi.org/10.1007/978-981-97-2144-3_30

and obligations of the evaluated subject [1]. At the same time, enhancing the development and construction of the financial performance evaluation database can achieve a scientific, standardized, unified, and reasonable financial fund expenditure indicator system, and establish corresponding databases to achieve the sharing of financial fund data. We need to further enhance the reasonable matching of decision-making and execution authority, vigorously promote the integration of financial resources, achieve deep integration of resources between different departments, and avoid the occurrence of financial waste [2]. At the same time, it is necessary to strengthen the constraints on fiscal expenditure, strengthen performance evaluation methods, and lay the foundation for future fiscal expenditure performance evaluation work.

The actual content of performance goals must be clear and detailed, not just broad scope and constraints. If the evaluation indicators are too macro, they cannot be quantified. The evaluation indicators should have quality and timeliness, cost targets, generated benefits, economic benefits, social benefits, ecological benefits, project revenue completion, and the cost resources required to achieve the established goals [3]. Accurately measuring its output and performance indicators, it is also necessary to determine the performance goals of each department, reasonably regulate the specific requirements for goal formulation, and thus make performance goal formulation more scientific and standardized [4, 5]. Government departments and units should set performance targets that meet the requirements of national economic development planning, and also match the functions of their own departments, highlighting the scope and effectiveness of fiscal expenditure. The goals of performance evaluation should be refined from multiple aspects such as quantity, quality, and cost. Quantifiable ones should be quantified, while those that cannot be quantified should adopt the basic principle of hierarchical classification and qualitative description. The diversity of financial performance evaluation methods and the scientific and reasonable design of financial performance evaluation methods will also determine the fairness of financial performance evaluation results to some extent. For different evaluation subjects, qualitative or quantitative evaluation methods can be selected, and for different performance goals, scoring and ranking methods can also be used to complete the evaluation.

30.2 Related Work

In recent years, the Chinese government has vigorously promoted the reform of the financial system and mechanism, strengthened the evaluation of fiscal expenditure, and established a public finance framework system. However, with the prominent contradiction between government units' income and expenditure, the problem of low efficiency in fund utilization has increased [6]. Therefore, deepening the performance evaluation of fiscal expenditure is a key point, and a complete fiscal expenditure performance evaluation system needs to be established [7]. These have become key issues that fiscal reform needs to face. This article discusses and analyzes the practical problems in the performance evaluation of domestic fiscal expenditure, in order to propose reasonable response measures and suggestions [8]. Optimize and improve financial performance evaluation rules and regulations, optimize the design of performance evaluation laws and regulations, establish a complete and scientific performance evaluation index system, expand the scope of performance evaluation work, vigorously promote innovation in performance work models, and further strengthen positive feedback and practical application of performance evaluation results [9].

We will vigorously implement performance evaluation indicators for fiscal expenditures, implement government fiscal policies, improve transparency in fiscal fund expenditures, establish a scientific fund utilization management system, and design fiscal performance evaluation indicators based on the principles of objectivity, practicality, and fairness [10]. Currently, there is still great room for improvement in the performance evaluation and management of fiscal expenditures, and it is necessary to strengthen the performance evaluation of fiscal fund expenditures, Helps to improve the efficiency of utilizing financial funds.

At present, the performance evaluation work of fiscal expenditure is not fully legal, and various provincial and municipal government departments in China have various performance evaluation activities for fiscal expenditure [11]. However, the use of evaluation standards is not very unified, and there is strong arbitrariness, which fails to fully play the true role of performance evaluation work. The investment department has proposed a process for investment decision-making, from project initiation to performance evaluation of financial funds expenditure, using a full process management model to avoid performance evaluation work becoming a formality. Moreover, due to the constraints of performance evaluation work, the guiding role of performance evaluation has not been fully demonstrated. At present, when designing evaluation indicators related to financial technology engineering management, most of the financial expenditure performance evaluations of relevant departments only focus on the compliance of the use of engineering funds, but there is no comprehensive evaluation of the benefits of fund use. In addition, there are many types of performance indicators for fiscal fund expenditures, coupled with the excessive functions of government departments, which have their own indicators in different departments, industries, and projects. This also leads to relatively cumbersome indicator design, and even creates overlapping problems, resulting in significant differences in evaluation results. The evaluation system for this type of financial expenditure lacks fairness and specificity, and designing too many indicators can also result in resource waste.

30.3 Research on Intelligent Analysis Model for Financial Management Performance Evaluation

The construction of a financial management performance evaluation model relies on the construction of a financial management performance evaluation index system and intelligent analysis methods. Therefore, the various indicators involved in the performance evaluation of financial management are first sorted out, and based on this, an evaluation model is designed to provide model support for financial management.

Financial Management Performance Evaluation Indicator System

This article mainly constructs an evaluation index system from eight dimensions: major policy implementation and response, financial system management, budget preparation management, budget execution management, budget performance management, asset management, local government debt management, and financial transparency management.

As shown in Table 30.1, the implementation and response of major policies include financial support for national strategic tasks, financial support for livelihood expenditures, and implementation of tax and fee reduction policies. Budgeting management includes government revenue and expenditure budgeting, departmental budgeting, and resource coordination management. Budget execution management includes the degree of budget execution, management of treasury funds, overall financial statements, and management of government procurement services. Among them, common performance indicators cover 12 categories: conference training, policy research, news promotion, large-scale activities, international exchange and cooperation, international organization membership fees, infrastructure engineering, equipment procurement, information system construction and operation maintenance, scientific research projects, supervision and inspection, and disciplinary inspection and supervision; The performance indicators by industry and field cover 18 industries, over 70 industry categories, over 360 fund uses, and over 3000 indicators. The industry category corresponds to the "category" level of the budget function account, the industry category corresponds to the "category" level, and the fund purpose corresponds to the "item" level. Output, benefits, and satisfaction are established as primary indicators. The output indicators are decomposed into four secondary indicators: quantity, quality, timeliness, and cost. The benefit indicators are decomposed into four secondary indicators: economic, social, ecological benefits, and sustainable impacts. Performance standards are established on three levels of indicators, covering plan, history, industry, international, and other standards. Performance indicators and their values have also been applied to construct a "traffic light monitoring system": a

Table 30.1 Financial management performance evaluation indicator system

Heading level	Example	Font size and style
Financial management performance evaluation indicator system	Major policy implementation and response	Financial support for national strategic tasks
		Financial guarantee for livelihood expenditure
		Implementation of tax and fee reduction policies
	Financial system management	Implementation of fiscal expenditure policies
	Budget preparation management	Preparation of government revenue and expenditure budget
		Departmental budgeting
		Resource coordination and management
	Budget execution management	Budget execution level
		Treasury fund management
		Financial final account
		Government procurement management
		Government procurement service management
	…	…

red light indicates a serious risk or inability to complete, a yellow light indicates difficulty or slow progress, a blue light indicates a clear unreasonable indicator setting at the beginning of the year, and a green light indicates completion or normal operation.

Intelligent Analysis Model for Financial Management Performance Evaluation

The intelligent analysis model for financial management performance evaluation consists of four parts. Firstly, it involves pre performance evaluation, which evaluates the necessity of project approval, investment economy, rationality of performance goals, feasibility of implementation plans, compliance with fundraising, and necessary conditions for budget application. Next is performance target management and evaluation, covering departmental and unit goals, policy and project goals, as well as output, cost, benefits, sustainable impact, and service object satisfaction. Then there is budget performance monitoring, covering the degree of achievement of performance goals and the progress of budget execution, as well as suspending or stopping budget

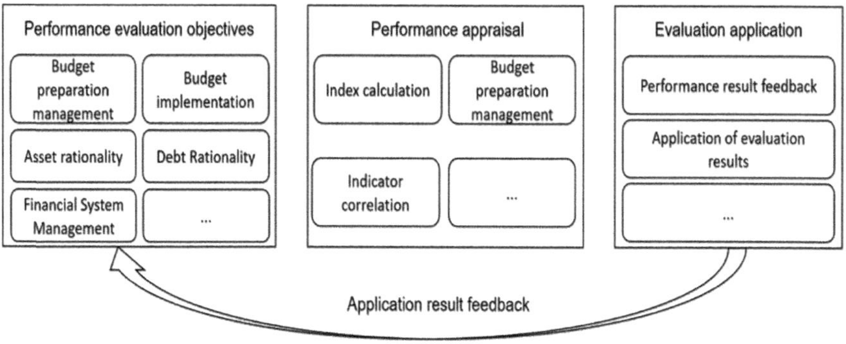

Fig. 30.1 Intelligent analysis model for financial management performance evaluation

appropriations for those with serious problems and promoting their rectification. Finally, there is performance evaluation and result application, covering third-party (financial, institutional, and third-party) evaluation, feedback on evaluation results, responsibility system for problem rectification, and application of evaluation results.

As shown in Fig. 30.1, the first step is to establish performance goals for financial management, so that the assessed units and departments can clarify the standards and rules for budgeting. Secondly, evaluate the collected indicators, including the formula for the evaluation results and the communication and analysis of the evaluation results. Finally, the application of evaluation results must be accurately and timely notified to the assessed object, so that the assessed object has a clear understanding of their assessment situation. The advantage of this is that it can help the evaluated department or unit clarify their shortcomings and improve their shortcomings or mistakes. The construction of this model provides the most basic business logic for the next step of system design. The system design can develop the core business logic of the system based on the construction ideas and framework of the model.

30.4 Design and Implementation of an Intelligent Analysis System for Financial Management Performance Evaluation

The design of an intelligent analysis system for financial management performance evaluation must fully consider the technical architecture and data foundation of the existing financial management budget system. The basic data of the current financial management budget system is stored in the data center, which integrates the management and business data of multiple business systems. There is a certain degree of independence in the data correlation between each business system. Therefore, the design and implementation of this intelligent system is based on the integration and integration of multi-source heterogeneous data between various business systems in financial management. This article has developed a data fusion plan through extensive research, and on this basis, data integration and access are carried out to complete the development and implementation of the system.

Design of Intelligent Analysis System for Financial Management Performance Evaluation

There are a total of 4 service modules in the existing financial management budget system, as well as two system backend management models, a data center, and a front-end service interface. There are a total of 3 services in the financial management project database system, including a management backend, a data center, and a front-end service interface. The financial management technology platform includes 9 services, including user center, authentication center, permission center, basic data center, function center, process center, report center, file service, etc.

As shown in Fig. 30.2, the design of an intelligent analysis system for financial management performance evaluation is mainly implemented from three aspects. Firstly, there should be a basic equipment aspect, including storage devices, management servers, network devices, and security devices. Secondly, in terms of data services, design metadata management, master data management, data governance management and control and data modeling function modules. Then, in terms of data services, functional modules such as budget analysis, project budget audit analysis, budget unit management, and data value mining analysis are provided. Provide users with functional modules such as budget involvement and application, budget execution progress monitoring, budget indicator calculation, and budget evaluation result feedback in application services.

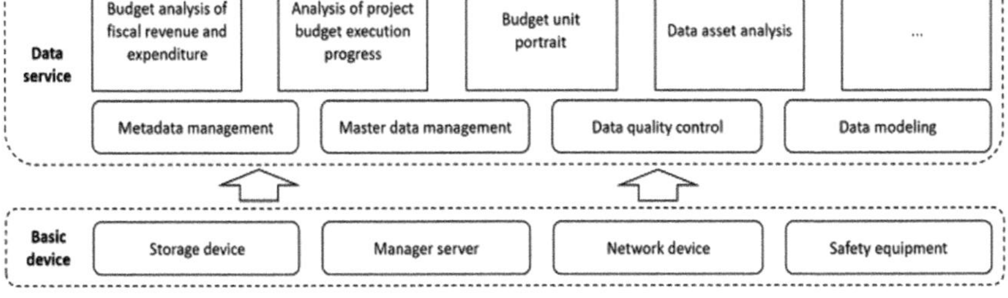

Fig. 30.2 Intelligent analysis system architecture for financial management performance evaluation

The storage device of this system stores the basic data from the budget management system. The data service layer is embedded with a unified data interface, which can be compatible with various types of database systems. By combining metadata management tools and data quality control module functions, the data is preliminary correlated. Next, define the storage objects associated with the main and reference data sources after establishing the data source association table. Then, taking time as the axis, the performance evaluation model for the entire process of budget management is constructed by connecting relevant data related to budget management across various time sections. The data service in the system providing users with functions such as budget analysis and financial performance evaluation.

Implementation of Intelligent Analysis System for Financial Management Performance Evaluation

The implementation of this system mainly includes four layers, namely software and hardware layers, data call layer, algorithm implementation layer, and application layer. These four levels support each other and provide progressive services, and the system provides users with visual services and operational interfaces through a browser on the client side. This system can meet users' concurrent access requirements of no more than 100 without significant performance degradation.

As is shown in Fig. 30.3, in the software and hardware layer, the database used in this paper includes relational database, graph database, HBase database and distributed file database. At the data invocation layer, basic data operations such as data invocation, table structure optimization, and data fusion are supported. At the algorithm implementation layer, functions such as data normalization, clustering analysis, indicator solving, and confidence calculation are provided. Then a unified budget business calculation model is formed, and the upper level business function functions can directly call and interact with data by calling the calculation model. Basic information management is the foundation and prerequisite for supporting the operation of intelligent budget management systems. It uniformly manages and maintains various basic information used in subsequent stages, facilitating the sharing and sharing of basic information among various business processes.

This system evaluates the efficiency and rationality of the entire process of provincial-level financial management. The data sources involve nine core financial business systems, including budget review, payment management, and government procurement. The performance evaluation results of the corresponding business processes of the above systems will be fed back to the financial management unit. When applying the system, it is necessary for the financial management department to uniformly set and maintain the first and second level indicators, as well as the common indicators in the third level indicators. Then set three-level indicators for each department and report them to each department and unit. According to the

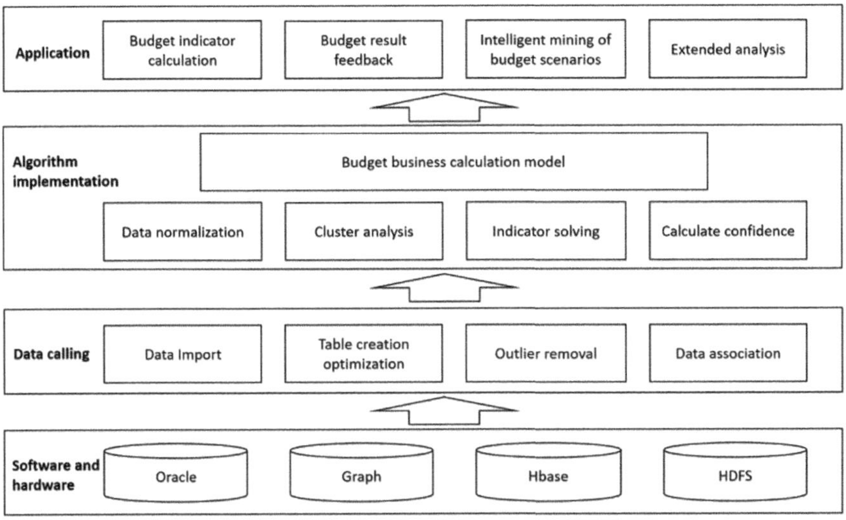

Fig. 30.3 Implementation of intelligent analysis system for financial management performance evaluation

needs of budget performance management, set three-level indicators for the department and report them to the same level of finance department. Finally, the reviewer will conduct the review, and after the review is passed, the budget report and related data will be stored in the database. Government financial management departments and units can dynamically update and maintain their performance indicators based on work needs or changes in the situation, and initiate adjustment applications to the same level of finance. The final financial management department will review the application for adjustment of indicators for departments and units in the system and update the performance indicators.

30.5 Application Results

At present, there are still seven main weak links in financial performance evaluation: performance measurement, total performance, allocation performance, cost performance, investment performance, financing performance, and incentive mechanism. One sided emphasis on operational performance evaluation is partly responsible for these weak links, especially project performance evaluation. Other reasons include unclear underlying logic and lagging basic theoretical research, making it difficult to provide accurate and effective guidance for performance practice. The intelligent analysis method for financial management performance evaluation proposed in this article starts from sorting out the underlying data and constructing a performance evaluation index system. An intelligent information system has been constructed using information technology, through which financial management units and personnel can intuitively monitor and manage the entire budget management process in real-time. At the same time, weaknesses in current financial management were identified through the application of the system. Strengthening the support of basic work is a fundamental prerequisite for promoting the full integration of financial performance evaluation and budget management, and also an important part of ensuring the standardization of financial fund use. Therefore, it is necessary to strengthen the investment in basic work and scientifically plan and design various links and processes of financial basic work. The next step is to adjust the current financial performance evaluation and budget management workflow to make it more in line with the needs of financial fund management, while also ensuring the standardization of each workflow to ensure the smooth progress of financial performance evaluation and budget management.

References

1. Zhejiang Provincial Department of Finance.: Case analysis and countermeasures for financial performance evaluation China finance **814**(17), 63–64 (2020). (In Chinese)
2. Yu, Z.X.: The application of fuzzy mathematics in financial performance evaluation. Pop. Sci. Technol. Innov. **2019**(12), 1 (2019). (In Chinese)
3. Bai, W.J.: Analysis of the connotation of financial expenditure performance evaluation. 42–46 (2021). (In Chinese)
4. Alimohammadlou, M., Bonyani, A.: A comparative analysis of dynamic and cross-sectional approaches for financial performance analysis[J]. Am. J. Finance Account. **5**(3) (2018)
5. Chen, Y.J.: Research on the performance evaluation index system of financial expenditure for urban greening construction. (5) (2022). (In Chinese)
6. Zan, W.Y.: Several thoughts on the problems in the performance evaluation of financial special projects. Enterprise Technol. Dev. **2**(2) (2019). (In Chinese)
7. Chen, K.: Challenges and countermeasures of accounting firms in financial performance evaluation. China Agric. Account. **4**(8) (2020). (In Chinese)
8. Yang, Y.: Reflection on promoting the combination of local financial performance evaluation and budget management. Business 2.0 (Economic Management) (18), 0117–0117 (2021). (In Chinese)
9. Li, S.R.: Research on financial performance evaluation by third party intermediaries: taking Xianyang city as an example. Western Finance Account. 3(1) (2023). (In Chinese)
10. Wang, Y.J.: Recent progress and prospects for financial performance evaluation and management. Financ. Sci. **9**, 14–31 (2022). (In Chinese)
11. Xue, G.: A brief discussion on the bottlenecks and solutions of grassroots financial performance evaluation. Econ. Manag. Abs. 2(11) (2021). (In Chinese)

Author Index

R. Kountchev (Deceased) et al. (eds.), *AI Methods and Applications in 3D Technologies*, Smart Innovation, Systems and Technologies 388,
https://doi.org/10.1007/978-981-97-2144-3

GPSR Compliance

The European Union's (EU) General Product Safety Regulation (GPSR) is a set of rules that requires consumer products to be safe and our obligations to ensure this.

If you have any concerns about our products, you can contact us on ProductSafety@springernature.com

In case Publisher is established outside the EU, the EU authorized representative is:

Springer Nature Customer Service Center GmbH
Europaplatz 3
69115 Heidelberg, Germany

Batch number: 10703867

Printed by Printforce, the Netherlands